Biomedical and social aspects of alcohol use

TNO Nutrition and Food Research

Biomedical and social aspects of alcohol use: a review of the literature

Dirk G. van der Heij & Gertjan Schaafsma (Editors)
TNO Nutrition and Food Research, Zeist, Netherlands

Pudoc Wageningen 1991

CIP-data Koninklijke Bibliotheek, Den Haag

Biomedical

Biomedical and social aspects of alcohol use: a review of the literature / Dirk G. van der Heij and Gertjan Schaafsma (eds.). – Wageningen : Pudoc [etc.]. – Ill.
Met reg.
ISBN 90-220-1051-1 geb.
NUGI 735
Trefw. alcoholgebruik.

ISBN 90-220-1051-1
NUGI 735

© Centre for Agricultural Publishing and Documentation (Pudoc), Wageningen, Netherlands, 1991.

All rights reserved. Nothing from this publication may be reproduced, stored in a computerized system or published in any form or in any manner, including electronic, mechanical, reprographic or photographic, without prior written permission from the publisher, Pudoc, P.O. Box 4, 6700 AA Wageningen, Netherlands.

The individual contributions in this publication and any liabilities arising from them remain the responsibility of the authors.

Insofar as photocopies from this publication are permitted by the Copyright Act 1912, Article 16B and Royal Netherlands Decree of 20 June 1974 (Staatsblad 351) as amended in Royal Netherlands Decree of 23 August 1985 (Staatsblad 471) and by Copyright Act 1912, Article 17, the legally defined copyright fee for any copies should be transferred to the Stichting Reprorecht (P.O. Box 882, 1180 AW Amstelveen, Netherlands). For reproduction of parts of this publication in compilations such as anthologies or readers (Copyright Act 1912, Article 16), permission must be obtained from the publisher.

Printed in the Netherlands.

Preface

The use of alcohol-containing beverages has been connected with both positive and negative social and health consequences. From time immemorial, wine and beer have belonged to the everyday menu in many countries and cultures. The use of distilled beverages is a comparatively recent phenomenon. In many cultures, the appreciation for alcohol has been high for centuries, and many artists and scientists still find support in alcoholic beverages as a source of inspiration. The role of alcoholic beverages in gastronomy is undisputed. Besides, many medicinal properties have been ascribed to these beverages. However, positive as the effects of alcohol consumption may be, it is undeniably true that some people misuse alcohol. This may explain, at least in part, the ambivalence in the attitude towards alcohol use in present-day society. On the one hand, this attitude is positive because of the sensory and aesthetic properties of alcoholic beverages and their highly valued role in a wide range of cultural expressions. On the other hand, society's attitude is negative because of the risks of misuse inherent in the use of any product.

Some governments have tried to influence the balance between positive and negative effects of alcohol use in the community by formulating alcohol control policies. These relate to price and excise measures, to restrictions imposed on advertising and availability, and to education and information. These policies are largely based on research data obtained by investigating the reasons for and consequences of excessive alcohol use. By far the majority – probably even as much as 99% – of scientific literature pertains to alcohol abuse and alcoholism.

As early as 1980, the Dutch alcohol industry and alcohol trade judged it desirable to have available an overview of scientific literature on moderate alcohol use and its social and biomedical implications and assigned the TNO Toxicology and Nutrition Institute at Zeist to carry out this work. This literature survey, entitled 'Biomedical and social aspects of the use of alcoholic beverages', was written by Ms Ir W.G. Vrij-Standhardt and Ms Ir E.L. Rasmussen-Conrad. It was completed in 1981 und updated in 1984. This state-of-the-art report prompted alcohol production and trade companies to embark on various activities in the fields of public education, training and education of staff in alcohol production and trade, starting an alcohol documentation centre and a current awareness newsletter 'Alcohol Selectief'. The report also provided the scientific basis for discussions with both the Dutch government and EC authorities on alcohol control measures. Finally, it resulted in the identification of the topic 'alcohol and cardiovascular disease protection' as an area for experimental research which has proved to be very fruitful.

From the very start in 1980, all these activities have been based at the TNO Toxicology and Nutrition Institute, under the guidance of the SAR ('Stuurgroep Alcohol Research', Steering Group for Alcohol Research). The SAR comprises representatives of the Dutch wine, spirits and beer industry and trade. Its present composition is: Mr H. Bergsma, chairman; Mr R.J.B. Wallast Groenewoud, secretary; Mr Th.G.Chr. Hooij, Drs N.J. Gratama, Drs J.J.M. Verhoek and Mr H.E. Jonker Roelants, members. The citical and constructive attitude of all SAR participants towards the subject matter has made the cooperation with SAR a most fruitful one.

In recent years, many new research data have become available. That is why both the SAR and TNO felt it necessary to update the 1984 literature survey. We feel pleasure and satisfaction in presenting this book, which is an updated and thoroughly revised edition of the 1984 report.

We are most grateful to the authors, Ms Ir W.G. Vrij-Standhardt, Ms Drs E. te Wierik and Dr J. Veenstra, who have spent a great deal of time and attention to writing the chapters. We also mention with gratitude the contributions made by the staff of the Alcohol Documention Centre, Ms M. van Ruyven and Ms E. Vermeulen-Hermans. Many of the drawings and schemes in this book have been aptly designed and lettered by Mr M. van der Vaart. We acknowledge with appreciation the SAR for generously providing the funds for this project. We are indebted to the publisher, Pudoc, for pleasant cooperation in producing this valuable review. The project has been coordinated by Dr G. Schaafsma. Finally, we wish to thank Mr D.G. van der Heij and Dr G. Schaafsma for undertaking the painstaking work involved in editing this book.

It is our hope that 'Biomedical and social aspects of alcohol use' will provide all those interested in the various aspects of alcohol consumption with the scientific facts underlying the implications of alcohol use. We also hope that a better understanding of all aspects of moderate alcohol use will help to formulate effective policies to prevent excessive use of alcohol.

Zeist, June 1991

Professor Dr R.J.J. Hermus, Director
TNO Toxicology and Nutrition Institute

Contents

Abbreviations	1
Introduction	3
1. Moderate and excessive drinking, problem drinking and alcoholism; Definitions / *E. te Wierik*	5
2. The biokinetics of alcohol / *W.G. Vrij-Standhardt*	14
3. Alcohol metabolism / *W.G. Vrij-Standhardt*	32
4. Alcohol and hormone metabolism / *W.G. Vrij-Standhardt*	47
5. Effects of alcohol on the metabolism of macronutrients / *W.G. Vrij-Standhardt*	62
6. Effects of alcohol on vitamin metabolism / *E. te Wierik and W.G. Vrij-Standhardt*	72
7. Effects of alcohol on water and mineral balance / *E. te Wierik and W.G. Vrij-Standhardt*	83
8. Effects of alcohol on the liver / *W.G. Vrij-Standhardt*	90
9. Effect of alcohol on the brain and the nervous system / *W.G. Vrij-Standhardt*	116
10. Effects of alcohol on the cardiovascular system / *J. Veenstra*	130
11. Effects of alcohol on various organs and tissues / *E. te Wierik and W.G. Vrij-Standhardt*	157
12. Alcohol and cancer / *E. te Wierik and W.G. Vrij-Standhardt*	169
13. Alcohol and pregnancy / *E. te Wierik*	180
14. Differences between alcoholic beverages / *W.G. Vrij-Standhardt*	190
15. Factors influencing behaviour in relation to alcohol / *E. te Wierik*	204
16. Alcohol and traffic / *E. te Wierik*	228
17. Prevention of alcohol abuse: models of control / *E. te Wierik*	238
18. Measures to control alcohol abuse / *E. te Wierik*	252
Subject index	263

Abbreviations

AA	Alcoholics Anonymous	HBV	hepatitis B virus
AAD	alcohol amnestic disorder	HDL	high-density lipoprotein
AC	adenylate cyclase	HLA	human lymphocyte antigen
ACTH	adrenocorticotropic hormone, corticotropin	HPA	hypothalamic-pituitary axis
ADH	alcohol dehydrogenase	IARC	International Agency for Research on Cancer
ALA	δ-aminolaevulinic acid	IBS	intermediate brain syndrome
ALAT	alanine aminotransferase	IQ	intelligence quotient
AMP	adenosine monophosphate	LD_{50}	median lethal dose
AP	alkaline phosphatase	LDL	low-density lipoprotein
Apo	apolipoprotein	LH	luteinizing hormone
ASAT	aspartate aminotransferase	LHRH	luteinizing hormone-releasing hormone
AST	aspartate aminotransferase		
ATPase	adenosine triphosphatase	MAO	monoamine oxidase
AUC	area under the (blood alcohol) curve	mAST	mitochondrial aspartate aminotransferase
AUDIT	Alcohol Use Disorders Identification Test (of the WHO)	MCV	mean corpuscular volume
		MEOS	microsomal ethanol-oxidizing system
AVP	arginine vasopressin	MLPA	minimum legal purchase age
BAC	blood alcohol concentration	NAD	nicotinamide adenine dinucleotide
cAMP	cyclic adenosine monophosphate	NADH	reduced form of NAD
CDAT	carbohydrate-deficient transferrin	NADP	nicotinamide adenine dinucleotide phosphate
CHD	coronary heart disease		
CNS	central nervous system	NADPH	reduced form of NADP
CoA	coenzyme A	PAI-1	plasminogen activator inhibitor 1
CRF	corticotropin-releasing factor	PGE1	prostaglandin E1
CT	computerized tomography	PLP	pyridoxal-5'-phosphate
CVD	cardiovascular disease	PSU	portal-systemic encephalopathy
DGLA	dihomogammalinolenic acid	PTU	propylthiouracil
DSMIII	Diagnostic and Statistical Manual of Mental Disorders	PUFA	polyunsaturated fatty acids
		RBP	retinol-binding protein
DWI	driving while intoxicated	RR	relative risk
EEG	electro-encephalogram	SCE	sister chromatid exchange
ERP	event-related potential	SES	socio-economic status
FAE	foetal alcohol effects	SHBG	sex hormone-binding globulin
FAS	foetal alcohol syndrome	t-PA	tissue pasminogen activator
FEV	forced expiratory volume	THBC	tetrahydrobetacarboline
FSH	follicle-stimulating hormone	TIQ	tetrahydroisoquinoline
GABA	γ-aminobutyric acid	TPP	thiamin pyrophosphate
GGT	γ-glutamyltransferase	TRH	thyrotropin-releasing hormone
GH	growth hormone	TSH	thyroid-stimulating hormone
GHRH	growth hormone-releasing hormone	UK-PA	urokinase-plasminogen activator
GOT	glutamic-oxaloacetic transaminase (aspartate transaminase, ASAT)	VLDL	very low-density lipoprotein
		WHO	World Health Organization
GPT	glutamic-pyruvic transaminase (alanine aminotransferase, ALAT)	WKS	Wernicke-Korsakoff's syndrome
GSH	glutathione		

Introduction

This book primarily offers a survey of literature relating to biomedical and social aspects of the consumption of alcoholic beverages.

The information contained in this book is aimed at those who are concerned with alcohol research, alcohol policies or the production of alcoholic beverages. Moreover, the book is intended to be a useful source of information for health educators and advisers, staff of alcohol and drugs centres, food scientists and nutritionists.

Relevant recent literature has been searched for in the Alcohol Documentation Centre of the TNO Toxicology and Nutrition Institute at Zeist, Netherlands. This Centre has the disposal of many journals, books and other publications dealing with the social and biomedical aspects of alcohol use and abuse.

Chapter 1 presents the various definitions of terms related to alcoholism, problem drinking and moderate and excessive alcohol consumption. It is followed by overviews of the biokinetics of alcohol (Chapter 2) and alcohol metabolism (Chapter 3). The subsequent eight chapters (Chapters 4–11) review the literature on the biomedical consequences of alcohol consumption and the effects of ethanol on various systems and organs. Studies into the relationship between the consumption of alcoholic beverages and the development of cancer are discussed in Chapter 12. The consequences of the consumption of alcoholic beverages during pregnancy are treated in Chapter 13. Differences between the three main categories of alcoholic beverages – beer, wine and spirits – are mentioned in Chapter 14.

The last four chapters are devoted to the social aspects of the use of alcohol. These chapters focus on factors that influence behaviour in relation to alcohol consumption (Chapter 15), alcohol and traffic (Chapter 16), prevention of alcohol abuse (Chapter 17) and measures to control abuse of alcoholic beverages.

Since the previous edition much progress has been made in many fields discussed in this book. Some of these will be mentioned briefly below.

Various studies have largely revealed the effects of alcohol on the brain and the way alcohol affects neurotransmission. A recent discovery is the presence of alcohol-metabolizing enzymes in the stomach which are claimed to metabolize a significant proportion of the alcohol ingested. The significance of this finding is not clear yet, but it could explain the fact that some alcohol is 'missing' when BAC curves are calculated. Research into the hepatotoxicity of alcohol has elucidated details on various mechanisms (with free-radical production being the most recent one) which could explain alcoholic liver injury. There is no such thing as a competition from one mechanism once will come out as *the* mechanism. Instead, various mechanisms act in concert to result in liver injury.

Another field in which much progress has been made in recent years is molecular biology and genetics. Identification of loci which code for enzymes has revealed wide genetically determined variation in enzyme activity. This may be why some people are more vulnerable to the adverse effects of alcohol abuse than others. Besides hereditary factors, other factors that influence behaviour in relation to alcohol use have been studied extensively. Interaction studies have been carried out to study several factors in one model.

A chapter on alcohol use in relation to traffic participation is new to this edition. Since many traffic accidents are related to alcohol use, many studies into the motives for drinking prior to driving have been performed and prevention strategies have been developed.

Various strategies have been designed to dissuade people from using alcohol excessively. The use of mass media for alcohol abuse prevention programmes is a novel phenomenon.

By reviewing the vast body of literature on the biomedical and social aspects of alcohol use the authors of this book have tried to give an outline of the present knowledge in this fascinating multidisciplinary field.

CHAPTER 1

Moderate and excessive drinking, problem drinking and alcoholism. Definitions

E. te Wierik

The word 'alcohol' stems from Arabic (*al* = the; *kuhl* = 'spirit' or 'refined'). Originally it referred to a fine powder used to darken the area around the eyes, i.e. eyeblack, and is related to the Hebrew *kahol*, a substance outlining the eyes (5, 10). In Latin pharmacology the term was used to indicate fine powdery substances or concentrations of distilled liquids. Paracelsus (1493–1541) used the term for the best or the spirit of wine.

The only alcohol suitable for consumption is ethyl alcohol, or ethanol. Ethanol is a clear, colourless liquid, generated by carbohydrate fermentation or by chemical synthesis, and constitutes the intoxication element of alcoholic beverages.

In its pure form ethanol may be used for several purposes, for example as a solvent, or – in a 70% solution – as a bactericide.

When administered as a medicine, alcohol may have a stimulating as well as a sedative, anaesthetic or narcotic effect. Its effectiveness depends on the concentration in the organism (13). The effects of alcohol may range from euphoria to intoxication and are responsible for the tremendous popularity of alcoholic beverages.

In fermented beverages the alcoholic concentration varies from 2 to 17%. In distilled drinks it may be considerably higher. Irrespective of the type of drink a standard drink (glass) contains about 10 g ethanol.

Research on drinking habits attempts to provide insight into patterns of alcohol consumption and in its consequences. Other motives are:
– to gain insight into the development of problem drinking and alcoholism by studying drinking habits,
– to collect data in order to establish aims for preventive action,
– to provide a basis for the development of theories regarding drinking habits,
– to lay a foundation for future research so that changes in drinking habits can be investigated in the long term.

Not only the motives of alcohol research vary, but also the way in which these drinking patterns are classified. Consequently, the results of these studies can be compared only to a limited extent. The quantities of alcohol drunk during normal and excessive alcohol consumption and in cases of alcoholism expressed in, for instance, grammes of pure alcohol per day vary widely. There are also differences in interpretation of terms 'alcoholism' and 'alcoholic'. Research on alcohol has developed into 'alcohology' in the course of time. As any science, alcohol research is characterized by a distinct terminology of its own. To define a set of concepts Keller et al. (10) have compiled a Dictionary of words about alcohol.

In this review, however, covering the literature up to 1991, terms are adopted from the original sources.

Since the meaning attached to terms may vary among authors, a number of frequently used terms, such as moderate drinking, excessive drinking, problem drinking and alcoholism, will be elucidated below.

1.1. Moderate and excessive drinking

It is by no means easy to give a straightforward answer to the question what quantities of alcohol should be considered moderate, reasonable or acceptable, nor to define what quantities represent light, moderate, safe, heavy and excessive drinking or alcoholism. There are several reasons why limits for 'safe' drinking are hard to give. First, it is not feasible to perform experiments with humans to investigate the chronic injuries incurred after different patterns of alcohol consumption. Second, some alcohol-related diseases occur rarely. Third, it is difficult to detect 'true' alcohol consumption by interviewing (12). Moreover, it is difficult to fix limits since there is a broad variation in sensitivity to alcohol. After the same amount of alcohol, women will have a higher blood alcohol concentration than men because of the differences in body composition. Besides, there is an intra-individual variation related to someone's state of health. Pregnancy, for example, goes with a much lower limit of safe drinking. In addition, under certain conditions (traffic, use of medicine) safe drinking levels are hard to give.

Although moderate alcohol consumption is hard to define many scientists have tried to do so. McDonald defines two drinks per day for the average woman and three drinks per day for the average man as moderate (14). Up to 13 drinks per week by women and up to 20 drinks per week by men is regarded as healthy drinking in another study (1). The Royal College of Physicians suggest 17 drinks per week for men and 11 drinks per week for women to be moderate (2).

Keller (10) defines the terms 'moderate drinking' and 'moderate drinker' as, respectively, a rate of alcohol consumption or someone whose alcohol consumption is:
– moderate in amount (i) in terms of daily caloric value of alcohol intake, and (ii) in terms of temporary effect (i.e., never or rarely involving intoxication); and
– moderate in long-term effects, i.e. there is no question of negative effects bearing on health, job, family or relatives, social live or community welfare.

According to Forrest (8), 'moderate drinkers' or 'social drinkers' may consume one or two alcoholic beverages per drinking occasion. Such individuals may drink once or twice a week or perhaps even far less frequently. A social drinker may consume only three or four alcoholic beverages each year, on specific festive occasions, but the definition also holds for a pattern of daily consumption that is limited to less than two drinks per day. Perhaps the key characteristics of social/moderate drinkers is their capacity to control their use of ethanol and the absence of behavioural, social, legal and medical problems associated with drinking.

Keller (10) describes an excessive drinker as someone whose alcohol consumption is:
– excessive in amount (i) in terms of daily caloric value of alcohol intake, and (ii) in terms of reversible effects, i.e., causing intoxication; or
– excessive in the course of time, accompanied with shorter or longer periods of intoxication; or
– excessive with regard to effects, e.g. in the eyes of people in the individual's social environment; or
– excessive with regard to aetiological (e.g. cultural, physical or psychological) factors.

Forrest (8) classifies heavy drinkers as those who consume three of four alcoholic beverages or more per drinking occasion and drink on three or more occasions a week. These people, like social drinkers, may drink to the point of intoxication on an infrequent basis.

1.2. Problem drinking

Rinaldi et al. (19) tried to reach a consensus on alcohol-related terms. They asked a multidisciplinary group of experts to cooperate. After four rounds of data gathering a list of 50 substance abuse terms was produced; the group reached consensus on most of these. In this list problem drinking has two definitions:
– a drinking pattern that has resulted in serious disturbances of health, work, social adjustment, or other areas of functioning, and
– a pattern of alcohol consumption that does not satisfy all the criteria of alcoholism, but that is characterized by a sufficiently large intake to have generated problems of health or social functioning.

A definition of a problem drinker is also presented by Keller (10): 'One whose heavy, deviant or implicative drinking causes private or public harm and who is seen to cause problems for himself and for others. The category includes the alcoholics.'

Problem drinking is defined by Forrest (8) as people who drink excessively, frequently drink to the point of intoxication, and experience family, interpersonal, legal, medical or economic problems as a direct result of their alcohol consumption.

1.3. Alcoholism

The term 'alcoholismus' was first coined by the physician Magnus Huss in 1849 (10). Over the years various attempts have been made to give general and universal definitions of the terms 'alcoholism' and 'alcoholic'.

In his 'The disease concept of alcoholism', Jellinek proposed a definition of alcoholism as 'any use of alcoholic beverages that causes damage to the individual or society or both' (9). After the proposal of the Disease Concept, alcoholism was redefined according to various professional (medical, social, psychological) interests.

At least ten definitions are commonly used. Boyd et al. (4) tested the following ten definitions of alcoholism used in the United States and Europe.

1. Alcohol abuse as defined by the Diagnostic and Statistical Manual of Mental Disorder, 3rd edition (DSM-III). The diagnostic criteria are:
– pattern of pathological alcohol use
– impairment in social or occupational functioning due to alcohol
– the disorder exists for at least one month (11, 21).

2. Alcohol dependence as defined by DSM-III:
– pattern of pathological alcohol use
– impairment in social or occupational functioning due to alcohol
– the disorder exists for at least one month
– tolerance
– withdrawal symptoms (11, 21).

3. Alcoholism as defined by the Research Diagnostic Criteria (RDC). This category is for subjects who have symptoms in at least three out of the following five groups for at least one month (20):
Group 1
– tremors, delirium tremens or a history of cirrhosis
– impotence associated with drinking
– alcoholic black-outs
– alcoholic binges or benders
Group 2
– drinking every day
– exceeding the equivalent, in terms of alcohol content, of 1½ litres of whisky per week
– being unable to answer questions concerning frequency or quantity of drinking
Group 3
– the subject had not been able to stop drinking when he wanted to
– the subject tried to control drinking by allowing himself to drink only under certain circumstances
– drinking before breakfast
– drinking non-beverage forms of alcohol
Group 4
– arrests for drinking
– traffic difficulties associated with drinking
– trouble at work owing to drinking
– fighting associated with drinking

Group 5
– the subject felt he drank too much
– his family objected to his drinking
– other people objected to his drinking
– he lost friends because of his drinking
– he felt guilty about his drinking

4. Alcoholism as defined by the Feighner criteria (6). A 'definite' diagnosis is made when symptoms occur in at least three of the four following groups. A 'probable' diagnosis is made when symptoms occur in only two groups.
Group 1
– any manifestation of alcohol withdrawal such as tremulousness, delirium tremens, convulsions or hallucinations
– history of medical complications
– alcoholic black-outs
– alcoholic binges or benders
Group 2
– the subject had not been able to stop drinking when he wanted to
– the subject tried to control drinking by allowing himself to drink only under certain circumstances
– drinking before breakfast
– drinking non-beverage forms of alcohol
Group 3
– arrests for drinking
– traffic difficulties associated with drinking
– trouble at work owing to drinking
– fighting associated with drinking
Group 4
– the subject thinks he drinks too much
– his family objected to his drinking
– other people objected to his drinking
– he lost friends because of his drinking
– he felt guilty about his drinking.

5. Alcoholism as defined by the National Council on Alcoholism. 'Alcoholism is a chronic, progressive and potentially fatal disease. It is characterized by: tolerance, physical dependency and/or pathological organ changes all of which are the direct or indirect consequences of the alcohol ingested' (3).

6. Jellinek's definition of alpha, beta, and gamma alcoholism (9):
– *Alpha alcoholism* is a purely psychological, continuous dependence on the capacity of alcohol to alleviate physical or emotional pain. Although this type of drinking exceeds conventional norms and has social consequences for the drinker and his relatives, it does not lead to uncontrolled drinking, nor does it cause an inability to

abstain from drinking. Alpha alcoholism may develop into gamma alcoholism (see below), i.e. it may be a stage development. On the other hand, there are persons who have maintained the drinking pattern characteristic of alpha alcoholism for thirty or forty years without showing any changes or progressive development.
– *Beta alcoholism* is the type of alcoholism in which physical complications occur such as polyneuropathy, gastritis or cirrhosis of the liver, but in which no physical or psychological dependence on alcohol is present. According to Jellinek, this type of drinking is found in a type of social setting in which nutritional deficiencies are rife and excessive drinking is habitual.
– *Gamma alcoholism* represents a serious form of drinking. An increase in tissue tolerance to alcohol, adaptation of cell metabolism, physical dependence, withdrawal symptoms, and loss of control over drinking may be observed. In gamma alcoholism there is a definite progression from psychological to physical dependence. The gamma alcoholic usually has alternating periods of intoxication and sobriety, and is not able to abstain from drinking for more than one or two days. Intra- and interpersonal changes in this type of drinking are considerable. General health damage and harmful effect on social standing are more manifestly present than other types of alcoholism.

7. The alcohol dependence syndrome as defined by the 9th Revision of the International Classification of Diseases (ICD-9): 'A state, psychic and usually also physical, resulting from taking alcohol, characterized by behavioral and other responses that always include a compulsion to take alcohol on a continuous or periodic basis in order to experience its psychic effects and sometimes to avoid the discomfort of absence; tolerance may or may not be present' (15, 23).

8. Alcoholism as defined by the Present State Examination (PSE) (22) holds if any item of the following check-list applies.
'During the past month, have you:
– had family problems because of drinking?
– missed work because drinking?
– had morning shakes or other withdrawal symptoms?
– had blackouts for several hours?
– heard voices or seen visions?'

9. The four different stages of alcoholism defined by the Iowa Alcoholism Stages Index (16). This diagnostic test contains four scales based on the four major types of signs that mark problem drinkers or alcoholics:
– trouble due to drinking
– drinking for personal effects
– preoccupied drinking
– uncontrolled drinking
A subject who qualifies on any one of the scales is classified as being in an early stage (stage I) of the process; if he qualifies on two scales he is in stage II, and so on.

10. Alcoholism as defined by the Michigan Alcohol Screening Test (MAST). This is a self-judgement scale comprising 25 items. Every item has its own weighed score. A score of 4 is said to be probably problematic and a score greater than 4 is indicative of alcohol dependence. Some of these items with their weighted scores are (11):

Item	Question	Score yes	no
4.	Is it possible for you to stop drinking after one or two glasses without difficulties?		2
16.	Do yo drink before the afternoon?	1	
19.	Have you ever asked help for your use of alcoholic beverages?	5	
23.	Have you ever been arrested for drunkenness?	2	

By testing these ten definitions on 23 alcoholic patients Boyd et al. (4) observed that all patients were defined as alcoholic by seven of the ten definitions (Nos. 1, 2, 3, 5, 6, 10), and at least 78 % of the patients were diagnosed as alcoholic by the other three definitions. The authors have developed the Alcohol Poly-Diagnostic Interview which is derived from the ten definitions mentioned. This interview was found to be sensitive for treated alcoholics, but has not yet been tested on untreated alcoholics (11).

In an experiment designed to find the lexical meanings of the term 'alcoholic' the conditions defined by DSM-III were investigated. The hypothesis that the presence of pathological alcohol use, impairment in social or occupational functioning, tolerance to alcohol, and withdrawal symptoms would each evoke 'alcoholic' with a greater frequency than the absence of these factors was corroborated. None of these factors was found to be analogous to a necessary definitional condition since the absence of any one did not result in a zero frequency. These variables were found to be supplementary (17).

Rinaldi et al. (19) tried to reach a consensus on alcohol-related terms. As described above, a multidisciplinary group of experts was asked to cooperate. Among the 50 definitions agreed upon are:
– *alcoholic*: a person who has experienced physical, psychological, social or occupational impairment as a consequence of habitual, excessive consumption of alcohol;
– *alcoholism*: a chronic, progressive and potentially fatal biogenetic and psychosocial disease characterized by tolerance and physical dependence manifested by a loss of control, as well as diverse personality changes and social consequences;
– *alcohol dependence*: chronic loss of control over the consumption of alcoholic beverages, despite obvious psychological or physical harm to the person. Increasing amounts are required over time, and abrupt discontinuance may precipitate a withdrawal syndrome. Following abstinence, relapse is frequent.

Although the results met with some criticism, this attempt to arrive at a consensus has been received favourably (7, 18).

Summary

A sharp distinction expressed in terms of amount of pure alcohol per day between moderate and excessive, or between excessive drinking and alcoholism, is hard to establish. Observations have been made with regard to physical, social and psychological damage.

One is supposed to be a moderate drinker when one is able to control his use of ethanol and when there are no behavioural, social, legal and medical problems associated with drinking.

Excessive/heavy drinking is associated with intoxication, whereas a problem drinker is characterized as one who's repeated heavy drinking has resulted in problems of health or social functioning. This category includes alcoholics. Moreover, in most definitions alcoholics are supposed to be physically dependent on alcohol.

Many definitions concerned with 'alcoholism' and other alcohol-related terms are commonly used, but an attempt to arrive at consensus has been made.

References

1. Anderson P. What is a safe level of alcohol consumption? Stress Medicine 1986;2: 99-101.
2. Anonymus. Alcohol and health. Lancet 1987;I: 876.
3. Babor TF, Kranzler HR, Kadden RM. Issues in the definition and diagnosis of alcoholism: implications for a reformulation. Prog Neuro- Psychopharmacol Biol Psychiat 1986;10: 113-28.
4. Boyd JH, Derr K, Grossman B et al. Different definitions on alcoholism. II. A pilot study on 10 definitions in a treatment setting. Am J Psychiatry 1983;140: 1314-17.
5. Caenegem RC et al. Grote Winkler Prins Encyclopedie. Amsterdam/Brussels: Elsevier Argus, 1980: 491.
6. Feigner JB et al. Diagnostic criteria for use in psychiatric research. Arch Gen Psychiatry 1972;26: 57-63.
7. Fleischhacker WW, Barnas C. Substance use terminology. JAMA 1988;260: 478.
8. Forrest GG. Guidelines for responsible drinking. Springfield, MA: Charles C. Thomas, 1989.
9. Jellinek EM. The disease concept of alcoholism. New Haven, CN: Hillhouse Press, 1960.
10. Keller M, McCormick M, Efron V. A dictionary of words about alcohol. New Brunswick, NJ: Rutgers Center of Alcohol Studies Publication Division, 1982.
11. van Limbeek J, Walburg JA. De vroege signalering van alcoholproblematiek. Lisse: Swetz & Zeitlinger, 1987.
12. De Lint J. Alkoholtrinken und gesundheitliche Schädigung: zur Diskussion über Sicherheitsgrenzen und Risikograde. Drogalkohol 1986;10: 39-51.
13. Majchrowitz E. The role of blood in alcohol intoxication and addiction. In: Lisansky Gomberg E, Raskin White H, Carpenter JA, eds. Alcohol science and society revisited. Ann Arbor, MI: The University of Michigan Press, 1982; 171-85.
14. McDonald J. Moderate amounts of alcoholic beverages and clinical nutrition. J Nutr Educ 1982;14: 58-60.
15. Meyer RE. What characterized addiction? Alcohol Health Res World 1989;13: 316-20.
16. Mulford HA. Stages in the alcoholic process: toward a cumulative, nonsequential index. J Stud Alcohol 1977;38: 563-83.
17. O'Donohue. Experimental semantics: the lexical definitions of 'prejudice' and 'alcoholic'. J Mind Behav 1989;10: 21-36.
18. Reder Sm. Substance use terminology. JAMA 1988;260: 477-8.

19. Rinaldi RC, Steindler EM, Wilford BB, Goodwin D. Clarification and standardisation of substance abuse terminology. JAMA 1988;259: 555-7.
20. Spitzer RL, Endicott J, Robins E. Research Diagnostic Criteria: rationale and reliability. Arch Gen Psychiatry 1978;35: 773-82.
21. Williams GD, Grant BF, harford TC, Noble J. Population projections using DSM-III criteria, alcohol use and dependence 1990-2000. Alcohol Health Res World 1989;13: 366-70.
22. Wing JC, Cooper JE, Sartorius N. Measurements and classification of psychiatric symptoms. New York: Cambridge University Press, 1974.
23. World Health Organization. Ninth revision of the international classification of diseases. Geneva: World Health Organization, 1979.

CHAPTER 2

Biokinetics of alcohol

W.G. Vrij-Standhardt

Biokinetics may be defined as the processing of substances by the body. In this chapter the absorption, distribution and elimination of alcohol will be discussed successively. The metabolism of alcohol (biotransformation) will be dealt with in Chapter 3.

2.1. Absorption

Alcohol is absorbed over almost the entire length of the digestive tract. Absorption from the mouth and oesophagus is minimal. A significant proportion of alcohol is absorbed from the stomach (ca. 20%) (37). Most of the alcohol is absorbed from the duodenum and small intestine. Some absorption may also occur in the large intestine.

The alcohol ingested is highly diluted by the admixture of saliva and digestive juices. After administration of 40 g alcohol, diluted to 10 and 20%, Israel (32) found a maximum concentration in the jejunum of 2.5 and 3%, respectively. The administration of 0.8 g alcohol per kg body weight (diluted to 25%) resulted in a maximum concentration of 1–5% in the jejunum. Its concentration in the ileum was equal to that in serum (0.2%). After two hours the alcohol concentrations in the entire digestive tract almost equalled the level in the serum (21). When alcohol (45 to 60 g, diluted to 20%) was administered directly to the stomach, peak concentrations of 6–10% and 5–7% were observed in the duodenum and the jejunum, respectively (53).

The alcohol molecules are absorbed through the intestinal wall by a process of passive diffusion (7); the quantity of alcohol absorbed per unit of time is directly proportional to the concentration gradient between intestinal lumen, epithelial cells, capillaries and the portal vein (4). This implies that alcohol absorption follows first-order kinetics. The presence of food in the stomach delays gastric emptying and consequently postpones the moment the alcohol is absorbed in the intestine. Considering the high rate of absorption in the intestinal tract, gastric emptying is a rate-limiting factor, thus affecting the rate of absorption and the overall absorption period.

Various models have been proposed to describe the processes of alcohol input and absorption (29, 67, 101). Pieters et al. (67) have developed the most extensive model with 6 parameters. It is a 3-compartment model with feed-back control of stomach emptying depending of actual stomach alcohol concentration.

It may take six hours or even longer until the stomach is entirely void of alcohol. Of the total amount ingested, however, 90% will be absorbed within 1 to 3 hours depending on various factors such as the quantity and concentration of alcohol as well as the amount and type of food present in the stomach (21). In normal circumstances a maximum blood alcohol concentration (BAC) will be reached 45 minutes after the consumption of a single dose (1).

2.1.1. Effects of food on alcohol absorption

Various experiments have been performed to establish the effect of the presence of food on the absorption rate of alcohol. As a rule, the influence of the consumption of food on the BAC or on the serum alcohol level was investigated. The BAC, however, is not only determined by the absorption of alcohol, but also by the distribution in the body and by the elimination rate. The results of these experiments are hard to compare because of differences in experimental design. The types and quantities of food ingested, the intervals between food intake and alcohol consumption, and the quantities of alcohol all tend to vary as well as the way of determining alcohol levels (in venous or capillary blood, serum or breath). Based on the results of many experiments the following conclusions may be drawn. The delay of gastric emptying caused by the presence of food results in a lower BAC peak: the alcohol will be absorbed into the blood over a longer period of time while elimination has already

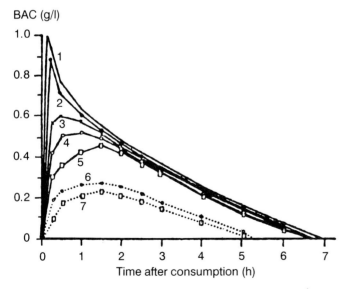

Fig. 2.1. BAC curves after the consumption of various kinds of spirits, wine and beer. Each of these beverages contained 0.6 g of alcohol per kg body weight. Types of beverage: 1, gin, vodka; 2, whisky; 3, dessert wine; 4, table wine; 5, beer; 6, table wine with meal; 7, beer with meal. After: Leake & Silverman (42).

set in. The effects of alcohol consumption are thus weakened by the intake of food. A larger amount of food results in a lower BAC peak (44). The effect of food consumption is maximal when alcohol is consumed with the meal or less than two hours after the meal (30, 79).

Haggard et al. (21) have devoted a great deal of research to the rate of alcohol absorption. A light meal, taken just before the consumption of 22–44 g of alcohol (whisky or martini-based cocktails) lowered the BAC peak by 50–80%. The BAC peak resulting from the consumption of 600–1200 ml beer was hardly affected by taking meals. Maximum BACs upon the consumption of beer, whether or not preceded by a meal, were comparable with a maximum BAC induced by the consumption of whisky or cocktails after a meal. Others (42), however, found a small decrease in BAC induced by beer when a meal had been taken (see Fig. 2.1).

Food, particularly food rich in fats and carbohydrates, delays gastric emptying, and thus makes alcohol reach the intestine more slowly. This explains the effect of the presence of food on the absorption rate of alcohol. The major factor accounting for the delay in absorption by solid food seems to be a slower delivery of alcohol to the small intestine, where it is rapidly absorbed, rather than the presence of solid

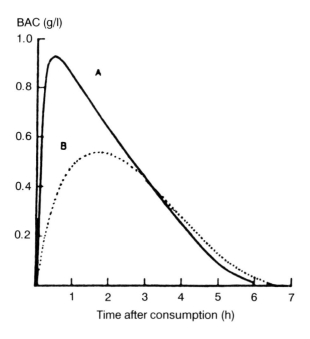

Fig. 2.2. Blood alcohol curves. Simulated data generated according to the one-compartment open model with first-order absorption and Michaelis-Menten elimination kinetics. For curve A, the absorption rate constant was assigned a value of 8/h, and for curve B the absorption rate constant was 1/h. After: Sedman et al. (78).

food acting as a mechanical barrier to the absorptive epithelium (30). It is assumed that the lower BAC peak in the presence of food is merely attributable to delayed absorption, not to a reduction of the total amount of alcohol absorbed. In experiments on dogs Harger et al. (23) and Southgate (82) found that 90% of the alcohol ingested is absorbed, irrespective of whether the stomach contains food. Welling et al. (99) based their assumption that less alcohol is absorbed if food is present in the stomach on the smaller area under the BAC curve (AUC). Sedman et al. (56), however, found that a delay in absorption per se resulted in a smaller AUC although the total amount of alcohol ingested remained equal (see Fig. 2.2). The maximum BAC is reduced more successfully by food rich in carbohydrates than by food with a high protein content. Fats have often been considered to exert a stronger effect than carbohydrates because of the stronger inhibitory effect of fat on gastric motility. In some experiments, however, fats were found to be less effective than carbohydrates (99), or roughly as effective as protein (78). In general, the various macronutrients may be said to delay absorption in the following order: carbohydrates > fats > proteins.

Clark et al. (12) distinguished various types of carbohydrate. They compared the effects of fructose, sorbitol and galactose. The administration of sorbitol resulted in the lowest BAC, with fructose as a close second, while even galactose still lowered the BAC. They explained the delaying action of these monosaccharides on alcohol absorption as follows. Sorbitol is absorbed extremely slowly, so large quantities of water will be retained in the intestine due to its osmotic effect. Not just water, but also the alcohol dissolved in it, is retained in the intestine for a longer period of time by the osmotic effect of sorbitol. Fructose is absorbed more rapidly than sorbitol, and galactose at still a faster rate. A quantity of sorbitol (125 g) as large as that administered in these trials causes fermentative diarrhoea because most of it eventually reaches the large intestine and is fermented by the microflora. A quantity of 40 g a day is the maximum amount deemed permissible (2). A similar quantity of fructose (125 g) also proved effective whereas half of this amount proved to have hardly any effect. Fructose not only affects the absorption of alcohol, but also accelerates its elimination from the blood (cf. Chapter 3). Consequently, fructose may also induce a lower BAC. Large quantities of glucose as well as of a mixture of glucose and fructose lowered the BAC through inhibition of gastric emptying (17, 78, 85). Broitman et al. (10), on the other hand, found a higher alcohol absorption rate in rats when glucose or other carbohydrates were present in the intestine. This held true when alcohol concentrations were low (0.23%) and carbohydrate contents were small (50 mmol/l). This phenomenon may be explained by an increased absorption of water.

Some milk and dairy products are considered to be particularly effective in lowering the BAC (5, 54); however, this has not been confirmed by other groups (33). Presumably, milk taken on an empty stomach has some effect by delaying gastric motility and diluting the alcohol. When the stomach is filled, milk has no additional BAC-lowering effect. In experiments with rats, caffeine proved to have a delaying effect on gastric emptying owing to relaxation of the gastric musculature resulting in a lower BAC, particularly when caffeine had been administered a quarter of an hour before alcohol was ingested (80).

2.1.2. Differences in absorption among alcoholic beverages

The quantity and type of alcoholic beverages consumed also affect alcohol absorption (68). The process is influenced by the alcohol concentration and by the occurrence of other substances in the beverage in question. Alcohol is absorbed relatively slowly from low-alcohol beverages, presumably as a result of dilution (the absorption rate in the intestine is directly proportional to the alcohol concentration present in the lumen). Large quantities of high-alcohol beverages, however, may slow down gastric motility and cause erosion of gastric mucosa (38). Pylorospasm, often followed by vomiting, may also occur. When alcohol consumption has become habitual, this will occur less frequently (21). Owing to these effects highly concentrated doses of alcohol will also be absorbed at a slower rate than doses with 15–30% alcohol. Alcohol from beer is absorbed rather slowly and results in a lower BAC than an equal quantity of alcohol from spirits (see Fig. 2.1). Alcohol from wine is also absorbed more slowly than alcohol from spirits (56).

The difference in absorption rate between these beverages is determined not only by differences in alcohol concentration but also by the presence of other components. The carbohydrate content of beer and sweet wine decreases gastric motility like meals do. Consequently, alcohol absorption from beer is delayed hardly any further by meals (21). Potential delaying effects on alcohol absorption have also been attributed to other qualities of alcoholic beverages (acidity, buffer capacity, alkaloid content) (41).

2.1.3. Other factors influencing absorption

Every physiological or pharmacological factor affecting gastric motility or circulation may basically modify alcohol absorption. An efficient blood circulation maintains the concentration gradient throughout the intestinal mucosa by rapidly removing the alcohol absorbed. Kalant (38) and Holford (29) have reviewed the factors affecting absorption. Alcohol absorption may be delayed not only by food ingestion, dilution of the alcohol and the presence of congeners, but may also result from a drop in body temperature, deep mental concentration or physical exercise. Aspirin, aminopyrine and anticholinergic drugs also slow down absorption whereas cholinergic drugs, insulin-induced hypoglycaemia and a raise in body temperature increase the absorption rate slightly.

Variation in hormonal status (e.g. stage of the menstrual cycle) affects ethanol absorption (36). The absorption rate shows a diurnal rhythmicity (45). Alcohol absorption seems also to be under genetic control (71).

2.2. Distribution

Once alcohol has been absorbed into the bloodstream it is rapidly transported throughout the body and distributed over the body fluids. The small electrically neutral alcohol molecules can diffuse easily through membranes. During absorption the arterial BAC tends to be higher than the venous BAC, peaking at a higher level and requiring less time to reach the peak, until an arterio-venous concentration equilibrium is reached, whereafter the venous BAC remains above the arterial BAC (51). About 1.0–1½ h after drinking an equilibrium is reached (63). This will take place sooner in organs highly perfused with blood (brain, lungs, kidneys, liver) than in other organs (96). It will take a relatively long time for this equilibrium to be reached in inactive skeletal muscular tissue because of the poor perfusion of this tissue. Once the equilibrium has been reached the amount of alcohol in the organs will be proportional to the organs' water content (63, 72).

Determination of alcohol in other body fluids as an alternative to blood alcohol determination has been proposed by several authors (20, 46, 76, 92). Blood sampling has various medical and legal drawbacks, whereas saliva, tear fluid en urine are easily obtainable. When the water content is considered, there is a close correlation between alcohol concentrations of blood and saliva in the elimination phase (from two hours after drinking) (20, 76). The tear/blood alcohol ratio was found to be 1.14, independently of BAC. The tear/blood water content ratio is 1.16, so the distribution of alcohol in tears follows closely its water content (46). The recent development of a simple and cheap dipstick based on alcohol oxidase holds promise for immediate determination of alcohol concentrations in saliva or other body fluids (27). Thorough studies to test the accuracy of these dipstick methods are to be recommended.

Sometimes urine is used for alcohol determinations (55), but these calculations often lack accuracy because the time of urine formation is usually unknown. In mutilated or partly decomposed bodies fluid of the eye, internal ear or brain can be used for forensic alcohol determination (92).

Hypothetically, physical exercise may lower the peak of the BAC curve. As the muscular tissue will be more richly perfused, the alcohol equilibrium between blood and muscular fluid may be reached sooner. Considering the large amount of muscular tissue (some 18 kg) and its high fluid content (79%) (81) the peak of the BAC curve might be flattened appreciably. Experiments by Pikaar et al. (69) did not prove BAC-lowering effects of physical exercise (cycling), although a lower breath alcohol content was found, presumably because an equilibrium in the lungs had not been reached due to panting. Body fat and skeletal mass absorb hardly any alcohol.

2.2.1. Breath alcohol analysis

The application of equipment to measure the alcohol content of breath from which the alcohol content of the blood may be deduced subsequently, serves various aims today. The principle is based on the facts that part of the alcohol present in the blood is excreted via the lungs and that the alcohol concentration in exhaled breath is proportional to the BAC. The partition coefficient between breath and blood has been estimated at 1:2100 to 1:2300 (57, 70), i.e. 2100–2300 ml of breath contains as much alcohol as 1 ml of blood once an equilibrium between blood and breath in the lungs has been reached. Determination of the alcohol content of expired breath has advantages as well as drawbacks as compared with determination of the alcohol content directly in the blood. The advantages lie mainly in the speed and ease of determination and in the fact that the method is non-invasive. One of the drawbacks is that the specimen cannot be stored for repeat analyses. Another problem is the lack of precision. The determination of breath alcohol content is an indirect way of determining the (arterial) BAC. It is only valid when a respiratory equilibrium has been established (34). Breathing techniques, temperature, humidity and other factors may interfere with the determinations. Gullberg (19) showed that variation in human breath by far exceeds the inaccuracy of breath alcohol analysis equipment. The first 7 to 20 minutes after alcohol consumption the alcohol still lingering in the mouth yields values that are too high, even after drastic rinsing and gargling with water (25, 83).

In the first 1½ hour high values tend to be found with breath tests as compared with blood tests. It should be noted that blood tests usually are performed in venous blood. The breath alcohol concentration, however, closely follows the pattern of arterial BAC during absorption. This discrepancy is in accord with the finding that the alcohol concentration in arterial blood is higher than that in venous blood during the first 1–1½ hour. Because alcohol is highly soluble in water it is easily transported via capillaries from the arteries to the tissues; later, it diffuses from the tissues to the veins with a reversed diffusion gradient (65).

The alcohol concentration in the brain also depends on the alcohol concentration in arterial blood. During elimination there is a close relationship between alcohol concentrations in breath, arterial blood and venous blood whereas during absorption breath alcohol concentration is a much better predictor of arterial BAC (51).

2.2.2. Sex differences in distribution

Men and women differ in bodily constitution. On average, women have less body water per kg body weight than males (500 and 600 ml/kg, respectively) (81). A certain quantity of alcohol per kg body weight will induce a higher BAC (considered as the degree of intoxication) in women than in men (36). Because the average body weight of women is lower, even larger mean differences in BAC between men and women will occur upon the consumption of the same quantity of alcohol. The 'reference man' weighs 70 kg, 60% of which is body water (42 litres), whereas the 'reference woman' weighs 58 kg, 50% of which is body water (29 litres). So, on average, the amount of body water over which the alcohol can be distributed is appreciably larger in men than in women (81). Kohlenberg (40) found alcohol distribution volumes of 51 and 38 litres, respectively, calculated with Michaelis-Menten kinetics from the BACs of men and women.

2.3. Elimination

The term 'elimination' is used in two meanings in the literature on alcohol. Sometimes it stands for the mere excretion of unconverted alcohol, sometimes for the elimination of alcohol from the body including the transformation of alcohol in the liver. In this paper the term will be used in the latter sense. Alcohol metabolism is discussed in Chapter 3.

Practically all alcohol is oxidized to CO_2 and H_2O. A small proportion of the alcohol is excreted as such with urine, breath, sweat and tears. At all times the excretion of alcohol is a function of the actual BAC, so the total amount of excreted alcohol can be seen as a function of the area under the BAC-time curve (AUC). There is a non-linear relationship between oral ethanol dose and AUC (101). An oral dose of 40 g alcohol for a person of 70 kg results in an AUC of about 3 g·h/l.

Assuming a breath/blood partition coefficient of 1:2100 and a lung ventilation rate (in rest) of 8 l/min, 720 mg of alcohol – i.e. 2% of the dose of 40 g – is expired. Elimination via the lungs depends on breath volume.

Alcohol is not concentrated in urine (57). The blood/urine alcohol concentration ratio is around 1 (0.5–1.0 in the absorption phase, 1.2–2.0 in the elimination phase (75). With an average urine production of 1.5 l/h 0.5% of a dose of 40 g alcohol is estimated to be excreted via the kidneys.

The amount of alcohol extreted in sweat is negligible except in case of excessive perspiration owing to extreme physical exertion (0.5% at most (50)).

In the literature (albeit, as a rule, not in primary sources) varying percentages are given for the various elimination routes (12, 14, 29, 50, 57, 63, 72, 96). Generally speaking, 90–98% may be said to be oxidized, 1–5% is excreted via the lungs, and 1–5% is excreted via other routes (urine, sweat).

2.3.1. The blood alcohol curve

The height of a BAC at any given moment will depend on the amount of alcohol absorbed at that particular moment, its distribution over the body, and the amount of alcohol already converted and excreted. A great deal of research has been devoted to the course of the BAC curve. Wilkinson (101) has produced quite an adequate state-of-the-art report in this field. The BAC curve first shows a steep increase, followed – sometimes after maintaining a plateau – by a slow decrease (see Figs. 2.1 and 2.2). The increase is often called the 'absorption phase' and the decrease the 'elimination phase'. A descending curve, however, only indicates that elimination exceeds absorption. It does not mean that absorption has finished yet. Absorption may take six hours or more, whereas the BAC curve peaks much earlier. Holford

(29) found a linear relationship between oral alcohol dose and BAC peak, which can be expressed by $C_{peak}(mg/l) = 20 \times dose\ (g/70\ kg)$.

Views on the way the curve descends have changed through the years. Around 1934, Widmark stated that elimination follows zero-order kinetics, that is to say, the amount of alcohol eliminated is not related to BAC values. Expressed as a graph this corresponds with a straight line:

$$C = C_0 - \beta t$$

in which C represents the concentration found at a given moment t, C_0 the initial concentration, and β the slope. The theoretical initial concentration C_0 is found by extrapolating the linear phase of the BAC curve to the ordinate according to Widmark. Multiplying C_0 with the volume in which alcohol is distributed results in the (theoretical) initial quantity of alcohol $r M_k C_0$ in which M_k is the body mass and r is the Widmark factor, which reduces the body mass to the alcohol distribution mass. This quantity proves to be smaller than the quantity of alcohol consumed, especially in case of delayed absorption due to food in the stomach. This discrepancy is called the 'alcohol deficiency' problem which has been the subject of many hypotheses (26, 50). According to Heifer (26) the difference is, at least to some extent, the result of an unallowed procedure of linear extrapolation. There is no proof for a chemical reaction (esterification with fatty acids or amino acids from food, as suggested by Widmark). Possibly the recently found 'first-pass metabolism' of alcohol in the stomach (16) can explain the 'alcohol deficiency'.

According to Widmark, the liver alcohol dehydrogenase (ADH) level is saturated even if alcohol concentrations are low. If this is true, the slope of the straight line (β) must not be related to the amount of alcohol. However this assumption often turns out to be incorrect: frequently higher values of the slope are found when initial BAC values are higher (15, 101). Presumably it is not ADH but the amount of NAD^+ that reduces the speed of this reaction (43).

The Michaelis-Menten equation is currently preferred as a representation of the descent of the BAC (47, 95):

$$C_0 - C + K_M \ln C_0/C = V_m t$$

in which K_M stands for the Michaelis constant and V_m for the maximum velocity (all other symbols represent the same elements as in the Widmark equation).

The Michaelis-Menten equation expresses the experimentally obtained course of the curve more adequately than Widmark's straight line does, particularly the tail of the curve (Fig. 2.2). At concentrations much greater than the K_M the elimination rate approaches V_m and elimination is approximately of a zero order, i.e. elimination is independent of alcohol concentration. At low concentrations elimination is concentration-dependent. Ueno (93) found V_m to be significantly correlated to β_{60} although the former value was slightly higher. When BAC values exceed 0.1 to 0.2 g/l virtually all experimentally obtained curves are (pseudo)linear. Recently new models have been developed, which account for Michaelis-Menten elimination kinetics of alcohol as well as kinetics of acetaldehyde and acetate, which are defined by first-order processes (39).

Although an approach based on the Michaelis-Menten equation is more correct, it remains feasible to approach elimination of alcohol from the blood by Widmark's straight line. Either approach may be used, dependent on the specific case it is to be applied to. The zero-order kinetics approach is simple, but less precise since it can be applied only to BACs exceeding 0.2 g/l. When applied to oral consumption of alcohol, when absorption and distribution are often disturbing factors, this method usually is sufficiently accurate. The Michaelis-Menten equation covers the whole of the elimination phase, even when concentrations are low. Its application, however, requires more complicated calculations.

2.3.2. Elimination rate

Both Wallgrenn & Berry (96) and Piendl (66) provide a survey of the literature on the rate of elimination of alcohol from the blood. The elimination rate corresponds to the slope of the decrease of a BAC curve (β). Since this slope is rarely linear in experimentally established curves (not even when the BAC is 0.2 g/l or higher) the elimination rate is not constant in time. Consequently it is not correct to speak about *the* elimination rate. The average elimination rate is 100 mg/kg body weight per hour.

Individual differences, however, may be large, due to differences in genetic constitution, body weight and habitual alcohol consumption. Depending on individual elimination rate and body weight, an amount of 4.5–12 g alcohol per pour (an average of 7 g/h) may be metabolized, calculated on the basis of the slope of the descending BAC curve (β). Back extrapolation of venous blood alcohol levels, based on a single known concentration, as a means of determining the BAC at the time of the accident, is legally valid in many countries, but can lead to estimates bordering on wild guesses (51, 91).

A significant difference, both in β_{60} and V_m, is seen between persons with normal aldehyde dehydrogenase and persons with the deficient type of it (93). Piendl (66) and Kohlenberg (39) found no consistent differences in elimination rate between men and women as suggested by Jones & Jones (36) and Sutker (88). Regarding the decline of the BAC, Jones & Jones found that alcohol is metabolized more slowly by women using oral contraceptives than by men. Women who do not use oral contraceptives show a higher elimination rate than men (expressed in mg kg^{-1} h^{-1}). According to Sutker (88) the elimination rate depends on the phase of the menstrual cycle, with the highest elimination rate during the mid-luteal phase (days 20–25).

The elimination rate is not the only factor in determining the duration of alcohol intoxication. The maximum BAC value also plays a role. For one particular amount of alcohol consumed, this value is usally higher in women than in men (cf. Section 2.2.2).

The individual differences reported by Piendl (66) and by Wallgren & Berry (96) may be ascribed to differences in study design (quantity of alcohol, method of calculation), differences in body composition, metabolic differences between light and heavy drinkers (see Chapter 3), and the time of day at which experiments are performed. A circadian rhythm in alcohol metabolism has been found (55, 86). Around 18.00 the elimination shows a dip (about 10% less then the 24-hour mean) (55).

The rate of elimination of alcohol from the blood is influenced by a number of factors. The effect of food components (vitamins, sugars, caffeine, oxygenated water), dietary patterns, thyroid hormones, physical exercise and specific sobering agents will be discussed consecutively.

Vitamins
Supplementation over a period of 5–10 days with a mixture of vitamins, at a dose several-fold higher than recommended, did not have a significant effect on the elimination rate of a quantity of alcohol amounting to 0.5 per kg body weight (3–4 glasses) (60, 63). According to Susick (87), ascorbic acid enhances alcohol metabolism. Possibly the alcohol metabolism is slowed down by serious vitamin deficiency. On the other hand, supplementation of vitamins to subjects with a normal dietary pattern appears pointless.

Sugar
The sugars most frequently tested are galactose, xylose, glucose and fructose. Galactose and xylose do not affect the elimination rate, but glucose has a small effect on alcohol elimination (52). Fructose clearly affects alcohol elimination (11, 13, 35, 60, 62, 63, 84, 94, 100). It is not easy to arrive at definite conclusions on the dose-effect relation which largely depends on the way alcohol and fructose are administered (orally or intravenous) and on the amount of alcohol administered. The effect of fructose also shows a large individual variation. So, it is not supprising that in some experiments fructose was found to have no effect at all (24). The effect of fructose is probably based on a reduction of the NADH/NAD ratio by one or more reactions of fructose or one of its metabolites for which NADH is required. Fructose consequently affects both absorption (see Section 2.1.1) and elimination of alcohol.

Caffeine
The ingestion of two cups of strong coffee supplemented with 50 mg of caffeine has no influence on the elimination rate of alcohol (62). On the other hand, large quantities of caffeine do influence the effect of alcohol on the central nervous system (in animal experiments), but this effect appears to be fortifying rather than otherwise (9, 59).

Oxygenated drinking water
Oxygenated water speeds up alcohol elimination in monkeys by 60% (31). However, in a comparable experiment with human subjects, no significant results were found (41).

Dietary composition
Carbohydrate-rich (73) and protein-rich (102) diets enhance the alcohol elimination rate. Pawan (61) found that both a 4.2 MJ diet in which fat accounted for 90% of energy intake and a diet containing only water and vitamins, over a period of one week, reduced the alcohol elimination rate. This may be caused by mineral deficiencies and other co-factors.

Thyroid hormones
The administration of thyroid hormones which speed up the entire metabolism does not affect the rate of alcohol metabolism (63).

Sobering agents
Quite a number of sobering agents are available now, especially in Germany where drug registration policies and prescription are rather less strict. Research, however, has demonstrated that these agents do not have any effect (8, 9).

Physical exercise
Physical exercise seems unlikely to influence the elimination rate of alcohol. Most scientists found no effects (3, 41, 48, 62, 64, 69, 74), whereas some others observed an increase in the rate of elimination (77).

Body temperature
The alcohol elimination rate decreases with decreasing body temperature. On the other hand, a large dose of alcohol (3.6 g/kg body weight) impairs thermoregulation. Alcohol-intoxicated mice placed in cold environments are not capable of maintaining normal body temperature (6).

2.4. An overview of factors influencing the BAC

In the preceding sections on absorption, distribution and elimination several factors affecting these processes have been mentioned. All these processes taken together define the course of the BAC curve. In Table 2.1 an overview is presented of the factors affecting the BAC curve by their effects on absorption, distribution and elimination. Some of these factors are under self-control (quantity and type of beverage, meals), whereas other ones can be modified only in the long run (body weight, nutritional deficiencies) or not at all (sex, stature). A distinction should also be made between direct (e.g. gastric motility) and indirect factors (e.g. presence of

Table 2.1. Overview of factors possibly influencing the concentration of alcohol in the blood through their effects on absorption, distribution and elimination. Numbers indicate references.

Factor	Absorption	Distribution	Elimination
Easily affected by self-control			
Amount of alcohol	21, 38		
Alcohol concentration	21, 38		
Type of beverage (congener levels)	42, 56		
Drinking rate	38		
Gastric motility	21		
Physical activity	38		69, 77
Presence of food in the stomach	30, 44, 99		
Dietary pattern	78, 99		61, 73, 102
Fructose	12		13, 35, 52, 63, 94
Various drugs	38	38	43, 97
Clofibrate			103
Caffeine	80		
Affected by self-control only indirectly or in the long term			
MEOS induction			43, 63
Circulation	38	96	
Body temperature	38		6
Oral contraceptives			37
Body weight		63, 71	97
Circadian rhythm	45		55, 86
Menstrual cycle	36		88, 97
Nutritional deficiencies	18		97
Alcoholism			97
Liver damage			97
Not affected by self-control			
Race			97
Sex		36	88, 89, 90
Stature (% body water)		63	

food in the stomach). Popular and pseudo-scientific literature use to publish nomograms, graphs and tables by means of which BAC values may be determined after the consumption of any given quantity of alcohol. As a rule, these values are means. The variability of BACs, however, is so large that the application of these nomograms and the like is most deceptive and may give rise to huge over- and underestimates (58). Even in controlled experiments, in which as many factors as possible were maintained at a constant level, the inter- and intra-individual variance found was large (58, 98).

Summary

A small quantity of the alcohol consumed is absorbed from the stomach, whereas most of it is absorbed from the small intestine following first-order kinetics. Of the total amount of alcohol consumed 90% is absorbed within 1 to 3 hours, depending on the type and quantity of the alcoholic beverage, and the presence of food in the stomach. A delayed gastric emptying extends the overall absorption period.

The maximum blood alcohol concentration resulting from the consumption of wine or spirits is consequently reduced by the intake of a solid meal immediately preceding or during alcohol consumption. Although there are slight differences in effectiveness among the various nutrients and foodstuffs in this respect, no substance can be said to have a particularly strong effect, exept possibly fructose. Carbohydrates present in alcoholic beverages (beer, wine, sweet liqueur) also slow down gastric emptying to some extent.

The absorption of alcohol in the intestine is directly proportional to the actual alcohol concentration gradient. This causes alcohol from weak alcoholic beverages to be absorbed more slowly than alcohol from more highly concentrated drinks (dilution effect). The consumption of large quantities of high-alcohol beverages may cause pylorospasm, thus delaying absorption. Once the alcohol has been absorbed into the bloodstream it is transported all through the body and distributed over the total amount of body fluid. The small alcohol molecules, soluble in water, easily diffuse through membranes. An equilibrium is reached about 1–1½ hour after drinking, depending on the degree of blood perfusion of the organs. Body fat and skeletal mass hardly absorb any alcohol.

Men and women differ in body constitution. Women have less body water per kg body weight than men (500 and 600 ml/kg respectively). An identical quantity of alcohol per kg body weight will induce a higher BAC (measure of intoxication) in women than in men. Because women usually also have a lower body weight than men, the differences in BAC between the 'reference woman' and the 'reference man' will be even greater after ingestion of an identical amount of alcohol. The amount of body water of the reference women (58 kg) is only 69% of that of the reference man (70 kg).

The alcohol concentration in blood can also be determined in an indirect manner by means of breath analysis. Once a respiratory equilibrium has been established, the alcohol concentration in exhaled breath will be proportional to the (arterial) BAC; consequently, the blood alcohol concentration can be calculated from the breath value. The breath test is not invasive, is both easy and fast, but is less precise than the direct determination of BACs.

Of the alcohol absorbed 90 – 98% is oxidized, 1–5% is excreted in an unaltered state in urine, and another 1–5% is expired via the lungs. Two mathematical functions are widely used to describe the elimination of alcohol from the blood: a straight line (zero-order kinetics according to Widmark), and the Michaelis-Menten equation. The Michaelis-Menten equation expresses more adequately the experimentally obtained course of the curve than Widmark's straight line approach does. Particularly the tail of the curve fits the Michaelis-Menten equation. Yet it remains feasible to approach the elimination of alcohol from the blood by means of a straight line, depending on the specific application required (balancing precision and ease). The mean elimination rate from blood is 100 mg/kg bodyweight per hour (range 88–132 mg/kg per hour. Depending on individual elimination rate and body weight, an amount varying between 4.5 g and as much as 12 g alcohol may be processed per hour (average 7 g/h). There are no marked differences in elimination rates (expressed in mg/kg per hour) between men and women. Differences in elimination rates relate to, among other factors, habituation, the use of large quantities of fructose, the time of the day and, for women, the use of oral contraceptives.

References

1. Anonymous. Defeating the breathalyser. Br Med J 1972;4: 744-5.
2. Anonymous. Informatorium voor diëtiek. 1976;2: IVa-11.
3. Barnes EW et al. Observations on the metabolism of alcohol in man. Br J Nutr 1965;19: 485-9.
4. Beck IT, Dinda PK. Acute exposure of small intestine to ethanol; effects on morphology and function. Dig Dis Sci 1981;26: 817-38.
5. Beckett AH, Mitchard M, Saunders A. Does drink men drunk? Autocar 1971;135: 34-5.
6. Bejanian M, Finn DA, Syapin PJ, Alkana RL. Body temperature and ethanol pharmacokinetics in temperature-challenged mice. Alcohol Int Biomed J 1990;7: 331-7.
7. Berggren SM, Goldberg L. The absorption of ethyl alcohol from the gastro-intestinal tract as a diffusion process. Acta Physiol Scand 1940;1: 246-70.
8. Besserer K, Springer E. Zur Wirksamkeit des 'Alkoholgegenmittels' SOBARO. Blutalkohol 1971;8: 122-32.
9. van der Brink G, de Gier JJ. Ontnuchteringsmiddelen. Voorburg: SWOV, 1979.
10. Broitman SA, Gottlieb LS, Vitale JJ. Augmentation of ethanol absorption by mono- and disaccharides. Gastroenterology 1976;70: 1101-7.
11. Brown SS, Forrest JAH, Roscoe P. A controlled trial of fructose in the treatment of acute alcoholic intoxication. Lancet 1972; ii: 898-9.
12. Clark ER, Hugnes IE, Letley E. The effect of oral administration of various sugars on blood ethanol concentrations in man. J Pharm Pharmacol 1973;25: 319-23.
13. Crownover BP, La Dine J, Bradford B, Glassman E, Forman D, Schneider H, Thurman RG. Activation of ethanol metabolism in humans by fructose: importance of experimental design. J Pharmacol Exp Ther 1986;236: 574-9.
14. Deurenberg P. Voeding en alcohol. Ned Tijdschr Diet 1980;35: 259-66.
15. Eggleton MG. Some factors affecting the metabolic rate of alcohol. J Physiol 1940;98: 239-54.
16. Frezza M, di Padova C, Pozzato G, Terpin M, Baraona E, Lieber CS. High blood alcohol levels in women: the role of decreased gastric alcohol dehydrogenase activity and first-pass metabolism. N Engl J Med 1990; 322: 95-9.
17. Goldberg L, Jones AW, Neri A. Effects of a sugar mixture on blood ethanol profiles and on ethanol metabolism in man. Blutalkohol 1979;16: 431-8.
18. Green PHR. Alcohol, nutrition and malabsorption. Clin Castroenterol 1983;12: 563-74.
19. Gullberg RG. Breath alcohol test precision: an in vivo vs. in vitro evaluation. Forensic Sci Int 1989;43: 247-55.
20. Haeckel R, Bucklitsch I. The comparability of ethanol concentrations in peripheral blood and saliva: the phenomenon of variation in saliva to blood concentration ratios. J Clin Chem Clin Biochem 1987;25: 199-204.
21. Haggard HW, Greenberg LA, Lolli G. The absorption of alcohol with special reference to its influence on the concentration of alcohol appearing in the blood. Q J Stud Alcohol 1941;1: 684-726.
22. Halsted CH, Robles EA, Mezey E. Distribution of ethanol in the human gastrointestinal tract. Am J Clin Nutr 1973;26: 831-4.
23. Harger RN, Hulpieu HR. Extent of absorption of alcohol at various intervals after oral administration. Proc Soc Exp Biol 1935;32: 1247-9.

24. Hartmann H, Förster H. Einfluss von Fruktose auf dem Alkoholumsatz. Ernährungsumschau 1980;27: 48.
25. Heifer U. Untersuchungen zur Differenz zwischen Atem- und Blutalkoholkonzentration in der Anflutungsphase. Blutalkohol 1982;19: 29-37.
26. Heifer U, Wehner HD. Zur Frage des Ethanol-'Resorptionsdefizits' = The problem of ethanol absorption deficiency. Blutalkohol 1988;25: 299-309.
27. Heller MB, Kaplan RM. Dipstick measurement of alcohol levels in body fluids. Ann Intern Med 1987;106: 328.
28. Hernández-Munoz R, Caballeria J, Baraona E, Uppal R, Greenstein R, Lieber CS. Human gastric alcohol dehydrogenase: its inhibition by H2-receptor antagonists, and its effect on the bioavailability of ethanol. Alcohol Clin Exp Res 1990;14: 946-50.
29. Holford NHG. Clinical pharmacokinetics of ethanol. Clin Pharmacokinet 1987;13: 273-92.
30. Horowitz M, Maddox A, Bochner M, Wishart J, Bratasiuk R, Collins P, Shearman D. Relationships between gastric emptying of solid and caloric liquid meals and alcohol absorption. Am J Physiol 1989;257: G291-G298.
31. Hyvärinen J, Laakso M, Sippel H, Roine R, Huopaniemi T, Leinonen L, Hytönen V. Alcohol detoxification accelerated by oxygenated drinking water. Life Sci 1978;22: 553-60.
32. Israel Y, Valenzuela JE, Salazar I, Ugarte G. Alcohol and amino acid transport in the human small intestine. J Nutr 1969;98: 222-4.
33. Janus ED, Sharman JR. Milk and blood alcohol. N Z Med J 1972;75: 339-42.
34. Jones AW. How breathing technique can influence the results of breath-alcohol analysis. Med Sci Law 1982;22: 275-80.
35. Jones AW. Effects of fructose, glucose, and mixed sugars on ethanol detoxification and blood glucose response in rats. Med Biol 1983;61: 319-23.
36. Jones BM, Jones MK. Women and alcohol: intoxication, metabolism and the menstrual cycle. In: Schuckit MA, Greenblatt, M, eds. Alcoholism problems in women and children. New York: Grune and Stratton, 1976: 103-36.
37. Jones MK, Jones BM. Ethanol metabolism in women taking oral contraceptives. Alcoholism Clin Exp Res 1984;8: 24-8.
38. Kalant H. Absorption, diffusion, distribution, and elimination of ethanol: effects on biological membranes. In: Kissin B, Begleiter H, eds. The biology of alcoholism I. New York: Plenum Press, 1971: 1-62.
39. Kohlenberg-Müller K, Bitsch I. Neue Methoden zur pharmakokinetischen Beschreibung des Alkohols und seiner Metaboliten bei weiblichen und männlichen Versuchspersonen = New methods for the pharmacokinetic description of alcohol and its metabolites in male and female volunteers. Blutalkohol 1990;27: 40-48
40. Laakso M, Huopaniemi T, Hyvärinen J, Lindros K, Roine R, Sippel H, Ylikahri R. Inefficacy of oxygenated drinking water in accelerating ethanol elimination in humans. Life Sci 1979;25: 1369-71.
41. Lang K. Biochemie der Ernährung. Darmstadt: Dietrich Steinkopff Verlag, 1974.
42. Leake CD, Silverman M. Alcohol beverages in clinical medicine. Chicago: Year Book Medical Publishers, 1965.
43. Lieber, CS. Ethanol metabolism and toxicity. Rev Biochem Toxicol 1983;5: 267-311.
44. Lin Y-J et al. Effects of solid food on blood levels of alcohol in man. Res Comm Chem Path Pharmacol 1976;13: 713.
45. Lötterle J, Husslein EM, Bolt J, Wirtz PM. Tageszeitliche Unterschiede der Alkoholresorption = Variations in alcohol absorption during the day. Blutalkohol 1989;26: 369-75.
46. Lund A. The secretion of alcohol in the tear fluid = Alkohol in Tränenflüssigkeit. Blutalkohol 1984;21: 51-4.
47. Lundquist F, Wolthers H. The kinetics of alcohol elimination in man. Acta Pharmacol Toxicol 1958;14: 265-89.
48. Mallach HJ. Alcohol und körperliche Belastung. Dt Med Wochenschr 1983;108: 958-9.
49. Mallach HJ, Hartman H, Schmidt V. Alkoholwirkung beim Menschen: Pathophysiologie, Nachweis, Intoxication, Wechselwirkungen. Stuttgart: Thieme Verlag, 1987.

50. Mallch HJ. Konzentrationsverlauf im Blut: Pharmakokinetik. In: Mallach HJ, Hartman H, Schmidt V. Alkoholwirkung beim Menschen: Pathophysiologie, Nachweis, Intoxication, Wechselwirkungen. Stuttgart: Thieme Verlag, 1987: 13-34.
51. Martin E, Moll W, Schmidt P, Dettli L. The pharmacokinetics of alcohol in human breath, venous and arterial blood after oral ingestion. Eur J Clin Pharmacol 1984;26: 619-26.
52. Mascord D, Smith J, Starmer GA, Whitfield JB. Effect of oral glucose in the rate of metabolism of ethanol in humans. Alcohol Alcoholism 1988; 23: 365-70.
53. Millan MS, Morris GP, Beck IT, Henson JT. Villous damage induced by suction biopsy and by acute ethanol intake in the normal human small intestine. Dig Dis Sci 1980;25: 513-25.
54. Miller DS, Stirling JL, Yudkin J. Effect of ingestion of milk on concentrations of blood alcohol. Nature 1966;212: 1051.
55. Minors DS, Waterhouse JM. Ethanol pharmacokinetics in healthy man : Michaelis-Menten parameters and the circadian rhythm. Chronobiologia 1985;12: 137-44.
56. Newman H, Abramson M. Absorption of various alcoholic beverages. Science 1942;96: 43-4.
57. Olson RE. Absorption, metabolism, and excretion of ethanol including effects on water balance and nutritional status. In: Gastineau CF, Darby WJ, Turner TB, eds. Fermented food beverages in nutrition. New York: Academic Press, 1979: 197-212.
58. O'Neill B, Williams AF, Dubowski KM. Variability in blood alcohol concentrations; implications for estimating individual results. J Stud Alcohol 1983;44: 222-30.
59. Osborne DJ, Rogers Y. Interactions of alcohol and caffeine on human reaction time. Aviat Space Environ Med 1983: 528-34.
60. Pawan GLS. Vitamins, sugars and ethanol metabolism in man. Nature 1968;220: 374-6.
61. Pawan GLS. Effects of the antecedent diet on the rate of metabolism of alcohol (ethanol) in man. Proc Nutr Soc 1968;27: 58A-59A.
62. Pawan GLS. Alcohol metabolism in man: acute effects of physical exercise, caffeine, fructose, and glucose on the rate of alcohol metabolism. Biochem J 1967;106.
63. Pawan GLS. Metabolism of alcohol (ethanol) in man. Proc Nutr Soc 1972;31: 83-9.
64. Pawan GLS. Physical exercise and alcohol metabolism in man. Nature 1968;218: 966-7.
65. Payne JP. Observations on the distribution of alcohol in blood, breath and urine. Br Med J 1966;i: 196-202.
66. Piendl A. Über die Grenze des Massvollen Verzehrs und des Überkonsums an Alkohol (I). Brauwelt 1979;119: 192-6.
67. Pieters JE, Wedel M, Schaafsma G. Parameter estimation in a three-compartment model for blood alcohol curves. Alcohol Alcoholism 1990; 25: 17-24.
68. Pihkanen TA. Neurological and physiological studies on distilled and brewed beverages. Dissertation, Helsinki, 1957.
69. Pikaar NA et al. Onderzoek naar de invloed van geslacht, inspanning, maaltijdgebruik, dosering, concentratie en tijdstip van consumptie op de verwerking van oraal toegediende doses alcohol bij 12 proefpersonen. Zeist: CIVO-Instituten TNO, 1984.
70. Raff G, Schmidt V, Staak M. Vergleichende Untersuchungen über den Verlauf von Atem- und Blutalkoholkurven. Blutalkohol 1980;17: 57-63.
71. Reed TE, Kalant H, Gibbins RJ, Khanna BM. Alcohol and acetaldehyde metabolism in Caucasians, Chinese and Amerinds. Can Med Assoc J 1976;115: 851-5.
72. Ritchie JM. The aliphatic alcohols. In: Goodman, Gilman, eds. The pharmacological basis of therapeutics. London: McMillan, 1970: 135-50.
73. Rogers J, Smith J, Starmer GA, Whitfield JB. Differing effects of carbohydrate, fat and protein on the rate of ethanol metabolism. Alcohol Alcoholism 1987;22: 345-53.
74. Sautier C et al. Anatomie et physiologie humaine: effet de l'exercise musculaire sur l'oxidation de l'éthanol chez l'homme. C R Acad Sci Paris 1982;295: 603-6.
75. Schmidt V. Alkoholnachweis im Körper. In: Mallach HJ, Hartmann H, Schmidt V. Alkoholwirkung beim Menschen: Pathophysiologie, Nachweis, Intoxikation, Wechselwirkungen. Stuttgart: Georg Thieme, 1987: 54-69.
76. Schulz E, Magerl H, Vock R. Der Alkoholgehalt des Speichels und seine Verwertbarkeit = Alcohol content of saliva and its utilization. Blutalkohol 1986;23: 55-63.
77. Schürch PM et al. The influence of moderate prolonged exercise and a low carbohydrate diet on ethaol elimination and on metabolism. Eur J Appl Physiol 1982;48: 407-14.

78. Sedman AJ et al. Food effects on absorption and metabolism of alcohol. J Stud Alcohol 1976;37: 1197-214.
79. Serianni E, Cannizzaro M, Mariani A. Blood alcohol concentrations resulting from wine drinking timed according to the dietary habits of Italians. Q J Stud Alcohol 1953;14: 165-73.
80. Siegers CP, Strubelt O, Back G. Inhibition by caffeine of ethanol absorption in rats. Eur J Pharmacol 1972;20: 181-7.
81. Snijder WS et al. Report of the task group on reference man. Oxford: Pergamon Press, 1975.
82. Southgate HW. The effect of alcohol, under varying conditions of diet, on man and animals, with some observations on the fate of alcohol in the body. Biochem J 1925;19: 737-45.
83. Spector NH. Alcohol breath tests: gross errors in current methods on measuring alveolar gas concentration. Science 1971;172: 57-9.
84. Sprandel U, Träger HD, Liebhardt EW, Zöllner N. Acceleration of ethanol elimination with fructose in man. Nutr Metab 1980;24: 324-30.
85. Stefenelli N, Priessnitz E, Klotz H, Hahn S. Die Beeinflussbarkeit der Alkoholresorption durch Änderung der Magenentleerung. Wien Klin Wochenschr 1977;89: 161-3.
86. Sturtevant FM, Sturtevant RP. Chronopharmacokinetics of ethanol. In: Majchrowicz E, Noble EP, eds. Biochemistry and pharmacology of ethanol, vol 1. New York: Plenum Press, 1979: 27-40.
87. Susick RL, Zannoni VG. Effect of ascorbic acid on the consequences of acute alcohol consumption in humans. Clin Pharm Ther 1987;41: 502-9.
88. Sutker PB, Goist KC, Allain AN, Bugg F. Acute alcohol intoxication: sex comparisons on pharmacokinetic and mood measures. Alcohol Clin Exp Res 1987;11: 507-12.
89. Teschke R, Wiese B. Sex-dependency of hepatic alcohol metabolizing enzymes. J Endocrinol Invest 1982;5: 243-50.
90. Teschke R, Heymann K. Effect of sex hormones on the activities of hepatic alcohol-metabolizing enzymes in male rats. Enzyme 1982;28: 268-77.
91. Taberner PV. Pharmacokinetics of alcohol and the law. Trends Pharmacol Sci 1988;9: 47-8.
92. Trela FM. Untersuchungen zur Äthanolverteilung in den Körperflüssigkeiten des Menschen unter rechtsmedizinischen Aspekten = Studies on the distribution of ethyl alcohol in human body in forensic aspect. Blutalkohol 1989;26: 305-18.
93. Ueno Y, Mizoi Y, Yamamoto K, Adachi J, Fujiwara S, Nakagawa K. Evaluation of ethanol elimination using pharmacokinetic models. Alcohol Clin Exp Res 1990;14: 347.
94. Wade A, Reynolds JEF. Martindale: the extra pharmacopoeia. London: Pharmaceutical Press, 1977;27: 36-42.
95. Wagner JG et al. Elimination of alcohol from human blood. J Pharm Sci 1976;65: 152-4.
96. Wallgren H, Berry H. Actions of alcohol. Amsterdam: Elsevier, 1970;1.
97. von Wartburg JP. The metabolism of alcohol in normals and alcoholics: enzymes. In: Kissin B, Begleiter H, eds. The biology of alcoholism, vol 1. New York: Plenum Press, 1971: 63-102.
98. Wedel M, Pikaar NA, van Dokkum W. Binnen- en tussenpersoons variabiliteit van de bloedalcoholcencentratie na consumptie van 3 glazen gesimuleerde wijn. Alcoholbelastingsproeven bij 12 proefpersonen. Zeist: CIVO-Instituten TNO, 1983.
99. Welling BG, Lyons LL, Elliot R, Amidon GL. Pharmacokinetics of alcohol following single low doses to fasted and nonfasted subjects. J Clin Pharmacol 1977;17: 199-206.
100. Whitfield J, Smith J, Mascord D, Starmer G. Effects of fructose and glucose on alcohol metabolism and redox state. Alcohol Clin Exp Res 1990;14: 352 (abstract 532).
101. Wilkinson PK. Pharmacokinetics of ethanol: a review. Alcohol Clin Exp Res 1980;4: 6-21.
102. Wissel PS. Dietary influences on ethanol metabolism. Drug Nutr Interact 1987;5: 161-8.
103. Yamauchi K et al. Effect of clofibrate and phenobarbiturate on clearance of ethanol and acetaldehyde in blood of rats. Jpn J Alcohol Stud Drug Depend 1981;16: 225-33.

CHAPTER 3

Alcohol metabolism

W.G. Vrij-Standhardt

3.1. Introduction

Many reviews have been devoted to alcohol metabolism (e.g. 4, 36, 39–41, 68, 92). Of the alcohol absorbed 90–98% is metabolized, the remainder is excreted as such (cf. Chapter 2). Most of the alcohol is oxidized completely to CO_2 and H_2O. A small proportion is first converted into acetate and then incorporated in body tissues as carbohydrates, proteins or lipids (see Fig. 3.1).

If completely converted via the ADH pathway, alcohol yields 30 kJ/g energy. In case of habituation to large quantities of alcohol, it yields less energy owing to a different metabolic mechanisms. In comparison, the energy yield of both carbohydrates and proteins is 17 kJ/g, and of fats 38 kJ/g.

This provides a direct answer to the question whether alcohol is a nutrient. Nutrients are defined as the chemical elements constituting a food (any consumable product) which can be used for growth, maintenance and/or functions of the human body, irrespective of whether these elements can be synthesized by the body to a sufficient extent. Ethanol fits in with this definition although it should be noted that it is also a psychotropic substance. Alcoholic beverages may be considered as foodstuffs which are neither useful nor indispensable to the anabolism of the human body. The same holds for sugar, but, unlike sugar, alcohol has psychotropic properties in addition to its nutritional value.

Alcohol is oxidized in several steps; it is oxidized first to acetaldehyde, which is

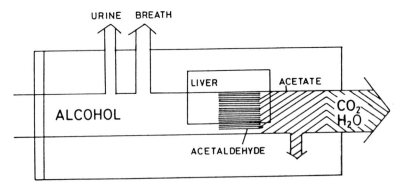

Fig. 3.1. The alcohol pathway.

subsequently converted into acetate. These reactions take place primarily in the liver (38, 68, 94). During oxidation to acetate ca. 30% of the total energy produced through the combustion of alcohol will be released. When ADH activity is maximal, this will be sufficient to provide 70–100% of the energy needed for the basal metabolism of the liver (92). The remaining energy will be released during extra-hepatic oxidation of acetate.

Although these fundamental pathways for alcohol metabolism are common to all human beings, the enzymes involved can take different forms – so-called isoenzymes or isozymes – which result from the substitution of one or more amino acids in the polypeptide chain. The isozymes have different catalytic characteristics which accounts for individual variation in rate of alcohol metabolism.

In the following sections will be discussed: the conversion of alcohol into acetaldehyde (first step); the intermediate product acetaldehyde; conversion of acetaldehyde into acetate (second step); further reactions of acetate (third step). Finally, the possibilities of influencing alcohol metabolism will be discussed.

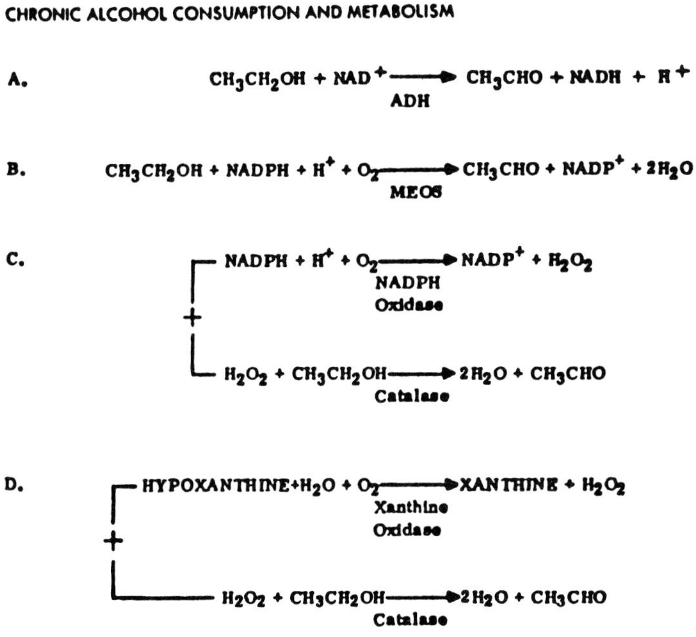

CHRONIC ALCOHOL CONSUMPTION AND METABOLISM

A. $CH_3CH_2OH + NAD^+ \xrightarrow{ADH} CH_3CHO + NADH + H^+$

B. $CH_3CH_2OH + NADPH + H^+ + O_2 \xrightarrow{MEOS} CH_3CHO + NADP^+ + 2H_2O$

C.
$\left[\begin{array}{l} NADPH + H^+ + O_2 \xrightarrow{\text{NADPH Oxidase}} NADP^+ + H_2O_2 \\ \\ H_2O_2 + CH_3CH_2OH \xrightarrow{\text{Catalase}} 2H_2O + CH_3CHO \end{array}\right.$

D.
$\left[\begin{array}{l} HYPOXANTHINE + H_2O + O_2 \xrightarrow{\text{Xanthine Oxidase}} XANTHINE + H_2O_2 \\ \\ H_2O_2 + CH_3CH_2OH \xrightarrow{\text{Catalase}} 2H_2O + CH_3CHO \end{array}\right.$

Fig. 3.2. Three enzyme systems converting ethanol into acetaldehyde. Ethanol is oxidized in the liver by: A, alcohol dehydrogenase (ADH), the oxidized form of nicotinamide adenine dinucleotide, NAD (NAD$^+$) and the reduced form of NAD (NADH); B, the microsomal ethanol-oxidizing system (MEOS), the reduced form of nicotinamide adenine dinucleotide phosphate, NADP (NADPH) and the oxidized form of NADP (NADP$^+$); C, a combination of NADPH oxidase and catalase; or D, xanthine oxidase and catalase. After: Lieber et al. (40).

3.2. First step: conversion of ethanol into acetaldehyde

Three enzyme systems are now known to catalyse the conversion of ethanol into acetaldehyde (see Fig. 3.2). The main reaction is the one in which ADH functions as the enzyme. The other two systems, i.e. the NADPH-dependent reaction catalysed by MEOS (microsomal ethanol-oxidizing system) and the reaction influenced by catalases, do not play an important part, although it has not been established to what extent they exactly contribute to the process in question.

The presence of three independent alcohol-metabolizing systems has been demonstrated convincingly (33, 40, 56, 83). They differ in intracellular localization, pH optimum, K-value, co-ferments and specific inhibitors.

3.2.1. ADH-dependent reaction

For the oxidation of alcohol to acetaldehyde by alcohol dehydrogenase (ADH), oxodized nicotinamide adenine dinucleotide (NAD^+), a hydrogen receptor, is needed. The chemical equation is:

$$\underset{\text{(ethanol)}}{CH_3CH_2OH} + NAD^+ \overset{ADH}{\rightleftharpoons} \underset{\text{(acetaldehyde)}}{CH_3CHO} + NADH + H^+$$

The NADH thus produced will be reoxidized to NAD^+ through various shuttle systems in the mitochondria. It can be calculated that the activity of rat liver ADH is limited to only 50–60% of its maximum velocity. Regeneration of NAD is considered the rate-limiting factor. Which step in this regeneration process exactly is rate-limiting, depends on the experimental design (species used, starved or fed animals). The dissociation of the ADH-NADH complex, the rate of which is controlled by the amino acid on position 47 of ADH, is seen as the rate-limiting factor (22). Re-oxidation of NADH to NAD^+ is not rate-limiting in man (54, 57). The importance of the respiratory chain activity and the transfer rate of reducing equivalents from the cytosol to the mitochondria by shuttle systems is not clear (28, 36).

ADH is not induced by chronic alcohol consumption (6). Norman et al. (59) even found a decrease in ADH activity in baboons fed alcohol for a long time. In a minority of people there is a swift and transient increase in alcohol metabolism (SIAM) of more than 40% when a second dose of alcohol is given a few hours after the first dose. SIAM most likely involves the increased reoxidation of NADH, as a result of activation of mitochondrial electron transport by enhanced adenosine diphosphate (ADP) supply (85).

The zinc-containing ADH has a low substrate specificity: it converts not only ethanol, but also other alcohols (69). There are several forms of ADH (isozymes), with different specific activities. The existence of five genes, coding for different variants of ADH, has been demonstrated. Three of these genes code for isozymes which are important for ethanol metabolism (class I enzymes). The other two genes code for other isozymes (class II and III enzymes) with a very low affinity for ethanol (22). New techniques have allowed to demonstrate a high degree of sequence homology between the 3 class I ADH genes and enabled the definition of ADH variants at the DNA sequence level (4a). On chromosome 4 an ADH gene cluster is described. All class I ADH genes are mapped in this specific region of chromosome 4 (9).

Individual variation in alcohol metabolism and tolerance can be partly explained by biochemical and molecular differences between the class I enzymes. Large individual differences in quantitative distribution of isozymes, and hence in ADH activity, have been demonstrated (93). Racial differences have been noted as well (2, 18, 19). Almost all Caucasians have the ADH_2^1 enzyme in addition to some other isozymes, whereas ca. 90% of the Japanese have an ADH isozyme with an activity at physiological pH values that is about a hundred times greater than the ADH_2^1 activity (101).

ADH occurs mainly in the cytoplasm of the liver, but several other tissues contain ADH isozymes as well (7, 67, 76). Local metabolism of alcohol in gastrointestinal tissue by ADH (called 'first-pass metabolism') has been reported. Calculated from the differences in AUC ('area under the blood alcohol curve') between intravenously and orally administered alcohol, first-pass metabolism could account for about 20% of alcohol oxidation (20). In women, first-pass metabolism, defined by AUC difference, is only 25% of the level in men (20). This could explain the higher blood alcohol concentration (BAC) in women after the same dosis of alcohol per kg body weight. In alcoholics gastric alcohol metabolism is only half that of non-alcoholics. However, the use of AUC differences for accurately determining the bioavailability of alcohol is not appropriate according to others (90). The gastric ADH activity in women has been found to be 70% of that in men. Ageing and, in particular, alcoholism affect gastric ADH activity (14).

3.2.2. MEOS

In 1965, Orme-Johnson & Ziegler (64) reported that ethanol may also be oxidized by an NADPH- and oxygen-dependent enzyme system in the microsomes of the liver, the so-called microsomal ethanol-oxidizing system (MEOS). This finding has been confirmed by others (45, 70). Its chemical equation is:

$$CH_3CH_2OH + NADPH + H^+ + O_2 \xrightleftharpoons{MEOS} CH_3CHO + 2H_2O + NADP^+$$

There is convincing evidence for the involvement of cytochrome P-450 in MEOS activity (83). The ethanol-oxidizing capacity is constructed in vitro by means of cytochrome P-450, lecithin (phospholipids) and NADPH-cytochrome P-450 reductase. Part of the P-450 system is inducible by ethanol. These isozymes are called P450IIE1. Since other cytochrome isozymes can also contribute to ethanol oxidation, the term MEOS is used as a generic term for the overall capacity of microsomes to oxidize ethanol rather than for the capacity of the P450IIE1 fraction (42). Many xenobiotics are metabolized by the cytochrome P-450 system. P450IIE1 has a high capacity of metabolizing not only ethanol, but also other aliphatic alcohols as well as other exogenous compounds. It can be induced by various compounds such as acetone, pyrazole and benzene (83).

Under normal conditions MEOS plays a minor role (less than 10% according to Wallgren & Berry (92), 20–25% according to Lieber (43), depending on the alcohol concentration). Chronic alcohol consumption results in enzyme induction and thus in an increased alcohol metabolism (43, 59, 66). Some suggest that MEOS, in combination with the catalase system, which will be discussed later, may be responsible for the catabolism of over 50% of the amount of alcohol consumed (5, 83).

The increased alcohol elimination in chronic alcohol consumption may be partly caused by a secondary increase in oxidation via ADH. The ADH pathway requires NAD^+, whereas NADPH is converted into $NADP^+$ via the MEOS pathway. The $NADPH/NADP^+$ and $NADH/NAD^+$ systems are linked by several shuttle systems. New NAD^+ may thus be generated by the MEOS reaction. In accordance with this hypothesis is the finding that acute administration of alcohol to rats that have become adapted to alcohol results in a smaller shift to the reduced state in the liver compared with the shift observed in unadapted control animals. In well-fed baboons to which alcohol was administered chronically, the elimination of alcohol from the blood increased progressively while the associated redox change decreased.

The conversion of alcohol into acetaldehyde by MEOS requires energy, whereas the formation of acetaldehyde by ADH (during which process NAD^+ is reduced) yields energy. If the first step proceeds via the MEOS pathway only 10 instead of 16 mol ATP per mol ethanol is formed during the complete conversion of alcohol. The MEOS pathway operates at subnormal efficiency (23% instead of 37%), producing less energy and more heat than the ADH pathway. Since the latter energy will not count, only 18.5 kJ becomes available per g alcohol instead of the 29.8 kJ/g usually believed to be the combustion energy of alcohol. The consumption of 180 g of pure alcohol per day (not uncommon in alcoholics) yields ca. (180×29.8 =) 5400 kJ/day via the ADH pathway; if half of it is converted via the MEOS pathway, the same quantity of alcohol yields only (90×29.8 + 90×18.5 =) 4400 kJ/day (6). It is, however, hard to explain why subjects who, over a period of one month, have received daily 8.4 MJ of energy derived from alcohol (282 g × 29.8 J/g) in addition to a basic diet supplying 9.2 MJ, gain hardly any weight, whereas an additional daily supply of 8.4 MJ in the form of chocolate results in a significant weight gain (44). Even if half of this huge dose of alcohol is metabolized via the MEOS pathway it should yield (141×29.8 J + 141×18.5 J =) 6.8 MJ.

A study of Reinus (72) suggests an alternative explanation for the absence of an effect of alcohol on body weight. Alcoholics receiving 60% of their energy supply in the form of alcohol administered nasogastrically lost weight as compared with the control period with an isoenergetic supply of carbohydrate (glucose). The results of the energy balance measurements indicated that this weight loss could not be attributed to a negative energy balance caused by excessive heat loss. Alcohol supplied only 5% less energy, relative to carbohydrate. This difference is primarily caused by respiratory and urinary losses of unmetabolized alcohol. A small and insignificant component of this difference was due to a higher heat release during ethanol feeding relative to carbohydrate infusion. Weight loss was due to a negative effect of alcohol on the N balance. Urinary analysis suggested that skeletal muscle breakdown contributed to the loss of N.

3.2.3. Catalase and other pathways

The third way via which alcohol may be converted into acetaldehyde is by means of catalase enzymes, which are present in peroxisomes of the liver. As early as 1936, Keilin and co-workers (see 88) pointed out that under certain conditions peroxidation of ethanol may take place, according to the following equation:

$$H_2O_2 + CH_3CH_2OH \rightleftharpoons 2H_2O + CH_3CHO$$

The peroxide needed is generated by NADPH or hypoxanthine oxidation under the influence of oxidases (39), or by β-oxidation of fatty acids (23, 85). The formation of H_2O_2 is considered to be the rate-limiting factor (40). To what extent the catalase system contributes to alcohol conversion is open to controversy. Most workers (10, 41, 44, 55) consider catalase to be of little significance, others mention values of up to 25% for the extent to which catalase accounts for alcohol conversion (23, 33). According to Thurman & Handler (85), MEOS is the combination of cytochrome P-450 (especially isozyme P450IIE1) and catalase, whereas cytochrome P450IIE1 plays a minor role in ethanol elimination in vivo. Earlier experiments designed to understand pathways of ethanol metabolism did not include fatty acids; therefore the role of catalase in alcohol metabolism has been underestimated. Thurman & Handler's recent experiments suggest that catalase plays an important role in ethanol oxidation in the presence of fatty acids in vivo (in rats). To arrive at that conclusion they suppose that methanol is not oxidized by ADH in the rat and that methanol oxidation occurs predominantly via catalase-H_2O_2. However, methanol is metabolized by both ADH and MEOS (83).

Several minor pathways of ethanol metabolism have been described (36). A non-oxidative pathway for alcohol metabolism is known, which results in the formation of fatty acid ethyl esters from ethanol and fatty acids. This reaction is catalysed by fatty acid ester synthase, which is found primarily in brain, pancreas and heart (34, 35, 52). Other reactions are conjugation reactions to form ethyl sulphate and ethyl glucoronide. These reactions appear to be quantitatively insignificant under normal conditions.

3.3. Acetaldehyde

The first intermediate product in alcohol conversion is acetaldehyde (CH_3CHO). It is a toxic substance held to be partly responsible for the harmful effects of excessive alcohol consumption (71, 95). Disulfiram, an antidipsotropic drug, inhibits acetaldehyde conversion, resulting in high concentrations of this substance. In patients taking disulfiram alcohol consumption actually has the effect of acute acetaldehyde poisoning. Acetaldehyde is supposed to play a role in hepatocellular injury (cf. Chapter 8 of this book) and in alcohol dependence, and possibly in intoxication and withdrawal symptoms. The cause of emotional and behavioural disturbances as a consequence of alcohol consumption is presumably the generation of morphine-like substances and/or 'false' neurotransmitters (12, 62, 71) – substances formed by condensation reactions of acetaldehyde and influencing the nervous system. According to others (50), these disturbances are caused by ethanol itself, which brings about a modification of the synaptic membranes (see also Chapter 9).

Following an intravenously administered dose of alcohol, alcoholics are found to have more acetaldehyde in their breath and blood than social drinkers (32, 46). There are several interpretations of this phenomenon (49). One interpretation is that the increased acetaldehyde level is a consequence of alcoholism; chronic alcohol consumption may have influenced metabolism to such an extent, either temporarily or permanently, that more acetaldehyde is formed. The type of transformation involved is not clear. It may be either a decrease in aldehyde dehydrogenase activity, owing to acetaldehyde-induced damage of the mitochondria, a decrease of the NAD^+-regenerating potential, or an accelerated alcohol oxidation (MEOS induction) (47, 96).

Others consider the increased acetaldehyde level as a cause of alcoholism: even before their addiction has become established, alcoholics would have a deviant metabolism, producing more acetaldehyde. This would be in accordance with Schuckit's finding (77) that first-degree relatives of alcoholics have a higher blood acetaldehyde level.

The reliability of the determination of acetaldehyde concentrations has been subject for debate (16, 47, 51). It has recently become evident that acetaldehyde concentrations determined by older methods are often artefacts brought about by the acetaldehyde generated during analysis.

3.4. Second step: conversion of acetaldehyde into acetate

During the second step in the oxidation process the acetaldehyde generated is converted into acetate by the NAD-dependent enzyme aldehyde dehydrogenase (AlDH).

Of the acetaldehyde 90–95% is converted in the liver (97). Almost all other tissues also contain AlDH; however, its extrahepatic activity is low (17).

Human AlDH is a tetrameric enzyme with a molecular weight of ca. 230 000. In the liver at least four different isozymes exist, which differ in structural and functional properties such as primary structure, molecular weight, electrophoretic migration, isoelectric point, catalytic constants and subcellular distribution. $AlDH_2$ has the highest affinity to alcohol (a low Michaelis constant, K_m). AlDH$_2$ is located in the mitochondria, the other AlDH isozymes in the cytosol. Cytosolic AlDH is more susceptible to inhibition by disulfiram (3). There appear to be interspecific differences in subcellular distribution and amounts of high- and low-K_m forms in the liver.

Whites have two predominant AlDH isozymes: AlDH$_1$ and AlDH$_2$. Many Asians do not have the active AlDH$_2$ isozyme, but an inactive form (21). Because one of the 500 amino acids differs (substitution of lysine for glutamine (80)), that form of AlDH$_2$ is inactive. This variant is found in about 50% of Japanese, 35% of Chinese and 40% of Vietnamese and Indonesians. The lack of the active form of this AlDH isozyme, causing an increased acetaldehyde level upon alcohol consumption, is considered the main cause of the flushing syndrome (1, 2, 24, 101) – a hypersensitivity to alcohol frequently seen in people belonging to the mongoloid race, manifesting itself in cardiac arrhytmia and flushing. These phenomena are comparable with the disulfiram reaction. They are considered to result from acute acetaldehyde intoxication. The atypical ADH isozyme demonstrable in ca. 90% of all orientals (2, 73) is now considered of little importance as a cause of flushing (47, 101).

The unpleasant reactions (flushing, cardiac arrhytmia) due to accumulation of toxic acetaldehyde discourage alcohol use and abuse. Almost all East Asian alcoholics (91%) have the active AlDH$_2$ form and are not flushers (37). Lack of the gene responsible for active AlDH$_2$ production is considered a genetic protection against alcoholism (79). So, a clear relation is seen between the genetic variant of an enzyme and drinking habits.

Some side-reactions of acetaldehyde take place in addition to its conversion to acetate, such as reactions with pyruvate, alpha-ketoglutaric acid (92), and various proteins (13). Condensation reactions between acetaldehyde and a number of neurotransmitters, such a catecholamines, dopamine and serotonine, during which morphine-like substances are formed, may also occur. These reactions are quantitatively unimportant in acetaldehyde conversion. The compounds formed, however, may exert an essential, but as yet not entirely elucidated, influence on the nervous system (15, 51, 87).

3.5. Third step: the conversion of acetate

All acetate reactions proceed via the activated form, acetyl-coenzyme A (acetyl-CoA), which is a metabolite central to the whole metabolism. It is also formed during catabolism of carbohydrates, fats and a number of amino acids.

It is not clear whether in the liver free acetate or acetyl-CoA is generated from alcohol (92). The acetate (or acetyl-CoA) formed in the liver is hardly oxidized during the citric acid cycle because of an NAD^+ shortage in the liver caused by previous oxidation of alcohol and acetaldehyde (92, 94). Most of the acetate is released into the circulation and oxidized extrahepatically to CO_2 and H_2O through acetyl-CoA in the citric acid cycle. In some tissues such as the heart, acetate serves as the preferred substrate over glucose and fatty acids. A small proportion of the acetate generated from alcohol is incorporated in tissue components by anabolic reactions. The endogenous acetate levels are rather low (0.1–0.3 mM). During alcohol oxidation blood acetate concentration rises in half an hour to about 0.5–1.2 mM and reaches a plateau which is maintained as long as alcohol is present in the blood. In alcoholics with an elevated rate of alcohol elimination, the acetate concentration in both the hepatic vein and peripheral veins is higher than in controls. There is a high correlation between the rate of alcohol elimination and the blood acetate concentration. It is suggested that an increased level of blood acetate during ethanol oxidation may be used as an indicator of enhanced ethanol oxidation (61). The higher blood acetate levels in alcoholics may have consequences for the heart and the liver, because acetate can have metabolic effects on the myocardium (30) and interferes with lipid metabolism (78, 81).

Extrahepic dissimilation of acetate yields about two thirds of the total combustion energy produced by alcohol. The remaining energy is released during the generation of acetate in the liver.

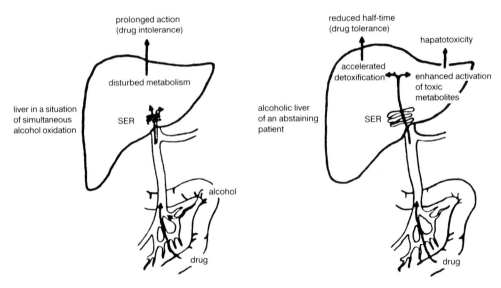

Fig. 3.3. Paradoxical interaction between alcohol and drugs. When liver cells come into contact with alcohol, drug metabolism will be inhibited competitively or non-competitively. This results in an extended effect of the drug in the organism (drug intolerance) (left-hand side of the figure). On the other hand, when the liver cell does not contain alcohol, but shows a proliferated smooth endoplasmatic reticulum (SER) due to chronic alcoholism (right-hand side of the figure), the organism may be either detoxified more rapidly (drug tolerance) or the drug may be converted to larger quantities of toxic metabolites and thus enhance liver intoxication. After: Hasumura & Takeuchi (25).

3.6. Interaction between alcohol and drugs

Several xenobiotics (for instance drugs) are converted through the cytochrome P-450 system. This partly explains the effect of alcohol on the metabolism of drugs, and vice versa. In alcoholism MEOS is induced and thus capable of metabolizing larger quantities. In sober alcoholics, drugs and other xenobiotics are converted more rapidly. These substances are consequently less effective, but the concentration of metabolites, which are quite often hepatotoxic, increases (41, 89). Following alcohol consumption, the system is blocked by ethanol, which causes a delay in elimination and an increased activity of drugs (25, 43, 75) (see Fig. 3.3).

MEOS induction may also result from the use of certain types of drugs such as barbiturates. Pre-treatment with these drugs may activate ethanol metabolism whereas the presence of drugs may have a inhibitory effect. Although acute and chronic forms of alcohol consumption have a strong effect on drug metabolism, the reverse effect is much smaller (41).

The interaction between alcohol and drugs is complex and does not take place via the MEOS pathway only (42). Some antimicrobial drugs inhibit AlDH, resulting in disulfiram-like reactions when combined with alcohol. The effect of drugs that are converted by ADH, such as digitalis, can be influenced by alcohol through direct competition. Gastric ADH is inhibited by H2-receptor antagonists such as cimetidine, resulting in higher blood alcohol levels (8). Alcohol consumption might alter drug absorption by increasing solubility (48) or by interfering with gastric emptying.

Liver injury is usually accompanied with a delay in biotransformation (31). Alcoholic cirrhosis causes reduced plasma binding of some drugs, resulting in toxic effects, even at therapeutic plasma concentrations. The metabolic rate of drugs metabolized through acetylation is increased by alcohol (63) due to enhanced acetyl-CoA production during alcohol metabolism. Theoretically, alcohol can interact with drugs via the hepatic blood flow and hepatic uptake, but there is no convincing evidence for this.

All effects mentioned so far fall under the heading of pharmacokinetic interactions (inhibition or stimulation of metabolism). No mention has been made as yet of pharmacodynamic interactions, in which the effects on the central nervous system are intensified. Noordhoek (60) has lucidly reviewed the interactions between drugs and alcohol.

3.7. Modification of alcohol metabolism

Theoretically, a number of nutrients may have the potential of influencing alcohol metabolism. Niacin and riboflavin, being part of the hydrogen-accepting co-factors NAD and FAD, may increase the metabolic rate, since unreduced NAD is often considered the rate-limiting factor in metabolism (53). Vitamins C and E as well as fructose might also affect metabolism through their effect on redox reactions (65). It is also possible to increase the metabolic rate of ethanol by induction of the MEOS.

Actually, however, alcohol metabolism proves as good as insusceptible to the influence of external agents. Pre-treatment with some drugs has been found to have some effect (via the MEOS pathway). Fructose (1–2 g/kg body weight) accelerates alcohol metabolism (11, 27, 98). The effect of glucose (54) is much smaller than the effect of fructose (27). A carbohydrate-rich diet (with carbohydrates including fructose accounting for 90% of energy intake; 74) or a protein-rich diet (99) is only slightly effective in this respect. These effects of fructose, glucose and other carbohydrates are often seen without a concomitant decrease of the alcohol-induced increase in lactate/pyruvate ratio (98), so the mechanism of action is not clear. Also a megadose of vitamin C (5 g daily over a period of two weeks) enhances alcohol metabolism without affecting the lactate/pyruvate ratio (82).

Other possible reactions have been discussed in Sections 2.3.2 and 2.4.

Summary

Ethanol may be considered both a psychotropic substance and a nutrient. Alcohol oxidation proceeds along several pathways. The first pathway causes acetaldehyde to be generated, which is subsequently converted into acetate. About 90% of these reactions take place in the liver. The acetate is for the greater part released into the circulation and may undergo several more reactions.

Three enzyme systems are now known to increase the conversion of alcohol into acetaldehyde. The most important reaction is the NAD-dependent one with ADH as an enzyme. There are large individual and racial differences in ADH activity (due to differences in quantitative and qualitative distribution of the isozymes). The other two systems, i.e. the NADPH-dependent MEOS reaction and the catalase-linked reaction, play no important role under normal conditions. Chronic alcohol consumption induces the cytochrome P-450, which is responsible for the MEOS activity (especially isozyme P450IIE1). So, more alcohol is converted via the MEOS pathway, but to what extent exactly this factor contributes to the process is not clear. Several other xenobiotics (such as drugs) are also converted via cytochrome P-450. This may partly explain the effect of alcohol on drugs and vice versa.

In the second oxidation step the acetaldehyde formed is converted into acetate by the NAD-dependent enzyme AlDH. The fact that one of the isozymes is inactive in ca. 50% of all orientals is responsible for the occurrence of the flushing syndrome (acute acetaldehyde poisoning upon alcohol consumption). A small part of the acetate formed is incorporated in issue through anabolic reactions, the remainder is broken down to CO_2 and H_2O extrahepatically.

Theoretically, many nutritional components might affect alcohol metabolism. Actually, a substantial enhancement of the rate of alcohol metabolism has only been proved for fructose.

References

1. Adachi J, Mizoi Y. Acetaldehyde-mediated alcohol sensitivity and elevation of plasma catecholamine in man. Jpn J Pharmacol 1983;33: 531-9.
2. Agarwal DP et al. Racial differences in biological sensitivity to ethanol: the role of alcohol dehydrogenase and aldehyde dehydrogenase isozymes. Alcohol Clin Exp Res 1981;5: 12-6.
3. Alderman J et al. Partial characterization of hepatic acetaldehyde dehydrogenase from the baboon. Prog Clin Biol Res 1982;114: 77-89.
4. Badaway AAB. The metabolism of alcohol. Clin Endocrinol Metab 1978;7: 247-71.
4a. von Bahr-Lindstrom H, Hoog JO, Heden LO, Vallee BL, Jornvall H. cDNA structures of class I human liver alcohol dehydrogenases. In: Lindros KO, Ylikahri R, Kiianmaa K. Advances in biomedical alcohol research: third congress of the International Society for Biomedical Research on Alcoholism (Helsinki, Finland, 8–13 June 1986. Oxford: Pergamon Press, 1987: 151-5.
5. Blanc B. Alkohol im Stoffwechsel des Menschen. In: Alkohol-Alkoholika. Zürich: Stiftung für zeitgemässe Ernährung SZE, 1980: 59-72.
6. Boeker EA. Metabolism of ethanol. J Am Diet Assoc 1980;76: 550-4.
7. Bühler R, Pestalozzi D, Hess M, Wartburg JP von. Immunohistochemical localization of alcohol dehydrogenase in human kidney, endocrine organs and brain. Pharmacol Biochem Behav 1983;18: 55-59.
8. Caballeria J, Baraona E, Rodamilans M, Lieber CS. Effects of cimetidine on gastric alcohol dehydrogenase activity and blood ethanol levels. Gastroenterology 96: 388-392
9. Cotton RW, Goldman D. Review of the molecular biology of the human alcohol dependence genes and gene products. In: Alcohol research from bench to bedside. New York: The Haworth Press, 1989: 171-82.
10. Crabb DW, Bosron WF, Li TK. Ethanol metabolism. Pharmacol Ther 1987; 34: 59-73.
11. Crownover BP, La Dine J, Bradford B, Glassman E, Forman D, Schneider H, Thurman RG. Activation of ethanol metabolism in humans by fructose: importance of experimental design. J Pharmacol Exp Ther 1986;236: 574-9.
12. Davis VE. Alcohol and aberrant metabolism of biogenic amines. In: Roach MK et al. Biological aspects of alcohol. Austin: University of Texas Press, 1971: 293-312.
13. Donohue TM et al. Acetaldehyde adducts with proteins: binding of (14c) acetaldehyde to serum albumin. Arch Biochem Biophys 1983;220: 239-46.
14. Egerer G, Simanowski UA, Schmier M, Chang G, Bogusz M, Seitz HK. Effect of age on the first pass metabolism of ethanol in man. Biomedical and social aspects of alcohol and alcoholism. In: Kuriyama K, Takada A, Ishii H, eds. Proceedings of the Fourth Congress of the International Society for Biomedical Research in Alcoholism (ISBRA) Kyoto, Japan, 26 June 26–2 July 1988. Amsterdam: Elsevier Science Publishers, 1988;115-8.
15. Eriksson CJP. The role of acetaldehyde in drinking behavior and tissue damage. Br J Alcohol Alcohol 1982;17: 57-69.
16. Eriksson CJP. Human blood acetaldehyde concentration during ethanol oxidation. Pharmacol Biochem Behav 1983;18: 141-50.
17. Erwin VG, Deitrich RA. heterogenicity of alcohol dehydrogenase enzymes in various tissues. Biochem Pharmacol 1974;21: 2915-24.
18. Ewing JR, Rouse BA, Pellizzari ED. Alcohol sensitivity and ethnic background. Am J Psychiatry 1974;131: 206-10.
19. Farris JJ, Jones BM. Ethanol metabolism in male American indians and whites. Alcoholism 1978;2: 77-81.
20. Frezza M, Padova C di, Pozzato G, Terpin M, Baraona E, Lieber CS. High blood alcohol levels in women: the role of decreased gastric alcohol dehydrogenase activity and first-pass metabolism. N Engl J Med 1990; 322: 95-9.
21. Goedde HW, Agarwal DP. Aldehyde dehydrogenase polymorphism: molecular basis and phenotypic relationship to alcohol sensitivity. In: Lindros KO, Ylikhari R, Kiianmaa K. eds. Advances in biomedical alcohol research: Third Congress of the International Society for

Biomedical Research on Alcoholism, Helsinki, Finland, 8—13 June 1986. Oxford: Pergamon Press, 1987: 47-54.
22. Grant DAW. Genetic polymorphism of the alcohol metabolising enzymes as a basis for alcoholic liver disease. Br J Addict 1988;83: 1255-9.
23. Handler JA, Thurman RG. Fatty acids supply H_2O_2 at high rates for the oxidation of ethanol by catalase. In: Lindros, KO, Ylikhari R, Kiianmaa, eds. Advances in biomedical alcohol research: Third Congress of the International Society for Biomedical Research on Alcoholism, Helsinki, Finland, 8—13 June 1986. Oxford: Pergamon Press, 1987: 225-9.
24. Harada S, Agarwal DP, Goedde HW. Aldehyde dehydrogenase deficiency as cause of facial flushing reaction to alcohol in Japanese. Lancet 1981; ii: 982.
25. Hasamura Y, Takeuchi J. Wechselbeziehungen zwischen Alkoholkonsum und Arzneimittelstoffwechsel in der Leber. Leber Magen Darm 1978;8: 286-90.
26. Jenkins WJ, Peters TJ. Selectively reduced hepatic acetaldehyde dehydrogenase in alcoholics. Lancet 1980;i: 628-9.
27. Jones AW. Effects of fructose, glucose, and mixed sugars on ethanol detoxification and blood glucose response in rats. Med Biol 1983;61: 319-23.
28. Kähönen MT, Ylikhari RH, Hassinen I. Ethanol metabolism in rats treated with ethyl-p-chlorophenoxyisobutyrate (clofibrate). Life Sci 1971;10: 661-70.
29. Khanna JM et al. Acute metabolic interaction of ethanol and drugs. In: Galanter M, ed. Currents in alcoholism. New York: Grune and Stratton, 1980;7.
30. Kiviluoma K, Hassinen I. Role of acetaldehyde and acetate in the development of ethanol induced cardiac lipidosis, studied in isolated perfused rat hearts. Alcohol Clin Exp Res 1983;7: 169-75.
31. Klinger W. Alkohol und Arzneimittelbiotransformation. Z Ges Inn Med 1981;36: 543-7.
32. Korsten MA, Matsuzaki S, Feinman L, Lieber CS. High blood acetaldehyde levels after ethanol administration. N Engl J Med 1975;292: 386-9.
33. Kürzinger R. Ethanol oxidation by hepatic microsomes: adaptive increase after ethanol feeding. Science 1968;162: 917-8.
34. Lange LG, Bergmann SR, Sobell BE. Identification of fatty acid ethyl esters as products of rabbit myocardial ethanol metabolism. J Biol Chem 1981;256: 12968-73.
35. Laposata E, Lange LG. Presence of nonoxidative metabolism in human organs commonly damaged by ethanol abuse. Science 1986;231: 497-9.
36. Li TK. The absorption, distribution and metabolism of ethanol and its effects on nutrition and hepatic function. In: Tabakoff B, Sutker PB, Randall PB, eds. Medical and social aspects of alcohol abuse. New York: Plenum Press, 1984; 47-77.
37. Li TK, Lockmuller JC. Why are some people more susceptible to alcoholism? Alcohol Health Res World 1989;13: 310-5.
38. Lieber CS. Alcohol, protein metabolism, and liver injury. Gastroenterology 1980;79: 373-90.
39. Lieber CS, Teschke R, Hasamura Y, DeCarli LM. Differences in hepatic and metabolic changes after acute and chronic alcohol consumption. Fed Proc 1975;34: 2060-74.
40. Lieber CS et al. Effect of chronic alcohol consumption on ethanol and acetaldehyde metabolism. Adv Exp Med Biol 1975;59: 185-227.
41. Lieber CS. Ethanol metabolism and toxicity. Rev Biochem Toxicol 1983;5: 267-311.
42. Lieber CS. Interaction of ethanol with drugs, hepatotoxic agents, carcinogens and vitamins. Alcohol Alcoholism 1990; 25: 157-176.
43. Lieber CS. Liver adaptation and injury in alcoholism. N Engl J Med 1973;288: 356-62.
44. Lieber CS. Microsomal ethanol oxidizing system (MEOS): interaction with ethanol, drugs and carcinogens. Pharmacol Biochem Behav 1983;18: 181-7.
45. Lieber CS, DeCarli LM. Ethanol oxidation by hepatic microsomes: adaptive increase after ethanol feeding. Science 1968;162: 917-8.
46. Lindros KO et al. Elevated blood acetaldehyde in alcoholics with accelerated ethanol elimination. Pharmacol Biochem Behav 1980;13: 119-24.
47. Lindros KO. Human blood acetaldehyde levels: with improved methods, a clearer picture emerges. Alcohol Clin Exp Res 1982;6: 70-5.
48. Linnoila M, Mattila MJ, Kitchell BS. Drug interactions with alcohol. Drugs 1979;18: 299-311.

49. Littleton JM. The biological basis of alcoholism: some recent experimental evidence. In: Edwards G, Grant M, eds. Alcoholism: new knowledge and new responses. London: Croom-Helm, 1977: 107-16.
50. Littleton JM. Neuropharmacological aspects of ethanol tolerance and dependence. In: Mendlewicz J, Praag HM van, eds. Alcoholism: a multidisciplinary approach. Basel: S Karger, 1979;3: 75-87.
51. Lundquist F. Acetaldehyde and aldehyde dehydrogenase – central problems in the study of alcoholism. Eur J Clin Invest 1983;13: 183-4.
52. Magelson S, Pieper SJ, Lange LG. Thermodynamic bases for fatty acid ethyl ester synthase catalyzed esterification of free fatty acid with ethanol and accumulation of fatty acid ethyl esters. Biochemistry 1984;23: 4082-7.
53. Marks V. Biochemical and metabolic basis of alcohol toxicity. In: Mendlewicz J, Praag HM van, eds. Alcoholism: a multidisciplinary approach. Basel: S Karger, 1979;3: 88-96
54. Mascord D, Rogers J, Smith J, Starmer GA, Whitfield JB. Effect of diet on [lactate]/[pyruvate] ratios during alcohol metabolism in man. Alcohol Alcoholism 1989;24: 189-91.
55. Matsuzaki S et al. Increased alcohol dehydrogenase independent ethanol oxidation at high ethanol concentrations in isolated rat hepatocytes: the effect of chronic ethanol feeding. J Pharmacol Exp Ther 1981;217: 133-7.
56. Mezey E, Potter JJ, Reed WD. Ethanol oxidation by a component of liver microsomes rich in cytochrome P-450. J Biol Chem 1973;248: 1183-7.
57. Morgan CJ, Badaway AAB, Thomas DR, Dando TG, Kirby A. The [lactate]/[pyruvate] ratio and alcohol metabolism: experiments with naloxone in fasting normal male volunteers. Alcohol Alcoholism 1989;24: 185-8.
58. Mungikar AM, Hetu C, Joly JG. Ethanol-inducible liver cytochrome P-450 in the rat: relative specificity for ethanol oxidation activity in vitro. Adv Exp Med Biol 1980; 132: 51-6.
59. Nomura F et al. Effect of ethanol administration on the metabolism of ethanol in baboons. J Pharm Exp Ther 1983;227: 78-83.
60. Noordhoek J. Interacties tussen alcohol en geneesmiddelen. Tijdschr Alcohol Drugs 1982;8: 100-4.
61. Nuutinen H, Lindros K, Hekali P, Salaspuro M. Elevated blood acetate as indicator of fast ethanol eliminiation in chronic alcoholics. Alcohol 1985: 623-6.
62. Offerhaus L. De farmacologie van alcohol. Ned Tijdschr Geneeskd 1979;123: 1243-6.
63. Olsen H, Morland J. Ethanol-induced increase in drug acetylation in man and isolated rat liver cells. Br Med J 1978;ii: 1260-2.
64. Orme-Johnson WH, Ziegler DM. Alcohol mixed function oxidase activity of mammalian liver microsomes. Biochem Biophys Res Comm 1965;21: 78-82.
65. Pawan GLS. Vitamins, sugars and ethanol metabolism in man. Nature 1968;220: 374-6.
66. Pawan GLS. Metabolism of alcohol (ethanol) in man. Proc Nutr Soc 1972;31: 83-9.
67. Pestalozzi DM, Bühler R, Wartburg JP von, Hess M. Immunohistochemical localization of alcohol dehydrogenase in the human gastrointestinal tract. Gastroenterology 1983;85: 1011-6.
68. Peters TJ. Ethanol metabolism. Br Med J 1982;i: 17-20.
69. Pietruszko R. Nonethanol substrates of alcohol dehydrogenase. In: Majchrowicz E, Noble EP. Biochemistry and pharmacology of ethanol, vol. 1. New York: Plenum Press, 1979: 87-106.
70. Pikkarainen PH, Lieber CS. Concentration dependency of ethanol elimination rates in baboons: effect of chronic alcohol consumption. Alcohol Clin Exp Res 1980;4: 40-3.
71. Rahwan RG. Toxic effects of ethanol: possible role of acetaldehyde, tetrahydroisoquinolines, and tetrahydro-beta-carbolines. Toxicol Appl Pharmacol 1975;34: 3-27.
72. Reinus JF, Heymsfield SB, Wiskind R, Casper K, Galambos JT. Ethanol: relative fuel value and metabolic effects in vivo. Metabolism 1989; 38: 125-35.
73. Ricciardi BR. Hepatic ADH and AlDH isoenzymes in different racial groups and in chronic alcoholism. Pharmacol Biochem Behav 1983;18: 61-5.
74. Rogers J, Smith J, Starmer GA, Whitfield JB. Differing effects of carbohydrate, fat and protein on the rate of ethanol metabolism. Alcohol Alcoholism 1987;22: 345-53.
75. Rubin E, Hutterer F. Ethanol increases hepatic smooth endoplasmatic reticulum and drug metabolizing enzymes. Science 1968;159: 1469-70.

76. Schmidt E, Schmidt FW. Enzym-Muster menschlicher Gewebe. Klin Wochenschr 1960;38: 957-62.
77. Schuckit MA, Rayses V. Ethanol ingestion: differences in blood acetaldehyde concentrations in relatives of alcoholics and controls. Science 1979;203: 54-5.
78. Snoswell AM, Trimble RP, Fishlock RC, Storer GB, Topping DL. Metabolic effects of acetate in perfused rat liver. Studies on ketogenesis, glucose output, lactate uptake and lipogenesis. Biochim Biophys Acta 1982;716: 290-7.
79. Stoil MJ. The case of the missing gene: hereditary protection against alcoholism. Alcohol Health Res World 1987/88;12: 130-6.
80. Suddendorf RF. Research on alcohol metabolism among Asians and its implications for understanding causes of alcoholism. Publ Health Rep 1989;104: 615-20.
81. Suokas A, Kupari M, Heikkila J, Lindros K, Ylikahri R. Acute cardiovascular and metabolic effects of acetate in men. Alcohol Clin Exp Res 1988;12: 52-8.
82. Susick RL, Zannoni VG. Effect of ascorbic acid on the consequences of acute alcohol consumption in humans. Clin Pharm Ther 1987;41: 502-9.
83. Teschke R, Gellert J. Hepatic microsomal ethanol-oxidizing system (MEOS): metabolic aspects and clinical implications. Alcohol Clin Exp Res 1986;10: S20-S32.
84. Thurman RG, McKenna WR, Brentzel HJ, Hesse S. Significant pathways of hepatic ethanol metabolism. Fed Proc 1975;34: 2075-81.
85. Thurman RG, Handler JA. New Perspectives in catalase-dependent ethanol metabolism. Drug Metab Rev 1989;20: 679-88.
86. Thurman RG, Hoffman PL, Eds. First congress of the international society for biomedical research on alcoholism. Fayetteville: ANKHO, 1983: 593 pp.
87. Topel H. Alkoholism – ein Komplex biochemischer Ereignisse. Drogalkohol 1983;7: 3-28.
88. Trémolières J, Lowy R, Griffaton G. Metabolic effects of ethanol. Proc Nutr Soc 1972;31: 107-15.
89. Tsutsumi R, Leo MA, Cho-il Kim, Tsutsumi M, Lasker J, Lowe N, Lieber CS. Interaction of ethanol with enflurane metabolism and toxicity: role of P450IIE1. Alcohol Clin Exp Res 1990;14: 174-9.
90. Ukita K, Fujimiya T, Fukui Y. Pharmacokinetic study of ethanol elimination: evaluation of first-pass effect. In: Kuriyama K, Takada A, Ishii H, eds. Biomedical and social aspects of alcohol and alcoholism: proceedings of the Fourth Congress of the International Society for Biomedical Research in Alcoholism (ISBRA), Kyoto, Japan, 26 June–2 July 1988. Amsterdam: Elsevier Science Publishers, 1988: 111-4.
91. Waldron HA, Cherry N, Johnston JD. The effects of alcohol on blood toluene concentrations. Int Arch Occup Environm Health 1983;51: 365-9.
92. Wallgren H, Berry H. Actions of alcohol vol. 1. Amsterdam: Elsevier, 1970.
93. von Wartburg JP, Papenberg J, Aebi H. An atypical human alcohol dehydrogenase. Can J Biochem Physiol 1965;43: 889-98.
94. von Wartburg JP. Biochemische Auswirkungen des Alkoholismus. Biblthca Nutr Dieta 1976;24: 7-16.
95. von Wartburg JP et al. The polymorphisms of alcohol and acetaldehyde dehydrogenase and their significance for acetaldehyde toxicity. Pharmacol Biochem Behav 1983;18: 123-5.
96. Weiner H. Aldehyde dehydrogenase: mechanism of action and possible physiological roles. In: Majchrowicz E, Noble EP. Biochemistry and pharmacology of ethanol, vol. 1. New York: Plenum Press, 1979: 107-24.
97. Weiner H. Acetaldehyde metabolism. In: Majchrowicz E, Noble EP. Biochemistry and pharmacology of ethanol, vol. 1. New York: Plenum Press, 1979: 125-44.
98. Whitfield J, Smith J, Mascord D, Starmer G. Effects of fructose and glucose on alcohol metabolism and redox state. Alcohol Clin Exp Res 1990; 14: 352 (abstract 532).
99. Wissel PS. Dietary influences on ethanol metabolism. Drug Nutr Interact 1987;5: 161-8.
100. Yamauchi K et al. Effect of clofibrate and phenobarbiturate on clearance of ethanol and acetaldehyde in blood of rats. Jap J Alc Stud Drug Depend 1981;16: 225-33.
101. Yoshida A. Differences in the isozymes involved in alcohol metabolism between Caucasians and Orientals. Isozymes Curr Top Biol Med Res 1983;8: 245-61.

CHAPTER 4

Alcohol and hormone metabolism

W.G. Vrij-Standhardt

In this chapter the effect of alcohol consumption on the various hormones and hormone systems will be considered. Attention will also be paid to the influence of some hormones on alcohol metabolism.

4.1. Vasopressin, renin and aldosterone

Extracellular water and sodium concentrations are regulated by two hormone systems. Vasopressin (antidiuretic hormone), secreted by the posterior lobe of the pituitary gland, stimulates water reabsorption in the kidney. Vasopressin is secreted upon stimulation of the osmoreceptors in the hypothalamus. The second hormone system is the renin-aldosterone axis. Under the influence of baroreceptors and sodium receptors in the juxtaglomerular apparatus in the kidney the secretion of renin is stimulated, thus causing a release of aldosterone by the angiotensin generated in the liver. Aldosterone is secreted by the adrenal cortex and is responsible for the renal reabsorption of sodium and chloride by stimulating the Na-K pump.

Besides the endocrinological action of the whole vasopressin molecule, there is evidence that certain fragments of vasopressin function as neuropeptides (99). These neuropeptides affect and modulate brain processes. One of these central actions is modulation of memory processes. Vasopressin represses the mild amnesia caused by a moderate dose of alcohol (2 ml wodka per kg body weight). Also in a sober state vasopressin improves human short-term memory processes (and interferes negatively with later learning)(71). This suggests that there is no interaction between vasopressin and alcohol as far as memory processes are concerned. There is some evidence that vasopressin plays a role in functional tolerance (46, 47) by affecting neurotransmission. Vasopressin attenuates the reinforcing action of some drugs (99).

One of the endocrinological effects of alcohol discovered earliest is its inhibitory effect on release of vasopressin (24). Consequently, the consumption of a single dose of alcohol causes diuresis. In case of repeated, orally administered doses of alcohol the diuretic response decreases to zero (98) in spite of a gradually increasing blood alcohol concentration (BAC). The mechanism involved in this diminished response to repeated doses of alcohol can be explained by a resetting of the osmoreceptors by alcohol. Alcoholics in the beginning of withdrawal show an increased vasopressin level, which decreases to normal during detoxification (27, 79). An increased vasopressin level leads to an antidiuretic state, and overhydration of the brain can occur (54, 63). It has been hypothesized that overhydration of the brain causes some of the neurological symptoms observed during withdrawal as well as some of the neuropathological abnormalities seen in alcoholics (54).

From several studies it has become apparent that alcohol also affects the renin-aldosterone axis (29, 30, 33, 38, 56). The results, however, are contradictory: unchanged, increased as well as decreased aldosterone levels have been found. Presumably the aldosterone response to alcohol consumption depends on the actual doses of alcohol consumed, the time after consumption (determined when the BAC either increases or decreases, or during a period of withdrawal or hangover), the individual involved (alcoholic or non-alcoholic), and the presence of liver disease. In patients with cirrhosis the rate of removal of aldosterone from the plasma is decelerated (18).

Although it has not been proved as yet, it seems likely that the release of renin which has been observed during alcohol intoxication and hangover, should be considered a normal response to ethanol-induced diuresis. A non-specific stress reaction and a direct influence of alcohol on aldosterone or renin synthesis or release have also been mentioned as possible causes.

4.2. Parathyroid hormone and calcitonin

The calcium concentration of blood is regulated by parathyroid hormone and calcitonin. Parathyroid hormone, secreted by the parathyroid gland, increases the blood calcium level by stimulating bone reabsorption (calcium and phosphate release), decreasing calcium secretion and increasing phosphate excretion in the kidney, and increasing calcium uptake in the intestine. Calcitonin is produced in the C-cells of the thyroid gland, and decreases the blood calcium concentration. Vitamin D regulates calcium absorption and calcium transport to and from the bone. administration of alcohol increases the rate of excretion of calcium and magnesium. Alcohol has a direct stimulating effect on calcitonin release (22, 104). In hospitalized alcoholics hypocalcaemia is frequently observed (10, 12, 28, 67); it often disappears spontaneously upon alcohol withdrawal and adherence to an adequate diet. In alcoholics suffering from both hypocalcaemia and hypomagnesaemia, the administration of parathyroid extract proved to be an insufficient remedy; hypocalcaemia did not disappear until magnesium had also been administered (28, 67). Hypocalcaemia probably occurs primarily as a result of hypomagnesaemia, since it has been suggested that magnesium is necessary for the action of parathyroid hormone in response to hypocalcaemia. So, hypomagnesaemia may result in parathyroid hormone resistance (28). Bjorneboe (10) and Chappard (12) report serum parathyroid levels in alcoholics within normal ranges. Consumption of 0.89 g alcohol per kg body weight has no significant effect on serum parathyroid hormone level (57).

In cirrhotic alcoholics the osteoblastic activity (bone formation) rate is reduced and extensive skeletal osteoporosis is found (12). It is possible that the activities of enzymes crucial in cholecalciferol (vitamin D-3) metabolism may be altered in alcoholics (10).

4.3. Thyroxine

Thyroid activity is regulated by the thyroid-stimulating hormone thyrotropin (TSH) secreted by the anterior pituitary. TSH is released under the influence of thyrotropin-releasing hormone (TRH) from the hypothalamus. The thyroid secretes mainly thyroxine (tetra-iodothyronine, T4), and in addition some tri-iodothyronine (T3). T4 is converted into the more active T3 in the liver. Thyroid hormones have numerous effects on the organism, such as an increase of basal metabolism, and an increased transmission of impulses in the peripheral nervous system.

In general, the thyrotropic axis will be hardly affected by alcohol, provided no liver injury is present. No deviant TSH levels were found in normal subjects after consumption of alcohol. The pituitary response to exogenous TRH was also within normal limits (55).

The results concerning serum T3, T4 and TSH levels in alcoholics are conflicting. Normal levels of T3, T4 and TSH have been reported (73), but also decreased T3 levels and raised T4 levels have been found (1, 73, 83, 87). It has been suggested that T4 is lowered in alcoholics, while during withdrawal the T4 level rapidly increases to normal or above (59).

The hypothalamic pituitary axis has been reported not to be influenced by alcohol (1), while others found an abnormal TSH response to exogenous TRH in alcoholics (58).

Both alcohol per se and alcoholic liver injury affect thyroxine levels. The liver plays a major role in peripheral metabolism of thyroxine. The conversion of T4 into T3 in the liver is diminished in case of liver injury. But also a direct toxic effect of alcohol on the thyroid gland, independent of the degree of liver damage, has been reported (45).

The mechanism of the abnormal TSH response to TRH in alcoholics with liver disease, and also in some other alcoholics, is unknown. Among the possible mechanisms suggested (73) are:
– a direct effect of alcohol on the hypothalamus;
– an increased concentration of circulating oestrogens observed in patients, since hyperoestrogenaemia leads to the same abnormal response of TSH to TRH (20);
– hepatic encephalopathy, associated with dopamine depletion, might facilitate TSH release (87);
– reduced liver function (7, 84).

Considering the prevalence of the abnormal response, a cause associated with liver injury appears most likely.

4.4. Growth hormone (somatotropin)

Growth hormone (GH) is secreted episodically by the pituitary anterior lobe with the highest peak values occurring shortly after sleep onset. It plays a central role in the regulation of growth. Further, it has an anabolic effect, it inhibits the insulin-mediated uptake of glucose, it delays the synthesis of fat from carbohydrates and it stimulates lipolysis. The regulation of its secretion is complex. Secretion is stimulated by the growth hormone-releasing hormone (GHRH) from the hypothalamus and is presumably also influenced by stress, hypoglycaemia, adrenergic stimuli, and non-esterified fatty acids. Secretion is inhibited by a release-inhibiting factor from the hypothalamus (somatostatin). Feedback by somatomedins also plays a role in this process.

Both in healthy subjects and in alcoholics with and without liver injury, the influence of alcohol on plasma growth hormone concentration and on the growth hormone response to hypoglycaemia has been investigated (see 52, 73 for reviews). Although the results are contradictory most results favour the conclusion that alcohol inhibits GH secretion. Owing to the pulsatile secretion of GH, drawing firm conclusions on the effects of alcohol on serum GH levels demands sampling over a long period. From the few studies based on such a strategy it is concluded that ethanol, even in small doses, abolishes both sleep- and non-sleep-associated peaks of GH secretion (55, 80, 93, 96). It has been suggested that this is not the result of inhibition at the hypophyseal level (97).

In cirrhotic alcoholics profound changes in GH metabolism are found. Basal GH levels are elevated owing to both increased GH secretion and reduced catabolism in the liver (96). However, Agner et al. (1) suggest that the increased growth hormone levels are due to decreased somatomedin levels.

4.5. Insulin and glucagon

Insulin is secreted by the β-cells of the islets of Langerhans in the pancreas. It decreases the blood sugar level by increasing glycogen synthesis, decreasing gluconeogenesis, increasing the glucose permeability of the cell wall, and decreasing the degradation of fat. Insulin secretion is stimulated by hyperglycaemia.

Glucagon, secreted by the α-cells of the islets of Langerhans, is an insulin antagonist. It increases the blood sugar level through the breakdown of hepatic glycogen, stimulation of gluconeogenesis and the breakdown of fat. It is released in case of hypoglycaemia.

In most investigations alcohol proved not to affect the basal plasma insulin level (8, 75). In only one study an increase has been found (89), in others a decrease has been observed (49). The latter may have been a response to the lowered blood glucose level induced by alcohol.

When administered to normal subjects, both orally and intravenously, alcohol increases the plasma insulin response to a glucose load (73, 75). The mechanism involved is unknown, although it has been suggested to be the result of a direct effect of alcohol on the β-cells (73).

In chronic liver disease, whether or not caused by alcohol, carbohydrate intolerance, usually without clinical symptoms, is frequently observed (73). Increase of plasma insulin levels is caused by diminished insulin breakdown in the liver as a result of hepatocellular failure (50).

A direct effect of alcohol on glucagon could not be demonstrated (77), but glucagon release may result from alcohol-induced hypoglycaemia.

Changes in plasma levels of insulin and glucagon, which may be common to alcoholics but may also be acute as a response to alcohol consumption, have no clinical consequences as a rule. Alcohol-induced reactive hypoglycaemia may be partly caused by a change in insulin response. According to Morgan (73), however, neither hormone plays an important part in the pathogenesis of alcohol-related hypo- or hyperglycaemia (see Chapter 5).

4.6. Glucocorticoids

Glucocorticoids, the most important of which is cortisol, are secreted by the adrenal cortex under the influence of ACTH (adrenocorticotropic hormone, corticotropin) released from the anterior pituitary. ACTH release is regulated by the corticotropin-releasing factor (CRF) in the hypothalamus. CRF release, and possibly ACTH release as well, may be influenced by the plasma cortisol concentration (negative feedback), and by the diurnal sleep-wake cycle. Plasma cortisol is partly bound to a cortisol-binding protein. The unbound fraction is biologically active. Cortisol is metabolized in the liver. It predominantly affects carbohydrate metabolism, thus increasing the blood sugar level, and has many other activities such as an anti-inflammatory effect.

The administration of intoxicating doses of alcohol to subjects usually results in an increase in plasma cortisol level. Both normal, increased and decreased values have been found in alcoholics (17, 73). Plasma ACTH is elevated in drinking chronic alcoholics. It has been suggested that chronic alcohol consumption might have a direct stimulatory effect on the adrenal cortex, leading to disorganization of the hypothalamic-pituitary axis (HPA) (61). The plasma cortisol response depends on the dose of alcohol (quantity, chronic or acute), the presence of liver disease and the time of measurement (during intoxication or after withdrawal).

There are indications that the increase in plasma cortisol levels following acute alcohol consumption in non-alcoholics is induced via the hypothalamic-pituitary axis (HPA) as well as by a direct effect on the adrenal cortex. It has been suggested that the latter effect is mediated by acetaldehyde (17). The indirect stimulation via the HPA may be the result of stress (77). The increase of plasma cortisol levels found in alcoholics – decreasing again after a moderate dose of alcohol had been administered (69, 70) – supports the stress hypothesis. This is also the case for the increase in plasma cortisol levels of subjects upon large doses of alcohol being associated with gastrointestinal illness (52). A third cause of an increased plasma cortisol level may be a diminished cortisol catabolism caused by hepatic dysfunction. In view of the sometimes conflicting experimental findings more definitive conclusions concerning aetiology can not be drawn.

In some cases alcohol-induced hypercortisolaemia gives rise to the pseudo-Cushing syndrome, which is reversible in contrast to the genuine Cushing syndrome (an overproduction of glucocorticoids due to dysfunction of the pituitary or the adrenal cortex) (73). Its symptoms are facial mooning, truncal obesity, and proximal muscle wasting. It is not clear as yet what causes the development of the pseudo-Cushing syndrome.

4.7. Adrenaline and noradrenaline

Adrenaline (epinephrine) and noradrenaline (norepinephrine) are produced in the adrenal medulla. They also function as neurotransmitters of the sympathetic nervous system. The secretion of (nor)adrenaline is controlled by the nervous system bypassing the pituitary. The adrenal medullar hormones play an important role in stress and are powerful vasopressors. (Nor)adrenaline breakdown follows both an oxidative and a reductive pathway.

Although some investigations produced contradictory results, it may be said that administration of alcohol to normal subjects as well as to alcoholics increases circulatory levels and urinary output of catecholamines (a collective term comprising adrenaline, noradrenaline and dopamine). Kakihana and Butte (52) and Paille et al. (77) have reviewed research in this field.

Acute alcohol consumption stimulates central noradrenergic functioning; however, in the long term, chronic intoxication induces noradrenergic hypofunctioning (76). Noradrenergic hypofunction is one of the factors involved in craving (3, 76). Other neurotransmitter systems (dopamine, opioid peptides) are also said to play a role (76). There is a great deal of controversy with regard to the relative importance of the diverse neurotransmitter systems in reward and reinforcement, resulting in craving (3). Withdrawal is accompanied with sudden noradrenergic hyperactivity (76).

According to Paille et al. (77) the decreased levels in chronic alcoholics may be caused by deficiencies of folate and vitamin B-6, which are necessary for (nor)adrenaline synthesis. In Pohorecky's view (78) catacholamine turnover is increased in alcoholics. The increased plasma (nor)adrenaline levels after acute alcohol consumption are said to be caused by reduction of its circulatory clearance rather than by an increased release (25). One of the ADH isoenzymes has a physiological role in the degradation of circulating adrenaline and noradrenaline (64).

Because of the excess of NADH during alcohol metabolism, reductive breakdown of catecholamines increases at the cost of oxidative breakdown. Upon condensation reactions with acetaldehyde the substances generated during this process (such as tetrahydropapaverolin) may play a part in alcohol dependence (77).

4.8. Male sex hormones

Two functions may be attributed to the testes:
- spermatogenesis under the influence of FSH (follicle-stimulating hormone) from the anterior pituitary, the androgens playing a 'permissive' role;
- production of androgens in the interstitial cells of Leydig under the influence of LH (luteinizing hormone) from the anterior pituitary. The androgens, the most important of which is testosterone, cause the development of the secondary sexual characteristics and have a strong metabolic effect.

The release of LH, and presumably also of FSH, is stimulated by LHRH (LH-releasing hormone), released from the hypothalamus and inhibited by negative feedback of the androgens (LH) and of inhibin (FSH).

The hypothalamic-pituitary-gonadal axis is heavily affected by alcohol, which may lead to regularly occurring clinical consequences. Mandell et al. (62) reported at least one form of sexual dysfunction in 84% of 44 alcoholics interviewed (erection disorders 59%, ejaculatory incompetence 48%). Further, hypogonadism (manifesting itself as testicular atrophy, impotence and diminished fertility) and hyperoestrogenism (manifested by gynaecomastia and loss of body hair) have been observed in alcoholics (34, 74).

These symptoms have been observed both in alcoholics with and alcoholics without liver disease, but also occasionally in non-alcoholic cirrhotics (74). Upon abstaining, improvement has been seen in alcoholics (35, 102, 103). Many alcoholics are affected by diminished fertility, the semen often showing decreased spermatozoal motility, a smaller number of spermatozoa per volume of semen, and abnormally shaped spermatozoa (74). The semen quality of moderate drinkers did not differ from that of abstainers (63). In vitro studies showed acrosomal loss in human spermatozoa incubated in alcohol (2). Part of the acrosome is necessary for egg penetration.

The symptoms seen in alcoholics suggest an imbalance of sexual hormones. In alcoholics often low levels of testosterone are found (43, 48, 74), although normal levels have also been observed (40, 65). According to some studies (see 41) acute alcohol administration does not affect plasma testosterone concentration, in other studies increased or decreased levels were found (14, 26, 31, 95, 96).

Most plasma testosterone is bound to albumin (55%) or sex hormone-binding globulin (SHBG) (40%). Only 2% circulates freely. SHBG-bound testosterone is less biologically active. Gluud et al. (40) found in cirrhotic alcoholics a normal total testosterone concentration, whereas the concentrations of the subfractions had changed (less non-SHBG-bound testosterone). Oestrogen concentrations in alcoholics are elevated or normal (40, 41).

The mechanisms of decreased testosterone and increased oestrogen levels are not completely understood. Some consider the liver disease often seen in alcoholics to be the major factor, but others blame alcohol itself. According to Bannister et al. (6) liver disease per se appears to cause sexual dysfunction and sex hormone changes, but these changes are amplified by ethanol. Gluud (41) concludes that decreasing liver function, but not ethanol consumption, is responsible for the low levels of testosterone in alcoholic cirrhotics.

There are indications that alcohol exerts multiple effects on the hypothalamic-pituitary-gonadal axis (13, 15, 16, 102). To what extent each of these possibilities contribute to sexual disturbances is not known as yet. Some suggest an effect of alcohol on LH release from the pituitary (86, 100), but according to other studies this site is unaffected by alcohol (26). Affection of the LHRH release from the hypothalamus (19) or LH receptors in the testicular membrane (9) are also suggested. Animal studies support the idea that alcohol or acetaldehyde has a direct inhibitory effect on testosterone synthesis in the testis (14, 35, 85, 96). An increased metabolic clearance rate of testosterone, due to liver injury, may also contribute to the low testosterone levels found in alcoholics (92).

In addition to direct effects of alcohol, increased adrenal secretion of cortisol, which is known to inhibit testosterone synthesis, may contribute to the decrease in serum testosterone (95).

Presumably alcohol has a direct influence on spermatogenesis as well. Retinal, generated from retinol via the alcohol hydrogenase (ADH) enzyme, is necessary in spermatogenesis. ADH activity has been demonstrated in the testes (11). Competitive inhibition by alcohol results in retinal deficiency and decreased levels of testosterone in the testes, thus diminishing spermatogenesis (49).

There are several mechanisms that can explain feminization as a result of hyperoestrogenism in chronic alcoholics. Oestrogenic activity can be increased by enhanced conversion of testosterone into estrogen due to alcohol-induced increment of hepatic aromatase activity (42), or by conversion of the less active oestrogen oestrone into the active oestradiol due to the hepatic redox reaction during alcohol metabolism (4). Oestrogen breakdown can be decreased in case of liver damage. Others consider an increase in oestrogen receptors, induced and maintained by low serum testosterone levels (23). This could explain feminization despite normal oestrogen levels. Recently, phyto-oestrogenic compounds have been identified in bourbon (82). These naturally occurring plant metabolites in bourbon possess weak oestrogenic activities, which can contribute to feminization in bourbon-drinking alcoholics (37).

It has been shown that sex hormones may in turn influence alcohol metabolism (90, 91). In female rats higher ADH activity and lower microsomal ethanol-oxidizing system (MEOS) and catalase activity have been demonstrated as compared with male rats. When male rats have been castrated, or when oestradiol has been administered to them, ADH activity increases and MEOS activity decreases. The effects of castration are totally annulled by administration of testosterone. In female rats testosterone decreases ADH activity and increases MEOS activity and catalase.

Administration of oestradiol and ovariectomy does not affect ADH activity in female rats. Whether these observations also apply to human beings and what consequences they may have, needs further investigation.

Therapy with testosterone or other anabolic-androgenic steroids is applied in the treatment of alcohol liver disease (39, 60). This therapy is based on the presumption that these anabolic hormones accelerate resolution of alcoholic fatty liver. Case-control studies suggest testosterone therapy to be ineffective, whereas with some other anabolic-androgenic steroids positive results are obtained (60).

4.9. Female sex hormones

4.9.1. Progesterone and oestrogen

The most important female sex hormones are progesterone and oestradiol. Progesterone is secreted by the corpus luteum under the influence of LH and, in pregnancy, by the placenta. Oestradiol is produced in the ovarian follicles under the influence of FSH from the pituitary. These hormones show a monthly cycle due to an intricate pattern of inhibition, stimulation and feedback, which complicates research on effects of alcohol.

Little research has been done as yet on the effect of alcohol on female sex hormones, possibly because alcoholism causes less striking sexual dysfunction in women than in men. Female alcoholics are often subject to menstrual disturbances such as irregular menses, menorrhagia and amenorrhoea (74). Ovulation is affected, resulting in diminished fertility. A decrease in gonadal mass has been observed due to the absence of corpora lutea (101). In alcoholic women with liver disease plasma concentrations of oestradiol and progesterone as well as urinary oestrogen excretion are reduced to roughly the same level as in post-menopausal women. The levels are significantly lower than those of age-matched menstruating women. Plasma gonadotropin levels are not elevated and their response to stimulation by LHRH is not enhanced. It has been suggested that disturbed regulation of gonadotropin secretion is an important factor in the genesis of oestrogen deficiency and amenorrhoea in chronic alcoholic women with liver disease, although ovarian function may also be directly impaired (94).

In pregnant alcoholics who had delivered babies with the foetal alcohol syndrome oestrogen concentrations had decreased throughout pregnancy, whereas alcoholics who had delivered healthy babies maintained normal levels (44).

In healthy women acute alcohol consumption (1.2 g alcohol per kg body weight) hardly affects plasma gonadal hormone concentrations (26, 93).

4.9.2. Prolactin

Prolactin, secreted by the anterior lobe of the pituitary, activates the corpus luteum and promotes lactation. The basal plasma prolactin level has been found to be either unaffected (21, 88, 105) or reduced (93) after acute alcohol intake, whereas the prolactin level after beer (but not after alcohol) was elevated (21). The response of prolactin to naloxone (an opioid antagonist) and TRH stimulation is exaggerated (1, 68, 105). Suckling-induced prolactin release and milk consumption by the pups were found to be strongly inhibited by alcohol in the rat (88).

By contrast to alcoholic men, hyperprolactinaemia is not among the hormonal abnormalities of chronic alcoholic women (94).

Nothing is known about possible consequences, although some suggestions have been made, such as feminization in men (16), breast cancer in women (16) and depression (72).

4.9.3. Oxytocin

Oxytocin is secreted by the posterior lobe of the pituitary. Oxytocin triggers the 'let down' reflex during lactation and stimulates uterine contractions in delivery. In animal experiments oxytocin release, determined from uterine motility, proved to be decreased by alcohol. When the BAC has reached values of ca. 2 g/l the intensity of labour decreases and delivery is delayed (52). This inhibition by oxytocin release has been clinically tested. BACs of ca. 1.2 g/l proved to prevent premature labour to some extent (33). However, considering its effect on the unborn child (see Chapter 13), this application of alcohol is debatable.

Summary

The nervous and endocrine systems coordinate and integrate a variety of physiological processes. From a number of glands hormones are released into the circulation, which transports them to the target organs. Many hormones are secreted under the influence of releasing hormones originating from the pituitary, which in their turn are released by commands from the hypothalamus.

Alcohol consumption and alcoholism affect the blood concentration of almost all hormones. Its aetiology has not, or not yet, been elucidated entirely. A number of more or less plausible hypotheses have been brought forward. The aetiology of growth hormones, insulin and female sex hormone changes is still entirely obscure. The direct and indirect effects of alcohol consumption on hormone metabolism may be manifested within a wide range of sites:
– the effect of alcohol on the hypothalamus and/or pituitary (thyroxine, cortisol, male sex hormones, oxytocin),
– the effect of alcohol on hormone production in or release from the gland (renin, male sex hormones),
– the effects on receptor level (male sex hormones),
– the effects of alcohol metabolites acetaldehyde and hydrogen (e.g. NADH) (cortisol, male sex hormones),
– the effects of liver injury (thyroxine, insulin, cortisol, male sex hormones),

- effects due to nutritional deficiencies accompanying alcoholism (parathyroid hormone, adrenaline),
- effects of stress reactions in alcoholism, or after an acute large dose of alcohol (renin, cortisol, adrenaline),
- effects caused by changes of other hormone concentrations as a reaction to alcohol (renin, thyroxine, glucagon).

The clinical consequences are often unknown. However, the diuretic effect of alcohol via vasopressin and the increased excretion of magnesium and calcium in alcoholism have been explained.

Two hormone axes (HPA- and HPG-axes) are known to be strongly affected by alcohol, leading to serious consequences. In alcoholics a disturbance of the cortisol metabolism can cause the pseudo-Cushing syndrome. Both upon a single dose of alcohol and in alcoholics disturbance of the hypothalamic-pituitary-gonadal axis has been observed in men. Most alcoholics have one or more symptoms of sexual dysfunctioning, due to hypogonadism and/or hyperoestrogenism. An acute dose of alcohol has a presumably reversible effect on spermatogenesis.

References

1. Agner T, Hagen C, Andersen BN, Hegedus L. Pituitary-thyroid function and thyrotropin prolactin and growth hormone responses to TRH in patients with chronic alcoholism. Acta Med Scand 1986;220: 57-62.
2. Alvarez JG, Lee MA, Iozzo RV, Lopez I, Touchstone JC, Storey BT. Ethanol accelerates acrosomal loss in human spermatoza. J Androl 1988;9: 357-66.
3. Amit Z, Brown ZW. Actions of drugs of abuse brain reward systems: a reconsideration with specific attention to alcohol. Pharmacol Biochem Behav 1982;17: 233-8.
4. Andersson SHG, Cronholm T, Sjovall J. Effects of ethanol on the levels of unconjugated and conjugated androgens and estrogens in plasma of men. J Steroid Biochem 1986;24: 1193-8.
5. Appleby J. Why 1992 may be disastrous for health. Br Med J 1988;296(6637): 1620.
6. Bannister P, Oakes J, Sheridan P, Losowsky MS. Sex hormone changes in chronic liver disease: a matched study of alcoholic versus non-alcoholic liver disease. Q J Med 1987;63(240): 305-13.
7. Becker U, Gluud C, Bennett P. Thyroid hormones and thyroxine-binding globulin in relation to liver function and serum testosterone in men with alcoholic cirrhosis. Acta Med Scand 1988;224: 367-73.
8. Bellet S, Joshimine N, De Castro OAP et al. Effects of alcohol ingestion on growth hormone levels: their relation to 11-hydroxycorticoid levels and serum FFA. Metabolism 1971;20: 762-9.
9. Bhalla VK, Rajan VP, Newman ME. Alcohol-induced luteinizing hormone receptor deficiency at the testicular level. Alcoholism Clin Exp Res 1983;7: 153-162.
10. Bjorneboe GEA, Bjorneboe A, Johnsen J et al. Calcium status and calcium-regulating hormones in alcoholics. Alcoholism Clin Exp Res 1988;12: 229-32.
11. Bühler, R, Pestalozzi D, Hess M. Immunohistochemical localization of alcohol dehydrogenase in human kidney, endocrine organs and brain. Pharmacol Biochem Behav 1983;18 (suppl 1): 55-9.
12. Chappard D, Plantard B, Fraisse H et al. Bone changes in alcoholic cirrhosis of the liver : a histomorphometric study. Path Res Pract 1989;184: 480-5.
13. Chiao Y-B, Van Thiel DH. Biochemical mechanisms that contribute to alcohol-induced hypogonadism in the male. Alcoholism Clin Exp Res 1983;7: 131-4.

14. Cicero TJ, Bell RD, Badger TM. Multiple effects of ethanol on the hypothalamic-pituitary gonadal axis in the male. In: Begleiter H, ed. biological effects of alcohol: Advances in experimental medicine and biology, vol 126. New York: Plenum, 1980: 463-78.
15. Cicero TJ. Alcohol-induced deficits in the hypothalamic-pituitary-luteinizing hormone axis in the male. Alcoholism Clin Exp Res 1982;6: 207-15.
16. Cicero T. Alcohol effects on the endocrine system. In: Biomedical processes and consequences of alcohol use. Alcohol and Health Monograph 2. Rockville: NIAAA, 1982: 53-91.
17. Cobb CF, Van Thiel DH. Mechanism of ethanol-induced adrenal stimulation. Alcoholism Clin Exp Res 1982;6: 202-6.
18. Coppage WS et al. The metabolism of aldosterone in normal subjects and in patients with hepatic cirrhosis. J Clin Invest 1962;41: 1672-80.
19. Dees WL et al. Effects of ethanol on rat hypothalamic luteinizing hormone releasing hormone. A study utilizing radioimmunoassay. Biol Reprod 1983;28: 1066-70.
20. Delean A, Labrie F. Sensitizing effect of treatment with estrogens on TSH response to TRH in male rats. Am J Physiol 1977;233: E235-E239.
21. de Rosa G, Corsello SM, Ruffilli MP et al. Prolactin secretion after beer. Lancet 1981;ii(October): 934.
22. Dymling JF et al. Whisky: A new provocative test for calcitonin secretion. Acta Endocrinol 1976;82: 500-9.
23. Eagon PK, Porter LE, Van Thiel DH. The role of estrogens and androgens in the feminization of the chronic alcoholic male. Alcoholism Clin Exp Res 1983;7: 140-3.
24. Edkins N, Murray MM. Sugar tolerance and alcohol. J Physiol 1931;71: 403-11.
25. Eisenhofer G, Lambie DG, Johnson RH. Effects of ethanol on plasma catecholamines and norepinephrine clearance. Clin Pharm Ther 1983;34: 143-7.
26. Ellingboe J. Acute effects of ethanol on sex hormones in non-alcoholic men and women. In: Lindros KO, Ylikahri R, Kiianmaa K, eds. Advances in biomedical alcohol research: third congress of the International Society for Biomedical Research on Alcoholism (Helsinki, 8–13 June 1986). Alcohol Alcoholism 1987(suppl 1): 109-16.
27. Emsley RA, Potgieter A, Taljaard JJF et al. Impaired water excretion and elevated plasma vasopressin in patients with alcohol-withdrawal symptoms. Q J Med 1987;64: 671-8.
28. Estep H et al. Hypocalcemia due to hypomagnesia and reversible parathyroid hormone unresponsiveness. J Clin Endocrinol Metab 1969;29: 842-8.
29. Fabre LF et al. Aldosterone secretion in pentobarbital-anesthetized ethanol infused dogs. Q J Stud Alcohol 1972;3: 476-84.
30. Fabre LF, Farmer RW, Davis HW. Effect of ethanol of adrenocortical steroid secretion. In: Roach MK et al. Biological aspects of alcohol. Austin: University of Texas Press, 1971: 418-40.
31. Falk O, Palonek E, Bjorkhem I. Effect of ethanol on the ratio between testosterone and epitestosterone in urine. Clin Chem 1988;34: 1462-64.
32. Farmer RW, Fabre LF. Some endocrine aspects of alcoholism. In: Majchrowicz E, ed. Biochemical pharmacology of ethanol. New York: Plenum, 1975: 277-89.
33. Fuchs F. et al. Effect of alcohol on threatened premature labour. Am J Obstet Gynecol 1967;99: 627-37.
34. Galvao-Teles A, Gonçalves L, Carvalho H et al. Alterations of testicular morphology in alcoholic disease. Alcoholism Clin Exp Res 1983;7: 144-9.
35. Gavaler JS, Gay VC, Egler K et al. Evaluation of the differential in vivo toxic effects of ethanol and acetaldehyde on the hypothalamic-pituitary-gonadal axis using 4-methylpyrazole. Alcoholism Clin Exp Res 1983;7: 332-6.
36. Gavaler JS et al. Ethanol: its adverse effects upon the hypothalamic-pituitary-gonadal axis. Subst Alcohol Actions/Misuse 1983;4: 97-110.
37. Gavaler JS, Imhoff AF, Pohl CR et al. Alcoholic beverages : a source of estrogenic substances?. In: Lindros KO, Ylikahri R, Kiianmaa K, eds. Advances in biomedical alcohol research: third congress of the International Society for Biomedical Research on Alcoholism (Helsinki, 8–13 June 1986). Alcohol Alcoholism 1987(suppl 1): 545-9.
38. Gleiberman L, Harburg E. Alcohol usage and blood pressure: a review. Hum Biol 1986;58: 1-31.

39. Gluud C, Henriksen JH. Liver haemodynamics and function in alcoholic cirrhosis : relation to testosterone treatment and ethanol consumption. J Hepatol 1987;4: 168-73.
40. Gluud C, Dejgaard A, Bennet P, Svenstrup B. Androgens and oestrogens before and following oral testosterone administration in male patients with and without alcoholic cirrhosis. Acta Endocrinol 1987;115: 385-91.
41. Gluud C. Testosterone and alcoholic cirrhosis : epidemiologic, pathophysiologic and therapeutic studies in men. Dan Med Bull 1988;35: 564-74.
42. Gordon GG, Southre AL, Vittek J et al. The effect of alcohol ingestion on hepatic aromatase activity and plasma steroid hormones in the rat. Metabolism 1979;28: 20-4.
43. Guechot J, Peigny N, Ballet F, Vaubourdolle M, Giboudeau J, Poupon R. Sex hormone imbalance in male alcoholic cirrhotic patients with and without hepatocellular carcinoma. Cancer 1988;62: 760-2.
44. Halmesmäki E, Autti I, Granstrom ML, Stenman UH, Ylikorkala O. Estradiol, estriol, progesterone, prolactin, and human chorionic gonadotropin in pregnant women with alcohol abuse. J Clin Endocrinol Metab 1987;64: 153-6.
45. Hegedus L, Rasmussen N, Ravn V, Kastrup J, Krogsgaard K, Aldershvile J. Independent effects of liver disease and chronic alcoholism on thyroid function and size : the possibility of a toxic effect of alcohol on the thyroid gland. Metabolism 1988;37: 229-33.
46. Hoffman PL. Structural requirements for neurohypophyseal peptide maintenance of ethanol tolerance. Pharmacol Biochem Behav 1982;17: 685-90.
47. Hoffman PL, Tabakoff B. Mechanisms of alcohol tolerance. In: Thirteenth Annual Symposium, The North Carolina Alcoholism Research Authority. Alcohol Alcoholism 1989;24: 251-2.
48. Irwin M, Dreyfus E, Baird S, Smith TL, Schuckit M. Testosterone in chronic alacoholic men. Br J Addict 1988;83: 949-53.
49. Joffre BI et al. Hormonal responses in ethanol induced hypoglycaemia. J Stud Alcohol 1975;36: 550-4.
50. Johnston DG et al. C-peptide and insulin in liver disease. Diabetes 1978;27 (suppl): 201-6.
51. Junge J, Bentsen KD, Christoffersen P, Orholm M, Sorensen TIA, Horn T. Fibronectin as predictor of cirrhosis in men who abuse alcohol. Br Med J 1988;296(6637): 1629-30.
52. Kakihana R, Butte JC. Ethanol and endocrine function. In: Majchrowicz E, Noble EP, eds. Biochemistry and pharmacology of ethanol, vol 2. New York: Plenum, 1979: 147-64.
53. Klingman GI, Goodall M. Urinary epinephrine and levarterenol excretion during acute sublethal alcohol intoxication in dogs. J Pharmacol 1957;121: 313-8.
54. Lambie DG. Alcoholic brain damage and neurological symptoms of alcohol withdrawal : manifestations of overhydration. Med Hypothese 1985;16: 377-88.
55. Leppäluoto J, Rapeli R, Ranta T. Secretion of anterior pituitary hormones in man: effects of ethyl alcohol. Acta Physiol Scand 1975;95: 400-6.
56. Linkola J. et al. Renin-aldosterone axis in ethanol intoxication and hangover. Eur J Clin Invest 1976;6: 191-4.
57. Ljunghall S, Lundin L, Wide L. Acute effects of ethanol intake on the serum concentrations of parathyroid hormone, calcium and phosphate. Exp Clin Endocrinol 1985;85: 365-8.
58. Loosen PT et al. Thyrotropin-releasing hormone (TRH) in abstinent alcoholic men. Am J Psychiatry 1983;140: 1145-9.
59. Loosen PT. Thyroid function in affective disorders and alcoholism. Endocrinol Metab 1988;17: 55-82.
60. Maddrey WC. Is therapy with testosterone or anabolic-androgenic steroids useful in the treatment of alcoholic liver disease? Hepatology 1986;6: 1033-5.
61. Majumdar SK, Shaw GK, Bridges PK. Relationship between plasma adrenocorticotropic hormone and cortisol concentrations in chronic alcoholic patients with depression. Drug Alcohol Depend 1989;23: 111-6.
62. Mandell W, Miller C. Male sexual dysfunction as related to alcohol consumption; a pilot study. Alcoholism Clin Exp Res 1983;7: 65-9.
63. Mander AJ, Smith MA, Kean DM et al. Brain water measured in volunteers after alcohol and vasopressin. Lancet 1985;ii(8463): 1075.

64. Mardh G, Dingley AL, Auld DS, Vallee BL. Human class II (pi) alcohol dehydrogenase has a redox-specific function in norepinephrine metabolism. Proc Natl Acad Sci USA 1986;83: 8909-12.
65. Markianos M, Moussas G, Lykouras LL. Normal testosterone plasma levels in non-abstinent alcoholics. Drug Alcohol Depend 1987;20: 81-5.
66. Marshburn PB, Sloan CS, Hammond MG. Semen quality and association with coffee drinking, cigarette smoking, and ethanol consumption. Fertil Steril 1989;52: 162-5.
67. Medalle R, Waterhouse C, Hahn TJ. Vitamin D resistance in magnesium deficiency. Am J Clin Nutr 1976;29: 854-8.
68. Mendelson JH, Mello NK, Cristofaro P et al. Alcohol effects on naloxone-stimulated luteinizing hormone, prolactin and estradiol in women. J Stud Alcohol 1987;48: 287-94.
69. Merry J, Marks V. The effect of alcohol, barbiturate, and diazepam on hypothalamic/pituitary/adrenal function in chronic alcoholics. Lancet 1972;ii: 990-2.
70. Merry J, Marks V. Hypothalamic pituitary-adrenal function in chronic alcoholics. Adv Exp Med Biol 1973;35: 167-79.
71. Millar K, Jeffcoate WJ, Walder CP. Vasopressin and memory: improvement in normal short-term recall and reduction of alcohol-induced amnesia. Psychol Med 1987;17: 335-41.
72. Miller F, Barasch A, Sacks M, Levitan J, Ashcroft L. Serum prolactin correlates with depressed mood during alcohol withdrawal. Drug Alcohol Depend 1986;17: 331-8.
73. Morgan MY. Alcohol and the endocrine system. Br Med Bull 1982;38: 35-42.
74. Morgan MY, Pratt OE. Sex, alcohol and the developing fetus. Br Med Bull 1982;38: 43-52.
75. Nikkilä EA, Taskinen MR. Ethanol induced alterations of glucose tolerance, post glucose hypoglycemia and insulin secretion in normal, obese and diabetic subjects. Diabetes 1975;24: 933-43.
76. Ollat H, Parvez H, Parvez S, Tipton KF. Alcohol and central neurotransmission. Neurochem Int 1988;13: 275-300.
77. Paille F, Baille N, Barrucand D. Principales conséquences biologiques de l'alcoolisation (2ième partie). Revue Alcoolisme 1983;29: 65-107.
78. Pohorecky LA. Effects of ethanol on central and peripheral noradrenergic neurons. J Pharmacol Exp Ther 1974;189: 380-91.
79. Potter JF, Bannan LT, Beevers DG. The effect of a non-selective lipophilic beta-blocker on the blood pressure and noradrenaline, vasopressin, cortisol and renin release during alcohol withdrawal. Clin Exp Hypertens Theor Pract 1984;6: 1147-60.
80. Prinz PN, Roehrs TA, Vitaliano PP et al. Effects of alcohol on sleep and nighttime plasma growth hormone and cortisol concentrations. J Clin Endocrinol Metabol 1980;51: 759-64.
81. Redmond GP. Effect of ethanol on endogenous rhythms of growth hormone secretion. Alcoholism Clin Exp Res 1980;4: 50-6.
82. Rosenblum ER, Van Thiel DH, Campbell IM, Eagon PK, Gavaler JS. Separation and identification of phytoestrogenic compounds isolated from bourbon. In: Lindros KO, Ylikahri R, Kiianmaa K, eds. Advances in biomedical alcohol research: third congress of the International Society for Biomedical Research on Alcoholism (Helsinki, 8—13 June 1986). Alcohol Alcoholism 1987(suppl 1): 551-5.
83. Rumilly F et al. Les hormones thyroïdiennes dans la cirrhose éthylique avant et après sevrage alcoolique. Semin Hop Paris 1983;59: 390-6.
84. Salata R, Klein I, Levey GS. Thyroid hormone homeostasis and the liver. Semen Liverdis 1985(5): 29-34.
85. Santucci L, Graham TJ, Van Thiel DH. Inhibition of testosterone production by rat Leydig cells with ethanol and acetaldehyde: prevention of ethanol toxicity with 4-methyl-pyrazole. Alcoholism Clin Exp Res 1983;7: 135-9.
86. Schade RR, Bonner G, Gay VC et al. Evidence for a direct inhibitory effect of ethanol upon gonadotropin secretion at the pituitary level. Alcoholism Clin Exp Res 1983;7: 150-2.
87. Schlienger JL. Thyroid status in fifty patients with alcoholic cirrhosis. Z Gastroenterol 1979;17: 452-61.
88. Subramanian MG, Abel EL. Alcohol inhibits suckling-induced prolactin release and milk yield. Alcohol: Int Biomed J 1988;5: 95-8.

89. Taskinen M-R, Nikkilä EA. Nocturnal hypertriglyceridemia and hyperinsulinemia following moderate evening intake of alcohol. Acta Med Scand 1977;202: 173-7.
90. Teschke R, Wiese B. Sex-dependency of hepatic alcohol metabolzing enzymes. J Endocrinol Invest 1982;5: 243-50.
91. Teschke R, Heymann K. Effect of sex hormones on the activities of hepatic alcohol-metabolizing enzymes in male rats. Enzyme 1982;28: 268-77.
92. Välimäki M, Salaspuro M, Härkönen M et al. Liver damage and sex hormones in chronic male alcoholics. Clin Endocrinol 1982;17: 469-77.
93. Välimäki M, Härkönen M, Ylikahri R. Acute effects of alcohol on female sex hormones. Alcoholism Clin Exp Res 1983;7: 289-93.
94. Välimäki M, Pelkonen R, Salaspuro M, Härkönen M, Hirvonen E, Ylikahri R. Sex hormones in amenorrheic women with alcoholic liver disease. J Clin Endocrinol Metab 1984:59: 133-8.
95. Välimäki MJ, Härkönen M, Eriksson CJP, Ylikahri RH. Sex hormones and adrenocortical steroids in men acutely intoxicated with ethanol. Alcohol 1984;1: 89-93.
96. Välimäki M, Ylikahri RH. Endocrine effects of alcohol. In: Parvez S, Burov Y, Parvez H, Burns E, eds. Alcohol nutrition and the nervous system. Utrecht: VNU Science Press, 1985: 265-86. (Progress in alcohol research vol 1.)
97. Välimäki M, Pelkonen R, Karonen SL, Ylikahri R. Effect of ethanol on serum concentrations of somatomedin C and the growth hormone (GH) secretion stimulated by the releasing hormone (GHRH). In: Lindros KO, Ylikahri R, Kiianmaa K, eds. Advances in biomedical alcohol research: third congress of the International Society for Biomedical Research on Alcoholism (Helsinki, 8–13 June 1986). Alcohol Alcoholism 1987(suppl 1): 557-9
98. Van Dyke HB, Ames RG. Alcohol diuresis. Acta Endocrinol 1951;7: 110-21.
99. Van Ree JM. Role of pituitary and related neuropeptides in alcoholism and pharmacodependence. Prog Neuropsychopharmacol Biol Psychiatry 1986;10: 219-28.
100. Van Thiel DH et al. Evidence for a defect in pituitary secretion of luteinizing hormone in chronic alcoholic men. J Clin Endocrinol Metab 1978;37: 499-507
101. Van Thiel DH, Gavaler JS. The adverse effects of ethanol upon hypothalamic-pituitary-gonadal function in males and females compared and contrasted. Alcoholism Clin Exp Res 1982;6: 179-85.
102. Van Thiel DH. Ethanol: its adverse effects upon the hypothalamic-pituitary-gonadal axis. J Lab Clin Med 1983;101: 21-33.
103. Van Thiel DH, Gavaler JS, Sanghvi A. Recovery of sexual function in abstinent alcoholic men. Gastroenterology 1983;84: 677-82.
104. Wells SA, Cooper CW, Ontjes DA. Stimulation of thyrocalcitonin secretion by ethanol in patients with medullary thyroid carcinoma – an effect not mediated by gastrin. Metabolism 1975;24: 1215-19.
105. Ylikahri RH, Huttunen MO, Härkönen M. Hormonal changes during alcohol intoxication and withdrawal. Pharmacol Biochem Behav 1980;13(suppl 1): 131-7.

CHAPTER 5

Effects of alcohol on the metabolism of macronutrients

W.G. Vrij-Standhardt

5.1. Introduction

The consumption of alcohol has various effects on the organism. These can be classified into:
- the effects of alcohol
- the effects of acetaldehyde
- the effect of the increase in NADH/NAD$^+$ ratio[1] in the liver.

The shifts in metabolism following alcohol consumption are mainly caused by the increase in NADH/NAD$^+$ ratio as a result of alcohol metabolization, whereas the effects of alcohol and acetaldehyde rather cause organ damage. The increase of NADH/NAD ratio resulting from ADH and ALDH reactions has several consequences (see Fig. 5.1). Basically, all reactions that need NAD$^+$ – or another oxidizing equivalent – are inhibited, whereas reactions in which NAD$^+$ is generated are stimulated. The increased NADH/NAD$^+$ ratio thus favours the conversion of pyruvate into lactate, fructose into sorbitol, acetoacetate into β-hydroxybutyrate, and oxaloacetate into malate. The activity of the citric acid cycle, the β-oxidation of fatty acids and the activation of acetate to acetyl-coenzyme A (acetyl-CoA) are inhibited.

Glycolysis yields pyruvate, which is normally oxidized in the citric acid cycle. During alcohol metabolism the citric acid cycle (which produces hydrogen as well) is inhibited because of the increase in NADH/NAD$^+$ ratio. Pyruvate will then be converted into lactate – for which process NADH is necessary – resulting in an increased lactate/pyruvate ratio (28, 38, 39). The extent of the shift in lactate/pyruvate ratio can be influenced by the diet. After a 5-day pre-treatment with a low-carbohydrate high-fat diet the lactate/pyruvate ratio during alcohol metabolism was significant higher than on a high-carbohydrate low-fat diet (39). There was no effect on alcohol metabolism rate. Persons consuming a low-carbohydrate high-fat diet are more vulnerable to negative metabolic effects of alcohol (such as increased fat synthesis and decreased fat catabolism, gluconeogenesis inhibition, decreased lactate clearance after exercise).

---

[1] NAD, nicotinamide adenine dinucleotide; NAD$^+$, oxidized form of NAD; NADH, reduced form of NAD.

As a rule, the increase in blood lactate concentration (hyperlactacidaemia) is moderate and the acidosis is mild. Severe lactic acidosis is uncommon in otherwise healthy intoxicated emergency patients (10). Usually clinical consequences are obviated by the buffer capacity of the blood. Occasionally, however, this buffer capacity may be exceeded, thus causing lactic acidosis, which may be fatal. This may particularly occur when there are already large amounts of lactate present in the blood, for example due to liver injury (resulting from alcoholism), or oxygen deficits attendant on pulmonary disease. Fructose is known to induce lactic acidosis and should therefore be avoided in cases of alcohol intoxication (32).

The overproduction of lactate may cause an increased level of uric acid in the blood (hyperuricaemia). This may lead to attacks of gout, well known in excessive drinkers. Hyperuricaemia is a consequence of decreased urinary excretion of uric acid secondary to hyperlactacidaemia owing to competitive inhibition of renal clearance by lactate (3, 25, 47), or to increased urate synthesis due to enhanced turnover of adenine nucleotides (8). Alcohol-induced ketosis (an abnormally elevated concentration of ketone bodies) may also promote hyperuricaemia (23).

Alcoholic ketosis is a syndrome occurring predominantly in chronic alcohol abusers who have had a recent binge, which is followed by severe abdominal pain and vomiting (due to gastritis, hepatitis or pancreatitis). This leads to interruption of food and alcohol consumption, causing dehydration, starvation and ketosis. However, according to Lefevre (23), ketonaemia can be induced by chronic alcohol administration in combination with calorically adequate diets.

The pathophysiology of alcoholic ketosis is reasonable well understood. In fact the term 'alcoholic ketoacidosis' is incorrect (60). Alcohol inhibits ketogenesis (41), probably due to the increased $NADH/NAD^+$ ratio. In response to starvation fat stores are mobilized, thus leading to a marked elevation of serum free fatty acids. When alcohol ingestion is stopped, due to abdominal complaints, the block in ketogenesis is released and the rapid conversion of free fatty acids leads to substantial production of ketone bodies in the liver. The normal peripheral metabolism of ketone bodies tend to decrease because levels of insulin – necessary for ketone body metabolism – are often low in this syndrome. In alcohol-associated ketosis the β-hydroxybutyrate/-acetoacetate ratio increases. This accounts for the often minimally positive serum ketones found in dipstick tests: this procedure only measures acetoacetate, not β-hydroxybutyrate.

Many, but not all, patients with lactic acidosis or ketoacidosis have a low blood pH (acidaemia). Acidaemia can be (over)compensated by hyperventilation or can occur together with metabolic alkalosis, caused by gastritis or pancreatitis, the final pH depending on whether acidosis or alkalosis predominates (11).

Another possible effect of hyperlactacidaemia is the stimulation of collagen production through increased activity of proline hydroxylase. In addition to hyperlactacidaemia, changes in the $NADH/NAD^+$ ratio may also cause changes in carbohydrate, fat and protein metabolism.

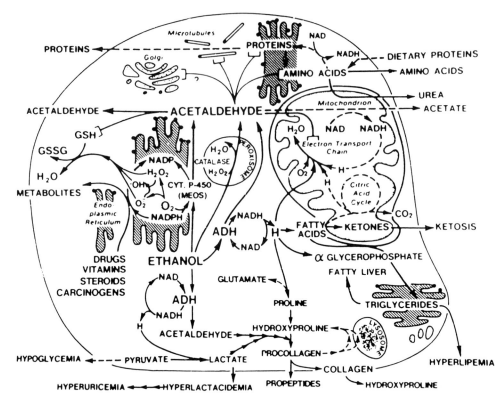

Fig. 5.1. Oxidation of alcohol in the hepatocyte and link of the two products (acetaldehyde and H) to disturbances in intermediary metabolism (fatty liver, hyperlipaemia, hyperuricaemia, hyperlactacidaemia, ketosis). NAD, nicotinamide adenine dinucleotide; NADH, reduced NAD; NADP, nicotinamide adenine dinucleotide phosphate; NADPH, reduced NADP; MEOS, microsomal ethanol-oxidizing system; ADH, alcohol dehydrogenase. The broken lines indicate pathways that are depressed by alcohol. The symbol —[denotes interference or binding. Arrowed lines denote stimulation or activation. After: Lieber (31).

5.2. Effects of alcohol on carbohydrate metabolism

Carbohydrate metabolism is a complex process in which at least five hormones are involved. Blood glucose level is maintained within certain limits (ca. 1 g/l) by these hormones. Dietary carbohydrates are digested and converted into glucose. Excess glucose is converted into glycogen, which is stored in the liver. Another source of glucose is synthesis from sources other than carbohydrates such as protein or fat (gluconeogenesis).

Alcohol affects carbohydrate metabolism by various mechanisms most of which are as yet not entirely understood. One possible consequence of alcohol consumption is a decrease of blood glucose concentration (hypoglycaemia), which may even cause sudden death (25, 37, 38, 43). The drop in blood glucose concentration may be so drastic that brain damage results. Brain damage in alcoholics must possibly be attributed to hypoglycaemia, rather than to alcohol itself (37, 38). The classic clinical features of hypoglycaemia are tachycardia, sweating, tremulousness, confusion, irritability and headache. Mild hypothermia may occur as well. Hypoglycaemia is brought about by several mechanisms.

The type of hypoglycaemia that is both the best known and the most serious form is alcohol-induced fasting hypoglycaemia. As a result of an increased $NADH/NAD^+$ ratio and an increased lactate/pyruvate ratio, gluconeogenesis in the liver is inhibited. In well-fed persons glucose is then released from the glycogen stores. When glycogen stores are depleted, normal blood glucose levels cannot be maintained and hypoglycaemia occurs. In starved alcoholics, but also in people consuming low-carbohydrate diets and fasting people (missing a meal or two while drinking), glycogen stores are depleted. Children are particularly susceptible, even without fasting, as their glycogen stores are readily depleted (51). Although alcohol-induced fasting hypoglycaemia is uncommon, it is important that it is diagnosed and treated early, because the outcome can be lethal.

Another type of hypoglycaemia is alcohol-induced reactive hypoglycaemia (37, 38). Following a carbohydrate-rich meal, a natural rise in blood glucose concentration takes place, resulting in an increased insulin release until a lower blood glucose level is reached. Alcohol seems to potentiate the insulin response to elevated blood glucose levels, resulting in reactive hypoglycaemia, about 2–3 hours after eating (44, 51). The exact mechanism is not known; possibly intracellular cyclic adenosine monophosphate (cyclic AMP, or cAMP) plays a part in it. Clinical tests have indicated that about 30 g alcohol plus 80 g carbohydrate are sufficient to induce reactive hypoglycaemia. The rapid decline in blood glucose often triggers a stress response (adrenaline release). In alcoholics the mechanism of alcohol-induced reactive hypoglycaemia seems to be different from that described for non-alcoholics, possibly due to disturbed hormonal processes (51).

In non-fasting animals and humans acute alcohol consumption has been found to result in glucose intolerance (prolonged elevated blood glucose level after glucose load) or hyperglycaemia (7, 35). However, other studies have not confirmed these effects (56, 57). Considerable variation exists in the reported results of glucose tolerance tests in alcoholics (48). Hyperglycaemia in relation to alcoholism has at times been observed, but its mechanism is not clear. Alcohol-related pancreatitis or adrenergic stimulation of glycogenolysis by acetaldehyde may play a role here (25, 43, 47). Pezzarossa et al. (49) suggest that glucose intolerance in alcoholics is primarily caused by an unadequate diet.

Oxidation of galactose is markedly inhibited by the increased redox state in the liver upon alcohol ingestion, but this has no clinically relevant consequences (18).

5.3. Effects of alcohol on fat metabolism

In the liver fat is synthesized from α-glycerophosphate and fatty acids originating from nutritional fats, adipose tissue or endogenous synthesis. The fat thus formed may be stored in the liver or released again into the bloodstream as very low-density lipoprotein (VLDL). Furthermore, fatty acids are used as a source of energy in the liver, through conversion into CO_2 and H_2O in the citric acid cycle after breakdown to 2-C fragments. There is a strong variation in either of these pathways prevailing at any given time. This is determined by the amount and type of nutrients a meal contains, the period of time elapsed since the meal has been taken, but also by hormonal factors and stress.

Alcohol also affects these mechanisms in various stages (see Fig. 5.2). The increase in $NADH/NAD^+$ ratio causes more α-glycerophosphate to be formed. It also leads to fatty acid synthesis from acetyl-CoA (partly derived from alcohol metabolism) (15) being promoted because hydrogen equivalents (NADPH) are consumed during the process. The oxidation of fatty acids is decreased (NAD^+-dependent reaction). Consequently, the mitochondria consume the H equivalents formed during alcohol metabolism, rather than the Hs obtained by the oxidation of 2-C fragments of fatty acids in the citric acid cycle. Mitochondrial injury due to chronic alcohol consumption also disturbs the citric acid cycle, even when actually no alcohol is metabolized. All these factors promote the hepatic formation of triglycerides causing fatty degeneration of the liver (25–27, 52). So, the fat present in a fatty liver is not generated from alcohol alone, as is sometimes thought. A direct effect of alcohol on adipose tissue has not been established (9, 61).

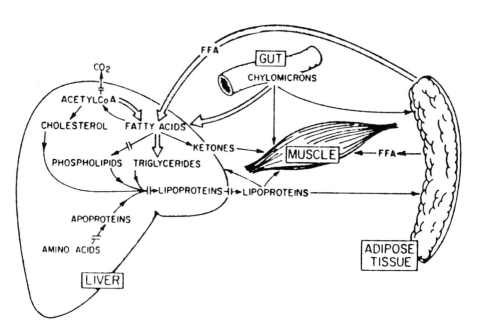

Fig. 5.2. Possible mechanisms of alcohol-induced fatty liver formation. Unbroken arrows: stimulation by alcohol; interrupted arrows: inhibition by alcohol. After: Lieber et al. (27).

The triglyceride concentration in blood is considered as a measure of the transport of lipids from the liver to the periphery (20). Even when blood alcohol concentrations are low, in the liver more fat is formed and less is broken down than without alcohol. This fat is partly released into the bloodstream as lipoprotein, as appears from an increase of the triglyceride level in the blood. In some investigations no change in the triglyceride level upon alcohol consumption was observed (14, 19). In most experiments, however, the triglyceride level proved to increase, mainly in the VLDL fraction (1, 4, 6, 12, 20, 21, 36, 46, 58). Lipoprotein release decreases at an alcohol concentration exceeding 2 g/l, possibly due to a disturbance of protein metabolism (21) or liver membranes. This leads to a lower plasma triglyceride level and an accumulation of fat in the liver. This fatty degeneration of the liver will disappear some time after alcohol has been eliminated from the body, if there is a time lag sufficient for recovery. (Fatty degeneration is discussed extensively in Chapter 8.)

Not just the total amount of triglycerides in the blood, but also the composition of the blood lipoproteins change during alcohol consumption. (These changes of the various lipoprotein fractions (HDL, VDL, VLDL), as well as cholesterol changes, are discussed in Chapter 10.)

Alcohol changes polyunsaturated fatty acid (PUFA) distribution (stimulation of eicosanoid formation, reduction of arachidonic acid). Results of most studies suggest that alcohol abuse leads to a PUFA deficiency in several tissues, affecting membrane fluidity. Dietary intervention may be effective, but recommendations for dietary supplementation cannot be given yet (53). Evidence is accumulating that part of the actions of alcohol is mediated by essential fatty acids and prostaglandins (13, 17). Alcohol affects essential fatty acid and prostagladin metabolism. The acute effect of alcohol is an increased production of prostaglandin E1 from dihomogammalinolenic acid (DGLA). Becease alcohol inhibits the production of DGLA from dietary linoleic acid, chronic alcohol consumption causes depletion of DGLA and hence of prostaglandin E1 (PGE1). Withdrawal will lead to a sudden drop of PGE1. PGE1 is known to have profound effects on the nervous system and behaviour. Horrobin (16, 17) suggests that disturbances in PGE1 and essential fatty acid metabolism mediate (via membrane changes) some of the acute behavioural effects of alcohol and some chronic effects of alcohol (alcoholic fibrosis, cardiomyopathy, pancreatitis, FAS and other alcohol-related disorders).

5.4. Effects of alcohol on protein metabolism

Until the 1980s, the changes in the amount of hepatic proteins in alcoholics used to be attributed to an indequate diet (alcoholic beverages may supply a lot of energy, but hardly any proteins or other essential nutrients), and to decreased protein absorption due to intestinal epithelial damage caused by excessive alcohol consumption. Alcohol proves, however, to have some effect on liver protein metabolism as well. The acetaldehyde formed, and the excess of NADH in particular, cause these effects. Lieber and co-workers are important investigators in this field (2, 5, 28, 33, 55).

Chronic alcohol feeding results in increased urinary excretion of nitrogen in rats and humans (40, 50). The increased excretion of nitrogen in man is associated with a negative nitrogen balance and weight loss. Chronic alcohol administration in rats decreases whole-body protein synthesis, due to a reduced efficiency in recycling nitrogen for protein synthesis (42). It seems clear that alcohol inhibits cardiac protein synthesis, be it by an as yet unidentified mechanism. The relationship between alterations in cardiac protein metabolism and the clinical and morphological consequences of excessive alcohol intake for the heart remains elusive (59).

The liver produces proteins which transport various compounds in the blood (transferrin, albumin, lipoproteins). Most (with the exception of albumin) are released into the bloodstream as glycoproteins. In vitro, alcohol decreases protein synthesis, but in vivo the acute effects of alcohol are less consistent (29). The release of liver proteins into the blood is inhibited by alcohol, in which process acetaldehyde-induced impairment of microtubular protein secretion or inhibition of the conversion of proteins and glucosamides into glycoproteins may play a part (2, 15, 33). Hepatomegaly (liver enlargement) due to alcohol is attributed to accumulation of fat for one half, and to an increase in proteins for the other (33).

Alcohol may, either directly or indirectly, influence both the synthesis and breakdown (catabolic reactions) of protein in the liver, as well as secretion of protein from the liver. So, both accumulation and a decrease in concentration of some amino acids and proteins may take place in the liver, as a net result. Plasma concentrations can be elevated or decreased too.

Plasma branched amino acid levels are increased by chronic alcohol consumption, but decreased by an inadequate diet and cirrhosis (34). In alcoholics increased plasma concentrations of the aromatic amino acids tyrosine and phenylalanine and of α-aminobutyric acid have been found (5, 55).

Tryptophan is decreased in blood of alcoholics with less severe liver injury, and increased in patients with severe liver injury and portal-systemic encephalopathy (PSE). Besides being a component of proteins, tryptophan is the precursor of the neurotransmitter serotonin and is possibly involved in the aetiology of PSE.

Methionine – a sulphur-containing amino acid – is a precursor of glutathione, a tripeptide protecting cells against oxidative and peroxidative damage. Chronic alcohol consumption is associated with a decrease in mitochondrial glutathione and increased peroxidation (24, 34). This may explain the partial protective effect of large amounts of methionine, which the liver can use to synthesize more glutathione (54).

In fibrosis liver cells proliferate and produce increased amounts of collagen. Lactate may be a factor in fibrogenesis by promoting collagen synthesis from proline. It is not clear whether increased collagen in fibrotic cells can be attributed to increased synthesis (demonstrable from the increased activity of proline hydroxylase) or decreased breakdown (22, 30, 32, 45).

From a hypothetical point of view, these modifications of protein metabolism may have far-reaching consequences for several organs (e.g. the brain) and body functions (such as the transport of iron and lipids by the blood). Hardly anything, however, is known about the clinical consequences.

Summary

Due to alcohol metabolism various changes in the metabolism of macronutrients (carbohydrates, proteins, fat) take place. The increase in $NADH/NAD^+$ ratio in the liver resulting from alcohol oxidation may have several consequences. Basically, all reactions for which NAD, or another oxidizing equivalent, is necessary will be inhibited, whereas reactions in which NAD is formed are stimulated. One of the first consequences is an increase in the level of lactic acid in the blood (hyperlactacidaemia). As a rule, the increased lactate level does not have any consequences because of the buffer capacity of the blood. If the buffer capacity is exceeded, lactate acidosis will follow. An increased level of uric acid in the blood (hyperuricaemia) may result from hyperlactacidaemia. This may cause attacks of gout, well known in excessive drinkers. The exact mechanism is not known. Alcohol-induced ketosis can be brought about by fasting after a binge.

The change in $NADH/NAD^+$ ratio does not only cause hyperlactacidaemia, but also changes in carbohydrate, protein and fat metabolism. Acute alcohol consumption may cause fasting and reactive hypoglycaemia, whereas in chronic alcoholics hyperglycemia and glucose intolerance are found.

Hyperlipidaemia and accumulation of fat in the liver are induced by a complex of factors. Both an increased production of fat in the liver and an increased supply of fat to the liver, as well as a decreased fatty acid oxidation, play a role here. At a high blood alcohol concentration the inhibition of lipoprotein release reinforces these effects. The decreased oxidation of fatty acids is a result of the liver using alcohol as an 'alternative fuel'.

The influence of alcohol on protein metabolism is an extremely complex matter, which as yet has hardly been clarified. Alcohol metabolism affects both protein synthesis and breakdown, as well as protein secretion from the liver.

References

1. Avogaro P, Cazzolato. Changes in the composition and physico-chemical characteristics of serum lipoproteins during ethanol-induced lipaemia in alcoholic subjects. Metabolism 1975;24: 1231-42.
2. Baraona E, Lieber CS. Effects of ethanol on hepatic protein metabolism. Am J Psychiatry 1982;139: 188-9.
3. Beck LH. Clinical disorders of uric acid metabolism. Med Clin N Am 1981;65: 401-11.
4. Böttiger LE, et al. Serum lipids in alcoholics. Acata Med Scand 1976;199: 357-61.
5. Branchey L, Lieber CS. Activation of tryptophan pyrrolase after chronic alcohol administration. Subst Alcohol Actions Misuse 1982;3: 225-9.
6. Castelli WP, et al. Alcohol and blood lipids. Lancet 1977;ii: 153-6.

7. Connelly DM, Harries EHL, Taberner PV. Differential effects of ethanol on the plasma glucose of non-alcoholic light and heavy social drinkers. Alcohol Alcoholism 1987;22: 23-9.
8. Faller J, Fox IH. Ethanol-induced hyperuricemia: evidence for increased urate production by activation of adenine nucleotide turnover. N Engl J Med 1982;307: 1598-1602.
9. Frain KN, Coppack SW, Walsh PE, Butterworth HC, Humphreys SM, Pedrosa HC. Metabolic responses of forearm and adipose tissues to acute ethanol ingestion. Metabolism 1990;39: 958-66.
10. Fulop M, Bock J, Ben-Ezra J, Antony M, Danzig J, Gage JS. Plasma lactate and 3-hydroxybutyrate levels in patients with acute ethanol intoxication. Am J Med 1986;80: 191-4.
11. Fulop M. Alcoholism, ketoacidosis, and lactic acidosis. Diabetes Metabolism Rev 1989;5(4): 365-78.
12. Ginsberg H, et al. Moderate alcohol ingestion and plasma triglyceride levels. Ann Internal Med 1974;80: 143-9.
13. Glen I, Skinner F, Glen E, MacDonell L. The role of essential fatty acids in alcohol dependence and tissue damage. Alcohol Clin Exp Res 1987;11: 37-41.
14. Glueck CJ, et al. Effects of alcohol ingestion on lipids and lipoproteins in normal men: isocaloric metabolic studies. Am J Clin Nutr 1980;33: 2287-93.
15. Haller H, et al. Alkohol und Fettstoffwechsel. Z Gesammte Inn Med 1981;36: 552-7.
16. Horrobin DF. A biochemical basis for alcoholism and alcohol-induced damage including the fetal alcohol syndrome and cirrhosis: interference with essential fatty acid and prostaglandin metabolism. Med Hypotheses 1980;6: 929-42.
17. Horrobin DF. Essential fatty acids, prostaglandins, and alcoholism: an overview. Alcohol Clin Exp Res 1987;11: 2-9.
18. Isselbacher KJ. Metabolic and hepatic effects of alcohol. N Engl J Med 1977;296: 612-6.
19. Johanson BG, Medhus A. Increase in plasma α-lipoproteins in chronic alcoholics after acute abuse. Acta Med Scand 1974;195: 273-7.
20. Jones DP, et al. Effects of ethanol on plasma lipids in man. J Lab Clin Med 1963;62: 675-82.
21. Käferstein H, et al. Einfluss von akutem Äthanolkonsum auf Gesamtmengen und Fettsäuremuster der Triglyceride und freien Fettsäuren im Serum. Blutalkohol 1982;19: 7-14.
22. Kowaloff EM, et al. Regulation of proline oxidase activity by lactate. Proc Natl Acad Sci USA 1977;74: 5368-71.
23. Lefèvre A, Adler H, Lieber CS. Effect of ethanol on ketone metabolism. J Clin Invest 1970;49: 1775.
24. Li T-K. The absorption, distribution, and metabolism of ethanol and its effects on nutrition and hepatic function. In: Tabakoff B, Sutker PB, Randall CL, eds. Medical and social aspects of alcohol abuse. New York: Plenum, 1984: 47-77.
25. Lieber CS. Liver adaptation and injury in alcoholism. N Engl J Med 1973;288: 356-62.
26. Lieber CS, et al. Effect of chronic alcohol consumption on ethanol and acetaldehyde metabolism. Adv Exp Med Biol 1975: 185-227.
27. Lieber CS, et al. Differences in hepatic and metabolic changes after acute and chronic alcohol consumption. Fedn Proc 1975(Abstr 34): 2060-74.
28. Lieber CS. Alcohol, protein metabolism, and liver injury. Gastroenterology 1980;79: 373-90.
29. Lieber CS. Alcohol, protein nutrition and liver injury. Curr Conc Nutr 1983;12: 49-71.
30. Lieber CS. Alcohol and the liver. In: Arias YM, Frenkel M, Wilson JHP, eds. Liver annual 3: A series of critical surveys of the international literature. Amsterdam: Elsevier, 1983: 106-48.
31. Lieber CS. Alcohol and the liver: 1984 update. Hepatology 1984;4: 1243-60.
32. Lieber CS. Alcohol and the liver: metabolism of ethanol, metabolic effects and pathogenesis of injury. Acata Med Scand 1987(Suppl 703): 11-55.
33. Lieber CS. Toxic and metabolic changes induced by ethanol. In: Goedde HW, Agarwal DP, eds. Alcoholism: biomedical and genetic aspects. New York: Pergamon Press, 1989: 57-83.
34. Lieber CS. Alcohol and nutrition: an overview. Alcohol Health Res World 1989;13: 197-205.
35. Lombrail P, Lang T, Degoulet P, et al. Alcohol consumption and impaired glycoregulation results in a population of 6665 salaried employees. Eur J Epidemiol 1988;4: 371-6.
36. Losowsky MS, et al. Studies of alcoholic hyperlipemia and its mechanism. Am J Med 1963;35: 794-803.
37. Marks V. Alcohol and carbohydrate metabolism. Clin Endocrinol Metabolism 1978;7: 273-96.

38. Marks V. Biochemical and metabolic basis of alcohol toxicity. In: Mendlowicz J, van Praag HM, eds. Alcoholism: A multidisciplinary approach, vol 3. Basel: Karger, 1979: 88-96.
39. Mascord D, Rogers J, Smith J, Starmer GA, Whitfield JB. Effect of diet on [lactate]/[pyruvate] ratios during alcohol metabolism in man. Alcohol Alcoholism 1989;24: 189-191.
40. McDonald JT, Margen S. Wine versus ethanol in human nutrition. 1. Nitrogen and calorie balance. Am J. Clin Nutr 1976;29: 1093-1103.
41. McGarry JD, Foster DW. The regulation of ketogenesis from oleic acid and the influence of antiketogenic agents. J Biol Chem 1971;246: 6247-53.
42. Mezey E. Metabolic effects of alcohol. Fedn Proc 1985;44: 134-8.
43. Morgan MY. Alcohol and the endocrine system. Br Med Bull 1982;38(1): 35-42.
44. O'Keefe SJD, Marks V. Lunchtime gin and tonic: A cause of reactive hypoglycaemia. Lancet 1977;i: 1286-8.
45. Oratz M. Alcohol and protein synthesis. In: Biomedical processes and consequences of alcohol use. Rockville, MD: NIAAA, 1982: 31-49. (Alcohol and health monograph 2.)
46. Ostrander LD, et al. Relationship of serum lipid concentrations to alcohol consumption. Arch Internal Med 1974;134: 451-6.
47. Paille F, Baille N, Barrucand D. Principales conséquences biologiques de l'alcoolisation (première partie). Revue Alcoolisme 1982;28: 129-61.
48. Patel DG. Effects of ethanol on carbohydrate metabolism and implications for the aging alcoholic. Alcohol Health Res World 1989;13: 240-6.
49. Pezzarossa A, Cervigni C, Ghinelli, F, Molina E, Gnudi A. Glucose tolerance in chronic alcoholics after alcohol withdrawal: effect of accompanying diet. Metabolism 1986;35: 984-8.
50. Reinus JF, Heymsfield SB, Wiskind R, Casper K, Galambos JT. Ethanol: relative fuel value and metabolic effects in vivo. Metabolism 1989;38: 125-35
51. Ryan K. Alcohol and blood sugar disorders: an overview. Alcohol Health Res World 1983;8: 3-7, 15.
52. Sabesin SM. Effects of alcohol on liver and blood lipids and lipoproteins. In: Biomedical processes and consequences of alcohol use. Rockville, MD: NIAAA, 1982: 131-70. (Alcohol and health monograph 2.)
53. Salem N. Alcohol, fatty acids, and diet. Alcohol Health Res World 1989;13: 211-8.
54. Shaw S, Jayatilleke E, Ross WA, Gordon ER, Lieber CS. Ethanol-induced lipid peroxidation: potentiation by long-term alcohol feeding and attenuation by methionine. J Lab Clin Med 1981;98: 417-24.
55. Shaw S, Lieber CS. Plasma amino acids in the alcoholic: nutritional aspects. Alcoholism Clin Exp Res 1983;7: 22-7.
56. Singh SP, Kumar Y, Snyder AK, Ellyin FE, Gilden JL. Effect of alcohol on glucose tolerance in normal and noninsulin-dependent diabetic subjects. Alcohol Clin Exp Res 1988;12: 727-30.
57. Shelmet JJ, Reichard GA, Skutches CL, Hoeldtke RD, Owen OE, Boden G. Ethanol causes acute inhibition of carbohydrate, fat, and protein oxidation and insulin resistance. J Clin Invest 1988;81: 1137-45.
58. Verdy M, Gattereau A. Ethanol, lipase activity and serum lipid level. Am J Clin Nutr 1967;20: 997-1003.
59. Ward LC. Ethanol and protein and amino acid metabolism in the rat. Int J Biochem 1987;19: 887-97.
60. Williams HE. Alcoholic hypoglycemia and ketoacidosis. Med Clin N Am 1984;68: 33-38.
61. World MJ, Ryle PR, Jones D, Shaw GK, Thomson AD. Differential effect of chronic alcohol intake and poor nutrition on body weight and fat stores. Alcohol Alcoholism 1984;19: 281-90.

CHAPTER 6

Effects of alcohol on vitamin metabolism

E. te Wierik and W.G. Vrij-Standhardt

Almost all research devoted to the effects of alcohol consumption on vitamin metabolism (and vice versa) has been done with alcoholics and animals. The effects of moderate alcohol consumption on vitamin metabolism, however, have hardly ever been investigated. In a series of papers Bonjour (10–16) describes the relation between vitamins and alcoholism. In this chapter various vitamins will be discussed in relation to alcohol use.

6.1. Vitamin A

In case of alcohol-induced cirrhosis of the liver symptoms of vitamin A deficiency may be observed, such as night blindness and testicular atrophy associated with decreased spermatogenesis (15, 61). The term 'vitamina A' is the generic descriptor for retinoids exhibiting qualitatively the biological activity of retinol.

Like ethanol retinol is an alcohol. A zinc-containing enzyme, alcohol dehydrogenase (ADH), is necessary to convert retinol into retinal, which plays an important role in vision and in spermatogenesis. By partly occupying the receptor site of retinol on the enzyme, ethanol may cause competitive inhibition of this conversion. Testicular atrophy is probably the result of enhanced peroxidation of testicular lipids that occurs upon ethanol exposure. Vitamin A stabilizes testicular membranes by acting as an antioxidant (62).

Decreased blood concentrations of vitamin A, retinol-binding protein (RBP), and often zinc, are frequently observed in alcoholics. This may be due to malnutrition since poor food dietary patterns have often been observed in alcoholics. Other causes of vitamin A deficiency may be malabsorption and a decreased storage of vitamin A in the injured liver (63).

Very low hepatic vitamin A levels have been found in patients with moderate liver disease such as fatty liver in the presence of normal serum concentrations of vitamin A and RBP (35). The reduction of hepatic vitamin A correlates with the degree of alcoholic liver injury in humans and in baboons (38).

A cause of the low vitamin A concentration is enhanced breakdown of vitamin A in the liver via cytochrome P-450, which is induced by alcohol consumption (65). Increased vitamin A metabolism as a consequence of chronic ethanol consumption probably results in the production of toxic vitamin A metabolites which may contribute to liver damage (35, 36, 37, 40). Therefore, although the alcoholic suffers from vitamin A deficiency, supplementation of the vitamin seems detrimental since vitamin A in excess can initiate hepatic injury. Furthermore, some patients do not respond to vitamin A when no supplementary zinc is administered as well.

6.2 Thiamin (formerly designated vitamin B-1)

Thiamin deficiency is frequently observed in chronic alcoholics; this may be attributed to (6, 17, 28, 56, 60, 69):
– inadequate thiamin intake,
– impairment of thiamin absorption due to an ethanol related nutritional deficiency state,
– decreased activation of thiamin to thiamin pyrophosphate (TPP) in the brain.

The inadequate dietary intake in many alcoholics, in particular a decreased thiamin absorption in the gastrointestinal tract, due to damage caused by alcohol and/or folate deficiency (23, 49), are presumably largely responsible for thiamin defiency in alcoholics. An altered thiamin requirement accompanying alcohol consumption has not been established. As a rule, thiamin requirement is related to energy intake since thiamin plays a role in catabolism, in particular in the conversion of pyruvate into acetyl-Coenzyme A (acetyl-CoA) and in the citric acid cycle. The breakdown of alcohol leads to the formation of acetyl-CoA acetaldehyde without pyruvate as an intermediate product. Consequently, no thiamin is necessary. The acetyl-CoA formed will be further broken down via the citric acid cycle, which process can actually do without thiamin. It should be pointed out, however, that alcohol consumption may increase daily energy intake, and thus thiamin requirement. Three important enzyme systems, all of which are involved in cerebral glucose utilization, are thiamin pyrophosphate (TPP)-dependent (17).

It has been noted that neurological disturbances caused by thiamin deficiency are enhanced by the consumption of ethanol in rats (74).

In some drinkers thiamin deficiency leads to Wernicke-Korsakoff's syndrome (WKS), cardiac beriberi and possibly polyneuropathy, which are not always distinguishable from the direct toxic effects of alcohol (6).

Some 90% of all WKS patients are alcoholics (18). Frequently stupor or coma is the dominant trait of WKS. Without treatment, mortality is estimated at 100% (54), but thiamin treatment can reduce this dramatic figure to 17% (18). Korsakoff's psychosis is generally believed to be the final stage of Wernicke's encephalopathy. It is characterized by severe impairment of the cognitive function (i.e. loss of past memories, inability to learn or form new memories, minor impairment of perceptual and conceptual functions, confabulation, and an apathetic loss of insight and initiative (30). Serious amnesia will occur, and presumably also irreversible brain damage, thus rendering thiamin treatment ineffective. The pathogenesis of the disease is still unknown. A possible mechanism is proposed by Butterworth (17) (Fig. 6.1).

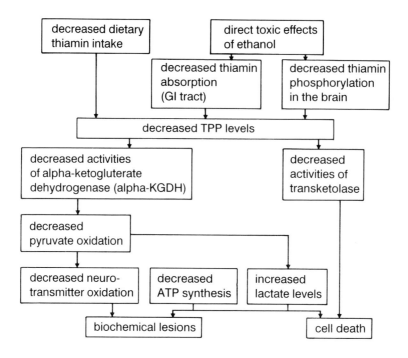

Fig. 6.1. Possible mechanism of thiamin deficiency and brain damage.

There is considerable variation not only in pathological changes as a response to thiamin deficiency, but also in susceptibility of the various organs. In some of the responders the brain is the target organ, in others it is the heart and in still others it may be both brain and heart. Cardiac beriberi was discovered at the start of the century in the Orient where many people subsisted largely on a diet of polished rice with its negligible thiamin but high carbohydrate contents. The occidental form of cardiac beriberi is usually associated with excessive alcohol intake, but is much rarer than WKS. Oedema may be a dominant feature and pulmonary congestion and dyspnoea is sometimes present as well (48, 72).

The recommended daily intake of thiamin is 0.12 mg/MJ. Alcoholic beverages do not contain thiamin (68) or only traces of it (7). Supplementation of thiamin to alcoholic beverages has been suggested. From a cost-benefit analysis of current conditions in the USA, Centerwall (18) concludes that the cost of supplementation is only 4–25% of the cost of (often lifelong) treatment of Korsakoff's psychosis. It would also decrease the number of cases of alcoholic polyneuropathy and cardiac beriberi (both caused by thiamin deficiency). Before embarking on thiamin supplementation, a great deal of research will have to be done on several aspects related to it. The addition of thiamin to beer does not entail technical problems (47).

However, fortification with a potentially harmful substance poses serious problems for health educators. Labelling of alcoholic beverages as 'vitamin-enriched' could result in changes in the community's attitude towards alcohol and hence in increased alcohol consumption. Besides, the necessity of modifying the often age-old procedures followed in production of these beverages would doubtlessly meet with resistance. It must also be questioned whether it is justifiable to add a substance to a product consumed by many in order to prevent the incidence of a serious disorder incurred by only few owing to abuse. Society's negative attitude towards fluoridation of drinking water, for instance, indicates that the possible effectuation of suppletion may well evoke political and social problems.

6.3. Riboflavin (formerly designated vitamin B-2)

Riboflavin is necessary for the electron transport system, which plays a part in alcohol metabolism. There are, however, no indications of an increased riboflavin requirement in alcoholics. A decreased riboflavin status in alcoholics presumably originates from a dietary deficiency to be attributed mainly to a low intake of milk and other dairy products. Ethanol does not affect the absorption of this vitamin (14). The riboflavin status of rats which had been administered alcohol in addition to a riboflavin-containing diet did not differ from that of controls (53). However, Kim & Roe (32) observed that chronic alcohol feeding can induce a biochemical riboflavin deficiency in hamsters fed a diet low in riboflavin as compared to hamsters fed the same riboflavin-poor diet without alcohol. Hietanen (27) has carried out experiments on the effect of riboflavin deficiency in rats consuming alcohol. Hardly any difference in liver function between deficient and non-deficient groups was found.

6.4. Vitamin B-6

Vitamin B-6 is the generic descriptor for compounds exhibiting qualitatively the biological activity of pyridoxine. Pyridoxine is transformed in the body mainly to pyridoxal-5'-phosphate (PLP) and is involved in amino acid metabolism, energy production, fat metabolism, central nervous system activity, and haemoglobin synthesis. On average, alcoholics have a decreased blood vitamin B-6 level. In about 50% of all cases some deficiency may be said to exist (i.e. PLP levels below the lower limit of the normal range) (12). In those with liver disease, the prevalence of reduced plasma PLP concentrations may be higher (34). Several possible causes have been suggested (12, 26, 29, 39, 49).
– Ethanol and/or acetaldehyde presumably have a direct influence on liver storage and vitamin B-6 metabolism, in particular the conversion of pyridoxine into its active form, PLP.
– Liver injury due to excessive alcohol use also causes a reduction of PLP, thus deteriorating the vitamin B-6 status even further.
– A decreased pyridoxine input, caused by an inadequate diet, may also play a role.

In contrast, chronic ethanol administration does not change intestinal pyridoxine absorption (51).
– Ethanol and/or alcoholic liver injury could promote vitamin B-6 deficiency by accelerating its degradation pathway.

No clear clinical symptoms of vitamin B-6 deficiency have been observed in alcoholics. Yet alcoholics are frequently treated with pyridoxine or PLP to re-establish the normal level.

Recently, Löwik et al. (42) have studied the effects of moderate alcohol consumption on vitamine B-6 indicators. No significant effect of moderate alcohol use on PLP level was observed, but there was a slight positive relationship between alcohol intake and PLP concentration. The authors suggest that that alcoholic beverages may contain much more vitamin B-6 than is indicated by food composition tables.

6.5. Folate

Folate is the most commonly deficient vitamin in malnourished alcoholics (4). Since beer contains considerably more folic acid (360 µg/l) than other alcoholic beverages (0–4 µg/l), higher serum folic acid concentrations are found in alcoholics drinking beer than those drinking predominantly spirits or wine (11).

Folate deficiency may cause megaloblastic anaemia (41). It also causes malabsorption of thiamin and vitamin B-12 as well as of folate itself (23). Deficiency frequently occurs, also among non-drinkers, but more often in alcoholics, particularly in combination with an inadequate diet (25) and liver disease (2). Alcoholics tend to have lower average blood folate levels than non-alcoholics (11, 24, 33). Megaloblastic anaemia is only found in individuals with both a small folate pool and a diet low in folate. If the body has an adequate folate pool, anaemia will usually not occur at once if the diet is lacking folate, not even in combination with alcohol use. However, a prolonged inadequate diet may deplete folate stores. In primates chronic alcohol consumption impairs folate coenzymes, increases haematological indices of megaloblastic anaemia, and may cause malabsorption of enterohepatically circulating folate in folate deficiency even when other essential nutrients are provided (9). Long-term experiments with monkeys showed that after a couple of years of chronic alcohol consumption a decrease of blood folate levels will occur while the plasma level will (as yet) remain unchanged (2). A direct toxic effect of alcohol on bone marrow, thus initiating megaloblastic anaemia, seems unlikely (11).

These low folate levels in alcoholics partly result from primary and secondary malnutrition (poor dietary intake and decreased absorption) (24). Besides, alcohol itself also directly affects folate metabolism. Both in rats and in man, administration of alcohol causes an acute decrease of the serum folate level, which returns to normal soon after termination of the alcohol administration (19, 55, 59). Alcohol consumption is attended by an increased urinary excretion of folate (45, 46, 64).

Ethanol also seems to block the enterohepatic circulation of folate and to interfere with intracellular folic acid utilization (1, 11, 57, 64).

6.6 Vitamin B-12

Vitamin B-12 deficiency, like folate deficiency, may cause megaloblastic anaemia. Besides its role in erythrocyte development, vitamin B-12 also plays a biochemical role, unrelated to folate, in the maintenance of myelin in the nervous system. In contrast with the vast body of literature on folate, literature on the vitamin B-12 status of alcoholics is scanty.

Chronic alcohol consumption in well-fed volunteers decreases the absorption of vitamin B-12 (33). Normal blood vitamin B-12 levels have been found in alcoholics with mild liver disease, but the vitamin B-12 level is elevated in severe alcoholic liver disease (5, 43). The normal or high serum vitamin B-12 levels may result from the failure of the damaged liver to take up several B-12 vitamers from the serum. This situation may be aggravated by a release of these compounds from the damaged liver into the serum (31). A high incidence of pathologically deficient levels of vitamin B-12 in cerebrospinal fluid, despite normal serum B-12 levels, has been observed in patients with alcohol-related brain damage (71).

Alcoholics rarely show vitamin B-12 deficiency (75); even if present, the deficiency probably reflects interference by alcohol with vitamin B-12 absorption (34, 66).

6.7 Vitamin C

In many alcoholics a decreased vitamin C status has been observed (3, 10). Although a negative correlation between alcohol consumption and vitamin C level has been found, there are no indications of a direct toxic effect of alcohol on vitamin C metabolism (10). Low vitamin C levels in alcoholics are presumably caused by alcohol-induced inhibition of the active transport system of vitamin C through the intestinal wall (21), an enhanced vitamin C excretion in the urine (20, 52), and/or poor dietary intake. Oral or intravenous administration of vitamin C to patients suffering from delirium tremens or from the alcohol withdrawal syndrome leads to an increase of the ascorbic acid concentration in the serum of these patients (58). This is contradictory to the presumed lower absorption of vitamin C in alcoholics.

No clinical symptoms resulting from vitamin C deficiency caused by excessive alcohol consumption have been reported.

6.8. Vitamin D

Vitamin D has an endocrine function in calcium homeostasis. Vitamin D from food (cholecalciferol, also called vitamin D_3) and vitamin D originating from the skin via ultraviolet irradiation is converted in the liver into 25-hydroxycholecalciferol (25(OH)D3), which is subsequently converted in the kidney into 1,25-dihydroxycholecalciferol (1,25(OH)2D3). The latter vitamer is the active form of vitamin D. It activates intestinal calcium absorption and calcium mobilization from the bone.

In several investigations decreased serum 25(OH)D3 levels have been found in alcoholics the causes of which have not been completely elucidated. Several factors can play a role (8, 16, 22):
– a poor diet,
– reduced exposure to sunshine,
– malabsorption, especially in case of fat malabsorption,
– an increased rate of vitamin D metabolism,
– disturbed hydroxylation in the liver,
– metabolic changes in the kidneys.

Opinions diverge as to whether disturbed hydroxylation in the liver, whether or not in conjunction with liver injury, may be a cause indeed (16, 33). It has been suggested that the alcohol-induced decrease of the metabolic effects of vitamin D may not take place in the liver, but in the kidney (50), or both (44). The activity of the renal enzyme regulating hydroxylation of 25(OH)D3 to 1,25(OH)2D3 decreases, while conversion into the presumably inactive 24,25-(OH)2D3 takes place. This may be caused by the effect of alcohol on the pituitary regulating the hydroxylase activity of the kidney (50).

Osteoporosis is frequently observed in alcoholics (67). In animals it has been observed that chronic alcohol administration results in disturbed vitamin D metabolism, a net increase in bone resorption and decreased mineralization of bone matrix (70). Changes in calcium and phosphate metabolism, whether or not of an endocrine nature, presumably also play a role.

6.9 Vitamin E

Vitamin E is a major lipid-soluble antioxidant in biomembranes. The vitamin E level is depressed in alcoholics. Reduced dietary intake and/or malabsorption may be the cause of vitamin E deficiency. It has also been suggested that a decreased serum vitamin E level is caused by liver disorders via a depression of low-density lipoprotein (LDL) formation (73).

Vitamin E deficiency may play a role in Zieve's syndrome because haemolysis and the transient haemolytic anaemia revert to normal upon vitamin E treatment.

Whether vitamin E treatment may prevent fatty liver in alcoholics, as has appeared from some animal experiments, is still to be further investigated (16).

Summary

The vitamin deficiencies frequently observed in alcoholics are presumably largely the result of a decreased vitamin supply to the organism. This is a consequence of the frequently insufficient diet attendant on alcoholism on the one hand, and of decreased vitamin absorption resulting from damage to the intestinal epithelial cells caused by excessive alcohol consumption and/or vitamin deficiency on the other. Vitamin deficiencies may be promoted by alcoholic liver disease (hepatitis, cirrhosis). The vitamin pool in the liver may be decreased as a result of the decreased capacity of the liver cells to store vitamins. Liver injury may also affect the metabolism of various vitamins. Moreover, a direct influence of alcohol on the metabolism of some vitamers (retinol, thiamin, pyridoxin, folic acid, cholecalciferol) seems plausible. Several disorders resulting from vitamin deficiency have been observed in alcoholics, the most important of which are:
- night blindness (caused by alcohol interacting with vitamin A and zinc deficiencies),
- decreased spermatogenesis (same causes),
- Wernicke-Korsakoff's syndrome (caused by thiamin deficiency),
- cardiac beriberi (caused by thiamin deficiency),
- various types of anaemia (caused by folate and vitamin B-12 deficiency).

Interaction between vitamin A and alcohol during drinking cannot be excluded, since the same enzymes participate in metabolism of both substances. Because the most important causes of vitamin deficiency (inadequate nutrition, malabsorption and liver injury) result from excessive alcohol consumption and accompany moderate drinking, the latter presumably affects vitamin metabolism only slightly. However, on the basis of our present knowledge, we cannot determine the intake level below which alcohol does not have any effect on vitamin metabolism.

References

1. Anonymous. Alcohol and the enterohepatic circulation of folate. Nutr Rev 1980;38: 220-3.
2. Anonymous. Long-term effects of ethanol consumption of folate status of monkeys. Nutr Rev 1983;41: 226-8.
3. Baines M. Vitamin C and exposure to alcohol. Int Z Vitamin Ernährungsforsch 1982:23: 287-93.
4. Baker H. A vitamin profile of alcoholism. In: Hank A, ed. Vitamins in medicine: recent therapeutic aspects. Bern: Hans Huber Publishers, 1983: 179-84.
5. Baker H, Frank O, DeAngelis B. VA Cooperative Study Group on Alcoholic Hepatitis. Plasma vitamin B12 titres as indicators of disease severity and mortality of patients with alcoholic hepatitis. Alcohol Alcoholism 1987;22: 1-5.
6. Basel HG. Alcohol intake and thiamin. Ernährung/Nutrition 1987;11: 329-33.
7. Bitsch R. Alkohol und Vitaminstoffwechsel. Ernährungs-Umschau 1987;34: 161-6.
8. Bjørnboe GEA, Johnson J, Bjørnboe A, et al. Effect of alcohol consumption on serum concentration of 25-hydroxyvitamin D3, retinol, and retinol binding protein. Am J Clin Nutr 1986;44: 678-82.
9. Blocker DE, Thenen SW. Intestinal absorption, liver intake, and excretion of 3H·folic acid in folic acid-deficient, alcohol consuming nonhuman primates. Am J Clin Nutr 1987;46: 503-10.
10. Bonjour JP. Vitamins and alcoholism I. Ascorbic acid. Int J Vitam Nutr Res 1979;49: 434-41.

11. Bonjour JP. Vitamins and alcoholism II. Folate and vitamin B12. Int J Vitam Nutr Res 1980;50: 96-121.
12. Bonjour JP. Vitamins and alcoholism III. Vitamin B6. Int J Vitam Nutr Res 1980;50: 215-30.
13. Bonjour JP. Vitamins and alcoholism IV. Thiamin. Int J Vitam Nutr Res 1980;50: 321-338.
14. Bonjour JP. Vitamins and alcoholism V. Riboflavin; VI. Niacin; VII. Pantothenic acid; and VIII. Biotin. Int J Vitam Nutr Res 1980;50: 425-40.
15. Bonjour JP. Vitamins and alcoholism IX. Vitamin A. Int J Vitam Nutr Res 1981;51: 166-77.
16. Bonjour JP. Vitamins and alcoholism X. Vitamin D; XI. Vitamin E; and XII. Vitamin K. Int J Vitam Nutr Res 1981;51: 307-18.
17. Butterworth RF. Effects of thiamine deficiency on brain metabolism, implications for the pathogenesis of the Wernicke-Korsakoff syndrome. Alcohol Alcoholism 1989;24: 271-9.
18. Centerwall BS, Crique MH. Prevention of the Wernicke-Korsakoff syndrome. N Eng J Med 1978;299: 285-9.
19. Eichner ER, Hillman RS. Effect of alcohol on serum folate level. J Clin Invest 1973;52: 584-91.
20. Faizallah R, Morris AI, Krasner N, Walker RJ. Alcohol enhances vitamin C excretion in the urine. Alcohol Alcoholism 1986;21: 81-4.
21. Fazio V, Flint DM, Wahlqvist ML. Acute effects of alcohol on plasma acorbic acid in healthy subjects. Am J Clin Nutr 1981;34: 2394-96.
22. Gascon-Barré M. Influence of chronic ethanol consumption on the metabolism and action of vitamin D, hypothesis and review. J Am Coll Nutr 1985;4: 565-74.
23. Green PH. Alcohol, nutrition, and malabsorption. Clin Gastroenterol 1983;12: 563-74.
24. Halsted CH, Robles EA, Mezey E. Intestinal malabsorption in folate-deficient alcoholics. Gastroenterology 1973;64: 526-32.
25. Halsted CH. Folate deficiency in alcoholism. Am J Clin Nutr 1980;33: 2735-40.
26. Halsted CH, Heise C. Ethanol and vitamin metabolism. Pharmac Ther 1987;34: 453-64.
27. Hietanen E, Koivusaari U, Laitinen M, Norling A. Hepatic drug metabolism during ethanol ingestion in riboflavin deficient rats. Toxicology 1980;16: 103-11.
28. Hoyumpa AM. Alcohol and thiamin metabolism. Alcoholism Clin Exp Res 1983;7: 11-4.
29. Hoyumpa AM. Mechanism of vitamin deficiencies in alcoholism. Alcoholism Clin Exp Res 1986;10: 573-81.
30. Jacobson RR, Acker CF, Lishman WA. Patterns of neuropsychological deficit on alcoholic Korsakoff's syndrome. Psychol Med 1990;20: 321-34.
31. Kanazawa S, Herbert V. Total corrinoid, cobalamin (vitamin B12), and cobalamin analogue levels may be normal in serum despite cobalamin in liver depletion in patients with alcoholism. Lab Invest 1985;53: 108-10.
32. Kim C, Roe DA. Development of riboflavin deficiency in alcohol-fed hamsters. Drug Nutr Interact 1985;3: 99-107.
33. Korsten MA, Lieber CS. Nutrition in the alcoholic. Med Clin North Am 1979;63: 963-72.
34. Leevy CM, Baker H, ten Hove W, Frank O, Cherrick GR. B-complex vitamins in liver disease of the alcoholic. Am J Clin Nutr 1965;16: 339-46.
35. Leo MA, Arai M, Sato M, Lieber CS. Hepatotoxity of vitamin A and ethanol in the rat. Gastroenterology 1982;82: 194-205.
36. Leo MA, Lieber CS. Hepatic vitamin A depletion in alcoholic liver injury. N Engl J Med 1982;307: 597-601.
37. Leo MA, Lieber CS. Hepatic fibrosis after long-term administration of ethanol and moderate vitamin A supplementation in the rat. Hepatology 1983;3: 1-11.
38. Leo MA, Sato M, Lieber CS. Effect of hepatic vitamin A depletion on the liver in humans and rats. Gastroenterology 1983;84: 562-72.
39. Li TK, Lumeng L. Vitamin B6 metabolism in alcoholism and alcoholic liver disease. In: Proceedings of a conference on vitamin B6 nutrition and metabolism held at the Barff Conference Centre, Barff, Alberta, Canada, October 8-10. New York: Alan R. Liss, 1984: 257-69.
40. Lieber CS. Alcohol and the lover, update 1984. Hepatology 1984;4: 1243-60.
41. Lindenbaum J, Roman MR. Nutritional anemia in alcoholism. Am J Clin Nutr 1980;33: 2727-35.

42. Löwik MRH, van Poppel G, Wedel M, van den Berg H, Schrijver J. Dependence of vitamin B status assessment of alcohol intake among elderly men and women. J Nutr 90;120: 1344-51.
43. Majumbar SK, Shaw GK, Thomson AD. Plasma vitamin E status in chronic alcoholic patients. Drug Alcohol Dependence 1983;12: 269-72.
44. Mawer EB, Klass HJ, Warnes TW, Berry JL. Metabolism of vitamin D in patients with primary biliary cirrhosis and alcoholic liver disease. Clin Sci 1985;69: 561-70.
45. McMartin KE. Increased urinary folate excretion and decreased plasma folate levels in the rat after acute ethanol treatment. Alcoholism Clin Exp Res 1984;8: 172-8.
46. McMartin KE, Shiao CQ, Collins TD, Redetzki HM. Acute ethanol ingestion by humans and subacute treatment of rats increase urinary folate excretion. Alcohol 1985;2: 473-7.
47. Meilgaard MC. Technical aspects of enrichment of beer with thiamine. J Stud Alc 1982;43: 427-33.
48. Meulders Q, Laterre PF, Corbeel MSL. Shoshin beriberi, a fulminant beriberi heart disease. Acta Clin Belg 1988;43: 115-9.
49. Mezey E, Potter JJ, Merchant CR. Effect of ethanol feeding on bone composition in the rat. Am J Clin nutr 1979;32: 25-9.
50. Mezey E. Alcohol and renal vitamin D metabolism. Gastroenterology 1980;78: 651.
51. Middleton HM, Mills LR, Singh M. Effect of ethanol on the uptake of pyridoxine·HCl in the rat jejunum. Am J Clin Nutr 1984;39: 54-61.
52. Mochizuki S, Yoshida A. Effects of dietary ethanol on ascorbic acid and lipid metabolism, and liver drug-metabolizing enzymes in rats. J Nutr Sci Vitaminol 1989;35: 431-40.
53. Mönks E, Potter JJ, Merchant CR. Effect of ethanol feeding on bone composition in the rat. Am J Clin Nutr 1979;32: 25-9.
54. Muller-Kobold MJP, Endtz LJ. Het syndroom van Wernicke-Korsakoff, een zeldzaam ziektebeeld? Ned Tijdschr Geneesk 1975;119: 991-7.
55. Paine CJ, Eichner ER, Dickson V. Concordance of radioassay and microbiological assay in the study of the ethanol induced fall in serum folate level. Am J Med Sci 1973;226: 135-8.
56. Paladin F, Perez GR. The haematic thiamine level in the course of alcoholic neuropathy. Eur Neurol 1987;26: 129-33.
57. Parkkinen E, Stowell L, Kohila T, Halkka O, Eriksson K. Synergistic effects of chronic ethanol consumption and folic acid deficient diet in rats, development of leucopenia and lack of effect on sister chromatid exchanges in bone marrow cells in vivo. Subst Alc Actions/Misuse 1981;1: 221-31.
58. Piatkowski J, Wiechert P, Ernst K. Ascorbic acid in chronic alcoholics. Int J Vitam Nutr Res 1987;57: 421.
59. Pikaar NA, Wedel M, van Dokkum W, van den Berg H, Schrijver J. Binnen- en tussenpersoonsvariabiliteit van de concentraties van melkzuur, vitamine B1, B2, B6, B12 en foliumzuur in van de γ-GT en ALAT-activiteit in bloed/plasma na consumptie van 3 glazen gesimuleerde wijn; alcoholbelastingsproeven bij 12 proefpersonen. Zeist: CIVO-Instituten TNO, 1983. (Rapport V 84.018/120224.)
60. Rindi G. Alcohol and thiamine of the brain. Alcohol Alcoholism 1989;24: 493-5.
61. Roe DA. Drug-induced nutritional deficiencies. Westport: AVI Publ. Comp., 1976.
62. Rosenblum ER, Gavaler JS, van Thiel DH. Lipid peroxidation, a mechanism for alcohol-injured testicular injury. Free Radical Biol Med 1989;7: 569-77.
63. Russel RM. Vitamin A and zinc metabolism in alcoholism. Am J Clin Nutr 1980;33: 2741-9.
64. Russel RM, Rosenberg IH, Wilson PD. Increased urinary excretion and prolonged turnover time of folic acid during ethanol ingestion. Am J Clin Nutr 1983;38: 64-70.
65. Sato M, Lieber CS. Increased metabolism of retinoic acid after chronic ethanol consumption in rat liver microsomes. Arch Biochem Biophys 1982;213: 557-64.
66. Sherlock S. Nutrition and the alcoholic. Lancet 1984;i: 436-9.
67. Spencer H, Rubio N, Rubio E, Indreika M, Seitam A. Chronic alcoholism, frequently overlooked cause of osteoporosis in men. Am J Med 1986;80: 393-7.
68. Stichting Nederlands Voedingsstoffenbestand, Voorlichtingsbureau voor de Voeding. NEVO Tabel 1989/1990. The Hague: Stichting NEVO, 1989.
69. Thomson AD, Jeyasingham MD, Pratt OE, Shaw GK. Nutrition and alcoholic encephalopathies. Acta Med Scand 1987(suppl 717): 55-65.

70. Turner RT, Aloila RC, Segel LD, Hannon KS, Bell NH. Chronic alcohol treatment results in disturbed vitamin D metabolism and skeletal abnormalities in rats. Alcoholism Clin Exp Res 1988;12: 159-62.
71. van Tiggelen CJM. Alzheimer disease/alcohol dementia, association with zinc deficiency and cerebral vitamin B12 deficiency. J Orthomol Psychiat 1983;13: 97-104.
72. Webster MWI, Ikram H. Myocardial function in alcoholic cardiac beriberi. Int J Cardiol 1987;17: 213-6.
73. Yoshikawa T, Takemura S, Kondo M. Alpha-tocopherol level in liver disease. Acta Vitaminol Enzymol 1982;4: 311-8.
74. Zimitat G, Kril J, Harper CG, Nixon PF. Progression of neurological disease in thiamin-deficient rats is enhanced by ethanol. Alcohol 1990;7: 493-501.
75. Zittoun J. Effects of some drugs and alcohol on folate and cobalamin metabolism. In: Zittoun J, Cooper BA, eds. Folates and cobalamins. Berlin: Springer-Verlag, 1989: 137-44.

CHAPTER 7

Effects of alcohol on water and mineral balance

E. te Wierik and W.G. Vrij-Standhardt

7.1. Water balance

The diuretic effect of alcohol consumption is well known. On the other hand, overhydration and ascites (also called hydroperitoneum) are frequently observed in alcoholics (32).

Experiments measuring the diuretic effect of acute large doses of alcohol are usually carried out over a short period of time. In these experiments an increase in urinary production appears to be caused by an inhibited vasopressin (antidiuretic hormone, ADH) release (3). Eiser (13), who has reviewed the studies concerning the effects of alcohol on renal function and excretion, concludes that both alcoholics and non-alcoholics retain sodium and water after a brief diuresis. In a study into the effect of ethanol on water balance and plasma vasopressin 75 ml alcohol was given to volunteers. It was been observed that ethanol inhibited vasopressin release by resetting the hypothalamic osmoreceptors, so that the rate of vasopressin release is reduced relative to plasma osmolality. The effect of this is a hydration state of increased plasma osmolality and reduced fluid volume which is induced via a transient diuresis (12). When drinking is discontinued the osmoreceptor sensitivity returns to normal, vasopressin secretion increases and normal volume is restored. Beard & Knott (2) studied chronic alcoholics and observed an expansion of the extracellular fluid volume. Moreover, they found that both total body water volume and calculated intracellular water volume were above normal values. During withdrawal total body water decreased significantly. In contrast to these findings, Mander et al. (31) observed fluid retention and increased plasma vasopressin levels in alcoholics during withdrawal. A decreased water excretion in alcoholic patients during withdrawal has also been observed by Emsley et al. (14).

It is well known that plasma volume may be expanded in alcoholics with a liver or cardiac disease. In alcohol-associated cardiomyopathy, with left ventricular dysfunction, plasma volume expands because low cardiac output leads to fluid and salt retention. In cases of liver cirrhosis volume expansion may occur because increased distal run-off (A-V shunting) decreases the effective (although not actual) plasma volume. Ascites may also occur in this setting (9).

7.2. Mineral balance

A close relationship exists between body water volume and mineral concentrations. The results of the various investigations into changes of mineral concentrations caused by acute or chronic alcohol consumption are difficult to interpret. With some minerals the effects during the ascending phase differ from those observed during the descending phase of the alcohol curve. Moreover, the differences in experimental design hamper a proper comparison.

The effects of acute or chronic alcohol consumption will be discussed only for sodium, potassium, chloride, calcium, phosphate, magnesium, iron and zinc.

7.2.1. Sodium

Sodium, together with potassium, assists in the maintenance of the body's electrolyte and water balance. In addition, sodium and potassium play an important role in neural conduction, muscle contraction and the transport of substances across membranes.

Some authors have observed an acute decrease of urinary sodium excretion after the consumption of alcohol (20, 43), whereas others have not found such an effect (26, 40). The decrease in sodium excretion is probably the result of an increase in elimination of 'free water' (water without salts) by the kidneys, resulting from inhibition of the release of vasopressin. A rise in plasma sodium level after alcohol consumption has been noted as well (24, 41, 44, 46). This plasma sodium increase is accompanied with water retention to compensate for the diuretic effect.

Hyponatraemia has frequently been observed in alcoholics. Several factors may contribute to the diminished concentration of sodium (32):
– the kidneys' ability to eliminate 'free water' is impaired,
– an elevated level of antidiuretic hormone enhances the kidneys' efficiency of water retention,
– the diminished total body potassium content indirectly affects the plasma sodium concentration,
– the use of medication such as diuretics or laxatives causes an increase in urinary and faecal elimination of sodium and water, which can result in low sodium concentrations when the excreted fluids are replaced with sodium-free water.

Besides these factors is has been mentioned that vagal neuropathy may contribute to the low serum concentration in patients with cirrhosis. In vagal neuropathy the baroreceptors might contribute to an inappropriate secretion of antidiuretic hormone, which might in turn induce hyponatraemia (11).

7.2.2. Potassium

The effects of alcohol on potassium metabolism are not unequivocal. A significant increase (19), a decrease (41), as well as no difference (20) in plasma potassium in response to the consumption of alcohol have been observed. Urinary potassium excretion seemed to decrease in some studies (20, 24, 26, 43, 44), and remained equal in other ones (46). Reduced potassium excretion during alcohol ingestion could result from metabolic acidosis. A mild metabolic acidosis is frequently observed during alcohol ingestion. An acute acidosis suppresses potassium ecretion in the distal tube (13).

Low serum concentrations of potassium are frequently observed in hospitalized alcoholics. They may reflect a deficit in total body potassium or merely a shift of potassium from the extracellular fluid to the cells (32).

7.2.3. Calcium

Calcium is required for blood clotting, muscle contraction and nerve transmission. Calcium is a major constituent of bones and teeth. Alcohol influences the neural calcium channels which may be linked to the behavioural effects of alcohol (18, 30). Acute alcohol consumption increases urinary excretion of this mineral (26) and decreases its plasma concentration (28). Chronic alcohol consumption decreases serum calcium concentration (5, 8). Several factors may account for the association between the occurrence of hypocalcaemia and severe alcoholism (6, 32).
– Calcium is bound to albumin, and a poor diet or liver disease results in diminished albumin levels (hypoalbuminaemia), thereby limiting the amount of calcium that can remain resolved in the blood.
– Hypomagnesaemia decreases the sensitivity of the parathyroid gland to low calcium levels.
– Vitamin D is necessary for adequate absorption of calcium from the gut. Vitamin D malabsorption impairs calcium absorption.
– In cases of rhabdomyolysis (destruction or necrosis of skeletal muscle) calcium may precipitate as calcium phosphate or calcium carbonate which leads to hypo-calcaemia.

7.2.4. Phosphate

Phosphate is a constituent of bones and teeth, and is an integral part of many molecules of the body. Hypophosphataemia has often been observed in alcoholics. It may be preceded by hyperphosphataemia. Hyperphosphaturia often results from the increased filtered load occurring when muscle releases phosphate (13). Other potential causes of hypophosphataemia in alcoholics are renal loss secondary to toxic injury, a dietary phosphorus deficiency, magnesium deficiency, cellular uptake of phosphorus following starvation, respiratory alkalosis, and insulin-stimulated phosphate uptake by cells (6, 32).

7.2.5. Magnesium

Magnesium is an important constituent of bones and teeth. It also assists enzymes involved in cellular energy production and protein digestion, and plays a role in the transmission of impulses in nerves.

Magnesium deficiency may accompany chronic alcoholism. Decreased magnesium concentrations in serum, lymphocytes and bone have been found. Alcohol-associated osteoporosis may be related to postmenopausal and senile osteoporosis (10). An increase in urinary excretion of magnesium after the consumption of alcohol has been observed by Kalbfleisch et al. (26). In his review Flink (16) concludes that alcohol induces excretion of magnesium. Besides magnesium loss in the urine, poor food intake, poor absorption (due to fat malabsorption), and diarrhoea may also result in magnesium deficiency.

A low magnesium concentration in the blood is mainly observed during withdrawal, when the alcohol has lost its effect and withdrawal symptoms occur (22, 24). This fall in magnesium level during withdrawal may be attributed to a rise in free fatty acids. Precipitation of magnesium with free fatty acids can contribute to the drop in plasma magnesium level (16). A negative association between serum magnesium concentration and alcohol intake has been found in drunkenness arrestees. This negative association could be expected since all drunkenness arrestees showed withdrawal symptoms (27, 38).

7.2.6. Iron

Iron is an important component of haemoglobin and is involved in a variety of enzyme systems. Several disturbances of iron metabolism are found in alcoholics. Serum iron and ferritin levels may be increased in some alcoholics, while various types of anaemia occur in others (29, 39, 48). The liver iron content is frequently elevated (7, 39). Iron overload due to genetic haemochromatosis (25, 37) or haemosiderosis (29) may occur.

A positive relation between drinking frequency and serum iron concentration has been observed in both male and female adolescents. Total iron-binding capacity, transferrin saturation and haemoglobin concentration were positively related to drinking frequency. These abnormalities are claimed to be precursors of liver iron overload and liver damage (17). Strain et al. (47) also observed that regular drinking increased some indicators of the iron status. No significant correlations between quantity of alcohol consumed and iron status have been observed (17, 47, 27).

The relatively large quantities of easily absorbable iron in some alcoholic beverages (e.g. wine) (21, 36, 45) may contribute to the iron overload. The effects of ethanol are moderated by other constituents of alcoholic beverages (15). An increased iron absorption caused by the presence of alcohol, a decreased utilization of iron in the bones (17), or interaction of iron with other nutrients mediated by alcohol (42) may also play a part.

Conversely, alcohol-induced diseases such as liver injury, pancreatitis, and gastrointestinal tract bleeding also cause disturbances in iron, ferritin and haemoglobin concentrations that can lead to anaemia (29, 48).

7.2.7. Zinc

Zinc is an essential trace element required for RNA and DNA synthesis and for the function of many enzymes. It also plays an important role in membrane stabilization. Zinc deficiency causes a progressive increase in membrane fluidity.

Zinc deficiency is a frequently observed phenomenon in alcoholics both with and without liver disease. A decreased dietary intake, decreased absorption, as well as increased urinary excretion (especially in liver injury) play a role in this (1, 4, 29, 33). Zinc is an essential element in a number of enzymes, such as alcohol dehydrogenase (ADH), so that zinc deficiency may manifest itself in several disturbances. Disturbances attendant on zinc deficiencies in alcoholics include skin lesions, night blindness, hypogonadism, anorexia, diarrhoea, changes in protein metabolism and wound healing, immune dysfunction, carcinogenesis, depressed mental function, and birth defects related to the foetal alcohol syndrome (32, 33, 34).

In the case of zinc it is not clear what the effects of moderate alcohol consumption may be. A single dose causes an increased urinary excretion of zinc, but the serum zinc concentration remains unchanged after consumption of 98 g alcohol a day over a period of 18 days (29, 35).

Not only may zinc deficiency occur with alcoholic liver disease, but there may also be an altered zinc metabolism. Alcoholic hepatitis patients appeared to have increased serum levels of the monokine interleukin 1, which is known to decrease the plasma zinc level and to change the internal distribution of zinc. It has therefore been suggested that patients with alcoholic liver disease are frequently faced with either zinc deficiency of altered zinc metabolism (34).

Summary

A direct consequence of the consumption of alcohol is an increase in urine production owing to an inhibited release of antidiuretic hormone. Moderate and incidental alcohol consumption causes some, mostly reversible, changes of the water and mineral balance, but does not lead to injury or other problems.

Chronic consumption of large quantities of alcohol may cause deficiencies of sodium, potassium, calcium, phosphate, magnesium, iron and zinc either as a result of altered mineral metabolism or consequent to inadequate intake. The effects of moderate alcohol consumption on mineral metabolism are less clear.

Calcium and phosphate deficiency decreases bone density. Magnesium deficiency plays a role in delirium tremens. Zinc deficiency may cause a wide range of disturbances.

References

1. Antonson DL, Vanderhoof JA. Effect of chronic ethanol ingestion on zinc absorption in rat small intestine. Dig Dis Sci 1983;28: 604-8.
2. Beard JD, Knott DH. Fluid and electrolyte balance during acute withdrawal in chronic alcoholic patients. JAMA 1968;204: 133-7.
3. Beard JD, Sargent WQ. Water and electrolyte metabolism following ethanol intake and during acute withdrawal. In: Majchrowicz E, Noble EP, eds. Biochemistry and pharmacology of ethanol, vol. 2. New York: Plenum Press, 1979: 3-16.
4. Beck IT, Dinda PK. Acute exposure of small intestine to ethanol. Dig Dis Sci 1981;26: 817-38.
5. Bjørnboe GEA, Bjørnboe A, Johnson J, et al. Calcium status and calcium- regulating hormones in alcoholics. Alcohol Clin Exp Res 1988;12: 229-32.
6. Blachley JD, Knochel JP. Ethanol and minerals. Pharmac Ther 1987;33: 435-48.
7. Chapman RW, Morgan MY, Boss AM, et al. Acute anc chronic effects of alcohol on iron absorption. Dig Dis Sci 1983;28: 321-7.
8. Chappard D, Plantard B, Fraisse H, Palle S, Alexandre C, Riffat G. Bone changes in alcoholic cirrhosis of the liver. Path Res Pract 1989;184: 480-5.
9. Clark LT. Role of electrolytes in the etiology of alcohol-induced hypertension. Magnesium 1989;8: 124-31.
10. Cohen L, Laor A, Kizes R. Lymphocyte and bone magnesium in alcohol- associated osteoporosis. Magnesium 1985;4: 148-52.
11. Decaux G, Cauchie P, Soupart A, Kruger M, Delwiche F. Role of vagal neuropathy in the hyponatraemia of alcoholic cirrhosis. Br Med J 1986;293: 1534-6.
12. Eisenhofer G, Johnson RH. Effect of ethanol ingestion on plasma vasopressin and water balance in humans. Am J Physiol 1982;242: R522-R527.
13. Eiser AR. The effects of alcohol on renal function and excretion. Alcohol Clin Exp Res 1987;11: 127-38.
14. Emsley RA, Potgieter A, Taljaard JJF, Coetzee D, Jouberts G, Gledhill RF. Impaired water excretion in patients with alcohol-withdrawal symptoms. Q J Med 1987;64: 671-8.
15. Fairweather-Tait SJ, Southon S, Piper Z. The effect of alcoholic beverages on iron and zinc metabolism in the rat. Br J Nutr 1988;60: 209-15.
16. Flink EB. Magnesium deficiency in alcoholism. Alcohol Clin Exp Res 1986;10: 590-4.
17. Friedman JM, Kraemer HC, Mendoza FS, Hammer LD. Elevated serum iron concentration in adolescent alcohol users. Am J Dis Child 1988;142: 156-9.
18. Gandhi CR, Ross DH. Influence of ethanol on calcium, inositol phospholipids and intracellular signalling mechanisms. Experientia 1989;45: 407-12.
19. Gill GV, Baylis PH, Flear CTG, Skillen AW, Diggle PH. Acute biochemical responses to moderate beer drinking. Br Med J 1982;285: 1770-3.
20. Green RJ, Baron DN. Lack of effect on acute ethanol consumption on active cation fluxes of leucocytes and erythrocytes in healthy humans. Clin Sci 1987;73: 387-93.
21. Hallberg L, Rossander L. Effect of different drinks on the absorption of non-heme iron form composite meals. Hum Nutr Appl Nutr 1982;36A: 116-23.
22. Harris RA. Metabolism of calcium and magnesium during ethanol intoxication and withdrawal. In: Majchrowicz E, Noble EP, eds. Biochemistry and pharmacology of ethanol, vol 2. New York: Plenum Press, 1979: 27-41.
23. Hemmingsen R, Kramp P. Effects of acute ethanol intoxification, chronic intoxication and ethanol withdrawal on magnesium and calcium metabolism in the rat. Psychopharmacology 1980;67: 255-9.
24. Howes LG, Reid JL. The effects of alcohol on local, neural and humoral cardiovascular regulation. Clin Sci 1986;71: 9-15.
25. Irving MG, Halliday JW, Powell LW. Association between alcoholism and increased hepatic iron stores. Alcohol Clin Exp Res 1988;12: 7-13.
26. Kalbfleisch JM, Lindeman RD, Ginn HE, Smith WO. Effects of ethanol administration on urinary excretion of magnesium and other electrolytes in alcoholic and normal subjects. J Clin Invest 1963;42: 1471-5.

27. Kärkkäinen P, Mussalo-Rauhamaa H, Poikolanen K, Lehto J. Alcohol intake correlated with serum trace elements. Alcohol Alcoholism 1988;23: 279-82.
28. Kischuk RP, Otten MD, Polimeni PI. Effect of acute alcoholic intoxication on myocardial electrolyte and water distributions. J Mol Cell Cardiol 1986;18: 197-205.
29. Kricka LJ, Clark PMS. Biochemistry of alcohol and alcoholism. Chichester: Ellis Horwood, 1979.
30. Leslie SW, Brown LM, Dildy JE, Sims JS. Ethanol and neuronal calcium channels. Alcohol 1990;7: 233-6.
31. Mander AJ, Young A, Merrick MV, Morton JJ. Fluid balance, vasopressin and withdrawal symptoms during detoxification from alcohol. Drug Alcohol Depend 1989;24: 233-7.
32. Marsano L, McClain GJ. Effects of alcohol on electrolytes and minerals. Alcohol Health Res World 1989;13: 255-60.
33. McClain CJ, Le-Chu S. Zinc deficiency in the alcoholic, a review. Alcohol Clin Exp Res 1983;7: 5-10.
34. McClain CJ, Antonow DR, Cohen DA, Shedlofsky SI. Zinc metabolism in alcoholic liver disease. Alcohol Clin Exp Res 1986;10: 582-9.
35. McDonald JT. Wine versus ethanol in human nutrition IV. Zinc balance. Am J Clin Nutr 1980;33: 1096-102.
36. McDonald JT. Wine and human nutrition. In: Wine, health and society, a symposium. San Francisco: Wine Institute, 1982: 107-18.
37. Muench KH. Hemochromatosis and infection: alcohol and iron, oysters and sepsis. Am J Med 1989;87: 40N-43N.
38. Mussalo-Rauhamaa H, Poikolanen K, Kärkkäinen P, Lehto J. Decreased serum selenium and magnesium levels in drunkenness arrestees. Drug Alcohol Depend 1987;20: 95-103.
39. Paille F, Baille N, Barrucand D. Principales conséquences biologiques de l'alcoolisation, 2ème partie. Rev Alcoolisme 1983;29: 65-107.
40. Potter JF, Watson RDS, Skan W, Beevers DG. The pressor and metabolic effects of alcohol in normotensive subjects. Hypertension 1986;81: 625-31.
41. Puddey IB, Vandongen R, Beilin LJ, Rouse IL. Alcohol stimulation of renin release in man, its relation to the hemodynamic, electrolyte, and sympatho-adrenal responses to drinking. J Clin Endocrin Metab 1985;61: 37-42.
42. Roque MC, DaCunha F. Interactions of alcohol and nutrition. Am J Dis Child 1989;143: 519.
43. Rubini ME, Kleeman CR, Lamdin E. Studies on alcohol diuresis. I. The effects of ethyl alcohol ingestion on water, electrolyte and acid-base metabolism. J Clin Invest 1955;34: 439-47.
44. Sargent WQ, Simpson JR, Beard JD. Twenty-four-hour fluid intake and renal handling of electrolytes after various doses of alcohol. Alcohol Clin Exp Res 1980;4: 74-83.
45. Stichting Nederlands Voedingsstoffenbestand, Voorlichtingsbureau voor de Voeding. NEVO Tabel 1989/1990. The Hague: Stichting NEVO, 1989.
46. Stott DJ, Ball SG, Inglis GC, et al. Effects of a single moderate dose of alcohol on blood pressure, heart rate and associated metabolic and endocrine changes. Clin Sci 1987;73: 411-6.
47. Strain JJ, Thompson KA, Barker ME, McKenna PG. Alcohol consumption and measurements of iron status. Proc Nutr Soc 1990;49: 23A.
48. Välimäki M, Härkönen M, Ylikahri R. Serum ferritin and iron levels in chronic male alcoholics before and after ethanol withdrawal. Alcohol Alcoholism 1983;18: 255-60.

CHAPTER 8

Effects of alcohol on the liver

W.G. Vrij-Standhardt

The organ most frequently affected by excessive alcohol consumption is the liver. Liver injury usually comprises one of three successive pathological conditions: fatty liver, hepatitis or cirrhosis. Fatty liver, also called steatosis, is a reversible disorder, in which fat accumulates in the liver cells. In the case of hepatitis much of the liver tissue is destroyed (necrosis) causing subsequent inflammation. Alcoholic cirrhosis, which is frequently equated with Laennec's cirrhosis, is characterized by hepatic fibrosis. The normal structure of the liver and liver functions are irreversibly disturbed. The assessment of alcoholic liver disease is complicated by the well known lack of correlation between the clinical syndrome and histopathological abnormalities. Patients without clinical or laboratory evidence of liver disease often suffer from histological damage as evidenced by biopsy.

This chapter starts with the clinical features and possible underlying mechanisms of these liver diseases, followed by options for treatment. Subsequently, threshold limits are dealt with as wll as sex differences in these limits. Next, epidemiological data on cirrhosis mortality and its temporal and regional relation with per capita alcohol consumption are discussed. Finally, biochemical markers of alcohol abuse – predominantly liver enzymes – are discussed.

8.1. Pathology and symptomatology

Fat accumulation in the liver cells is the earliest and most common response to alcohol consumption. Normally, the liver contains 2–4% (w/w) of lipids which cannot be discerned by light microscopy. The term 'fatty liver' is used when the lipid content of the liver exceeds the level of 5% (w/w). In this condition, lipid drops can be detected by light microscopy. A real fatty liver contains more than 10% (w/w) lipids, and fat drops as big as the liver cell nucleus can be observed in more than half of the liver cells (113).

Fatty liver is usually not accompanied with clinical symptoms. Occasionally there are complaints about a heavy feeling in the belly. Severe pains may occur occasionally, especially after a period of heavy drinking in which the liver is enlarged very rapidly and capsule stretching occurs (13). In laboratory tests highly increased serum γ-glutamyl transferase (GGT) values are often found in cases of fatty liver as well as slightly increased serum aspratate aminotransferase (ASAT or GOT) and serum alanine aminotransferase (ALAT or GPT) levels.

Fatty liver is not reversible immediately. After some days of abstinence fatty degeneration decreases and after 2 to 6 weeks the disorder usually has disappeared entirely (13, 113). Generally speaking, fatty liver is fairly harmless, also because its attendant functional and clinical symptoms are limited. There are, however, indications that fatty degeneration of the liver may be the cause of sudden death, particularly after excessive alcohol consumption (4, 87).

In more than 90% of alcoholics with fatty liver who continue to drink, alcoholic hepatitis develops (85). Alcoholic hepatitis is a toxic, degenerative and inflammatory lesion of the liver, characterized by necrosis, inflammation and, in some instances, acidophilic hyalin (Mallory bodies). Alcoholic hepatitis is the most difficult alcoholic liver syndrome to diagnose, because of its inconstant manifestations – varying from a total asymptomatic process to an acute, life-threatening liver failure (108). Usually, however, there are clear symptoms of illness, such as nausea, vomiting, pain in the stomach, loss of weight, anorexia and fever. The illness is often accompanied with jaundice (13, 33). Classically, alcoholic hepatitis develops over weeks, in contrast to viral hepatitis in which the illness evolves over days. Necrosis, and later fibrosis, will first be observed in the perivenular zones of the cells. These are the areas most remote from the vessels with oxygen-saturated blood, and thus having the lowest oxygen tension in the liver.

Alcoholic cirrhosis represents the end-stage lesion of alcoholic liver injury. The onset can be insidious or may follow acute alcoholic hepatitis. The patient may be entirely asymptomatic, have non-specific complaints (fatigue, anorexia, weight loss, nausea, abdominal discomfort), or may suffer specific syndromes relating to liver cell failure or portal hypertension (108).

Cirrhosis is described as sclerosis of the active liver tissue and overgrowth caused by fibrous cell proliferation (fibrosis). The necrosis and inflammation in the central lobular zone in cases of hepatitis stimulate the development of fibrosis. This results in so-called hyalin sclerosis (70). The strands of fibrous tissue thus formed may expand and eventually cause cirrhosis. Cirrhosis has, however, also been observed following alcohol-induced fibrosis unattended with inflammation (73, 77). Myofibroblasts have been recognized in the normal liver, and their proliferation was found to accompany and to precede the earliest signs of fibrosis. In alcohol patients and baboons that were found to have these perivenular fibrotic lesions, cirrhosis proved to develop more often than in controls (76, 77).

Hepatomegaly appears to be a common clinical finding in alcoholic liver injury. The enlargement of the liver is primarily due to an increase in size, not in number, of hepatocytes. Israel et al. (44, 46) suggest that this cell enlargement is accompanied with a reduction of the extracellular volume resulting in an increased portal pressure. In alcoholics liver enlargement is strongly associated with portal hypertension (45).

Cirrhosis may entail other complications, such as ascites and hepatic encephalopathy (13, 33). Alcoholic liver disease patients with ascites may have a normal body weight despite smaller skinfold thickness and malnutrition, because of the accumulation of litres of fluid in the abdominal cavity (90).

Liver cancer, which may develop after alcoholic liver disease, is discussed in Chapter 12.

8.2. Mechanisms of alcoholic liver disease

8.2.1. Fat accumulation

Alcohol consumption may bring about an accumulation of fat (triglycerides) in the liver cells. On the one hand, this is caused by decreased fat oxidation since the liver is prompted to use alcohol instead of fat as fuel. On the other hand, the supply of fatty acids to the liver and the synthesis of triglycerides from fatty acids are increased, mainly due to the increased $NADH/NAD^+$ ratio during alcohol breakdown (see Fig. 8.1 and Chapter 5). The simultaneous administration of alcohol and pyrazole, the latter of which inhibits metabolic handling of alcohol, does not result in fatty liver, in contrast to the same dose of alcohol without pyrazole, which induces fat accumulation (83).

Chronic alcohol abuse results in functional and structural changes in liver cell mitochondria (70). This in turn leads to a decrease in lipid oxidation and thus promotes accumulation of fat.

The influence of alcohol metabolism on protein synthesis in the liver also adds to fat accumulation, namely by a decreased production or decreased release of lipoproteins (56, 72). Liver fattening has also been observed in individuals with diabetes or obesity and as a consequence of pharmacotherapy.

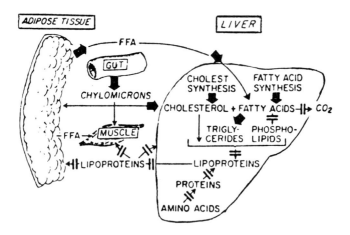

Fig. 8.1. Influence of alcohol in lipid metabolism. After: Lieber et al. (40).

8.2.2. Mechanisms of structural damage

Susceptibility to the development of alcoholic liver disease is higly variable. Only 8–30% of alcohol abusers actually develop structural liver damage (15, 100). If alcohol were directly hepatotoxic, it could be expected that liver damage occurs predictably and is dose-related. The variable response to alcohol can be better understood if alcoholic liver injury is seen as a result of a complex of factors, some of which are alcohol-mediated, and some other ones are genetically determined. Several mechanisms underlying the development of hepatitis and cirrhosis seem to be involved:
– hypoxia (24, 27, 42, 43, 48, 50, 156);
– increased peroxidation (1, 19, 40, 54, 55, 59, 83, 113, 133);
– abnormal cellular immunological activity and disturbed lymphocyte activity
  (6, 10, 20, 36, 48, 86, 89, 146);
– a change in collagen production (27, 38, 45, 51, 73, 78, 99, 152);
– mitochondrial injury caused by alcohol itself or by one of its metabolites
  acetaldehyde or hydrogen (73);
– interference with essential fatty acid and prostaglandin metabolism (32, 38, 39;
  see also Chapter 5);
– an antitubular effect of alcohol or acetaldehyde, causing the formation of
  Mallory bodies (48).

Hypoxia, peroxidation, immunological factors and collagen production will be extensively discussed below.

As opposed to some thirty years ago, nutritional deficiencies are now not regarded as the primary cause. Alcohol consumption itself, rather than the attendant malnutrition in alcoholics, plays a key aetiological role in the development of liver injury (2, 74). There is no solid evidence that dietary deficiency alone results in cirrhosis. Nor has a consistent deviating picture been found in the nutritional status of alcoholics with and without cirrhosis (15, 135). Both animal experiments with baboons and epidemiological investigations have shown that cirrhosis may also develop in cases of adequate dietary intake (in addition to alcohol consumption) (74).

However, nutritional factors may exert a modifying influence on the incidence of liver injury (73). In rodents, steatosis in relation to a given alcohol consumption was less frequently observed when the animals were fed low-fat diets than when they received diets with a normal fat content (75). Livers from nourished rats may be perfused in the absence of oxygen for up to 2 hours without detectable damage, whereas fasted livers are damaged irreversibly in 20–40 minutes (12). A diet low in vitamin E seems to enhance the effects of lipid peroxidation (54).

8.2.3. Hypoxia

Hypoxia (an inadequate supply of oxygen to the cells) seems to be an important factor in hepatic damage, because necrosis (cell death), and later fibrosis (formation of fibrous tissue as a replacement for the damaged tissue), will first be observed in the perivenular zones of the cells. These are the areas most remote from the vessels with oxygen-saturated blood, which thus have the lowest oxygen tension in the liver (125).

Israel et al. (42, 43) consider necrosis in these zones as caused by a disturbance of the equilibrium between the availability of oxygen and the liver's oxygen requirement. Oxygen is indirectly necessary to metabolize alcohol. It has been demonstrated that liver cells have an increased oxygen demand during alcohol metabolization (43, 48) as well as an elevated lobular oxygen gradient in isolated perfused livers from alcohol-treated rats (145). However, in experiments with baboons stimulation of oxygen consumption seemed to be fully offset by a parallel increase in bloodflow. Despite the absence of hypoxia, ethanol increases the lactate/pyruvate ratio 15-fold in hepatic venous blood and only 3-fold in liver tissue. Jauhonen et al. (50) and Baraona et al. (9) proposed an alternative mechanism for the selective perivenular injury, namely, that the low pO_2 at the perivenular site aggravates the redox-linked toxicity of ethanol.

In rat experiments a combination of alcohol consumption and low oxygen tension in the ambient air proved to cause a high degree of liver necrosis whereas the same quantity of alcohol in combination with a normal oxygen tension, or an alcohol-free diet in combination with a low oxygen tension, did not cause necrosis (42, 43). An in vitro experiment showed marked liver damage due to the combination of alcohol exposure and low blood supply, whereas alcohol exposure combined with normal blood supply resulted in moderate damage (156).

Hypoxia may manifest itself in alcoholics as a consequence of respiratory depression caused by severe alcohol intoxication, or as a consequence of pneumonia, an illness which is common among alcoholics. Anaemia, which is also frequently observed among alcoholics as a result of gastrointestinal haemorrhage, proved to promote the development of necrosis in the experiments of Israel and co-workers. The administration of propylthiouracil (PTU) reduces the oxygen consumption of the liver and thus protects alcohol-fed rats from liver injury (42, 43).

According to French (24) chronic alcohol ingestion leads to hepatocellular injury only if multiple factors are operative in combination to induce centrilobular hypoxia. Possible factors are a shift in redox state, MEOS induction, a high blood alcohol concentration (BAC), a diet rich in polyunsaturated fatty acids (PUFA), and episodic decreased oxygen supply to the liver.

8.2.4. Free radicals

Evidence is accumulating for a major role of free radicals in the pathogenesis of alcohol-induced liver damage. Free radicals are molecules containing an unpaired electron, which gives the free radical a high chemical reactivity. They can damage cells and tissues in many different ways (see Fig. 8.2).

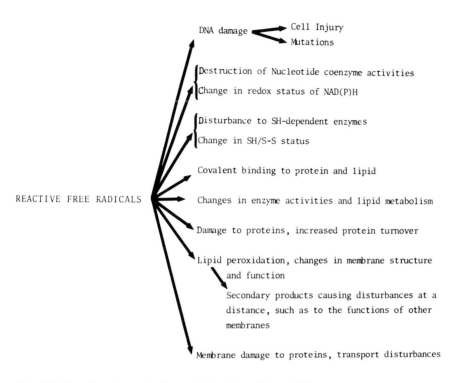

Fig. 8.2. Reactions of reactive free radicals. After: Slater (137).

The susceptibility of a tissue to peroxidation is a function of the balance between pro-oxidant and antioxidant systems. The pro-oxidant free radicals can originate from lipids, acetaldehyde and even alcohol itself. Iron is a powerful catalyst of free radical mechanisms. Glutathione (GSH), ascorbic acid, selenium and α-toco-pherol have antioxidant properties. Alcohol seems to influence both the pro-oxidant and the antioxidant systems at various sites.

Recently the in vivo formation of a free radical metabolite of alcohol (α-hydroxy-ethyl radical) has been demonstrated (55). Acetaldehyde is also a substrate of free radical production (113). In vitro peroxidation occurs after the addition of acetaldehyde in the presence of xanthine oxidase. In vitro studies of free radical generation with acetaldehyde have revealed dependence of and stimulation by iron. Administration of iron to rats potentiates hepatic radical production due to acute alcohol administration.

Another source of free radicals is lipid peroxidation. Lipid peroxidation is dependent of cytochrome P-450 reductase activity, particularly the type induced by chronic alcohol consumption (MEOS) (40). Chronic alcohol administration increases microsomal lipid peroxidation in the presence of added iron. Dicker & Cederbaum (19) found that chronic alcohol consumption leads to enhanced generation of reactive oxygen intermediates and increased inactivation of enzymes.

So, both acute alcohol consumption (direct and via acetaldehyde) and chronic alcohol consumption (by inducing MEOS) promote free radical production, which may contribute to the hepatotoxic effects of alcohol.

In a number of studies impairment of the antioxidant systems – which are important in the protection of cells against free radicals – have been reported in alcoholics. Hepatic GSH is depressed after chronic alcohol administration (133). Acute alcohol administration lowers GSH both in baboons chronically fed alcohol and in baboons not pre-treated with alcohol (133). Alcoholics, regardless of whether they suffer from hepatitis, have lower liver glutathione levels than non-alcoholic liver disease patients, and alcoholics (without liver disease) have lower plasma glutathione levels (a reflection of liver concentration) than healthy controls (59).

In cirrhotics decreased selenium concentrations have been found (1). In alcoholics without, and especially in alcoholics with, hepatic injury decreased vitamin E and selenium levels have been found (142). Alcohol induces a marked alteration of vitamin E metabolism in the liver (54).

Younes & Strubelt (155) suggest a model which links hypoxia and free radical production. Hypoxia should convert xanthine dehydrogenase into xanthine oxidase, which produces reactive oxygen radicals as a result of both acetaldehyde conversion and purine degradation. The reactive oxygen radicals mediate the inhibition of glycolysis and the direct toxic effects towards liver cells. Kato et al. (53) present a partly different hypothesis. They find it unlikely that acetaldehyde can serve as a substrate for xanthine oxidase in vivo, but do mention radical formation resulting from purine degradation.

8.2.5. Immunity

There is also increasing evidence of immunologically mediated hepatic damage (10, 20, 36, 86, 146). It has been suggested that acetaldehyde is capable of inducing liver damage not only by free radical production, but also by affecting immunity. However, its role in alcoholic liver disease remains speculative (10).

Acetaldehyde binds to free amino groups (lysine, valine and tyrosine) of the proteins in the liver cell, thus forming protein-acetaldehyde adducts (10, 79). This results in altered surface and structural conformation of the protein molecules, changing the immunogenicity. Antibodies against these protein-acetaldehyde adducts can be detected in mice chronically fed alcohol (47) and in alcoholics (96). Alcoholics with human lymphocyte antigen (HLA) B35 (86) and HLA B8 (127) develop cirrhosis faster and/or after less alcohol than other alcoholics, which suggests that a genetically determined immunological response to alcohol-induced antigens is responsible for the difference in susceptibility.

8.2.6. Collagen

Collagens are a heterogeneous group of proteins. In fibrosis liver cells proliferate and produce increased amounts of collagen. An increase in collagen content plays a direct role in the pathophysiology of alcoholic liver disease through its effect on both hepatic structure and function. It is not clear whether increased collagen in fibrotic cells can be attributed to increased synthesis (demonstrable from the increased activity of the proline hydroxylase enzyme) or decreased breakdown (27, 38, 45, 78).

Lactate, the level of which increases during alcohol metabolism, promotes collagen synthesis from proline. It is not clear whether alcohol itself affects collagen synthesis, in view of contradictory results (51, 99).

Genetic factors may also play a role in enhanced collagen production. In cirrhotic alcoholics a specific pattern in one of the genes coding for collagen has been determined (152). This suggests the existence of a genetic predisposition for hepatic fibrinogenesis.

8.3 Treatment

In this section the treatment of alcoholic liver disease will be discussed, rather than the treatment of the underlying drinking problem – which is just as well essential. An important factor in the treatment of alcoholic liver disease is abstinence. After some days of abstinence fat accumulation in fatty liver decreases and after 2 to 6 weeks the disorder usually has disappeared entirely (13, 114). In hepatitis and cirrhosis abstinence is essential too, but sometimes cirrhosis develops from hepatitis despite abstinence.

Nutritional support in alcoholic liver disease therapy is very important. Patients are often malnourished or have subclinical nutritional deficiencies. In cirrhosis dietary support has to be given carefully, because an oversupply of nutrients can give rise to complications, due to the disturbed metabolism (49).

The only effective treatment of end-stage cirrhosis is liver transplantation (130, 150). This 'start of a new life' seems to stimulate the patient to remain abstinent. However, in some alcoholics there are medical objections to liver transplantation, such as multiple organ damage or malnourishment.

Quite a number of drugs have been proposed to modify the disease process. Sometimes in vitro experiments yield promising results, whereas the results of clinical trials are disappointing. Saunders (128) has reviewed the drugs used in alcoholic liver disease therapy. The thyroid inhibitor propylthiouracil (PTU) appears now to be the most promising treatment available for treating the basic disease process. PTU reduces the oxygen consumption of the liver and should thus protect against liver injury (42, 43).

Saunders (128) states that corticosteroids should be regarded as having no role in the treatment of alcoholic liver disease. The effect of anabolic steroids should be examined further. Colchicine inhibits collagen synthesis and increases collagenase activity. It has also anti-inflammatory properties. Its therapeutic potential seems to be considerable, but has not yet been established unequivocally. Propanolol, a β-adrenergic blocking drug, reduces portal venous pressure, thus preventing bleeding of oesophageal varices (128).

8.4. Threshold values for alcohol consumption relative to liver injury

A vast body of literature on threshold values for alcoholic liver diseases has been published in the 1970s (see Table 8.1). The opinion has taken root since that a large number of factors, both of an environmental and a genetic nature, modify the association between alcohol consumption and alcoholic liver disease, so threshold values are not a very actual topic now.

The advances in modern molecular biology have revealed that every individual metabolism is a unique concert of enzymes, in which the activity of each enzyme is (partly) genetically determined. Processes can be more or less promoted or inhibited, resulting in individual differences in vulnerability for injury. Immunity probably is involved in the development of alcoholic liver injury. In alcoholics different genes coding for some antigens are found, suggesting genetically determined vulnerability. These can be seen as internal factors modifying the relation between alcohol and injury.

Life-style and eating and drinking habits could be seen as external, environmental modifying factors. After a period in which malnutrition was seen as the most important aetiological factor in alcoholic liver injury, followed by a period in which dietary factors were neglected, it is assumed now that the diet can have a modifying effect. Excess of fat promotes fatty liver (110), vitamin E deficiency enhances free radical injury (54) and the dietary composition of polyunsaturated fatty acids affects the PUFA composition in the body, possibly attenuating alcoholic injury (123).

However, alcohol is of course indispensable for the development of alcoholic liver

Table 8.1. Threshold values for alcohol consumption according to various authors. After: Piendl (106, 107).

Ref.	Alcohol use (g/day)	Results
61–67	a) < 160	patients suffering from the consequences of excessive alcohol consumption were not found to have incurred cirrhosis
	b) > 160	after 12 years of alcohol abuse 25% of the patients were found to have incurred cirrhosis, and after 20 years 50%
95	> 66 up to > 110 (88 on average)	danger limit for the onset of liver cirrhosis
41	160–200 over > 10 years	danger zone for the pathogenesis of liver cirrhosis
22	> 125 over	increased risk for alcoholic liver cirrhosis
102, 104	a) < 80	limit of harmlessness
	b) 80–160	danger zone ('Gefährdungsbereich')
	c) > 160	increased risk of developing liver cirrhosis
7	100	critical limit
5	a) < 80	alcohol consumption generally harmless
	b) > 100	danger limit
	c) 160	critical limit

disease. Interviews with patients suffering from alcoholic hepatitis or cirrhosis (or their relatives) about the quantity and duration of alcohol consumption provide global information about threshold values, and so do animal experiments. In the subsequent sections these limits with respect to fatty liver, hepatitis and cirrhosis will be discussed successively. The limits presented, however, should be interpreted with caution. Increasing alcohol consumption increases the risk for injury gradually. 'Safe' limits cannot be given because of the large variation in individual vulnerability.

8.4.1. Fatty liver

Glouberman (33) mentions (without referring to a particular source) an alcohol intake accounting for 20% of total energy intake (20 en%) as a conceivable limit below which fatty liver will not develop. This level corresponds with ca. 70 g alcohol (7 glasses) per day at a daily energy intake of 10 MJ. In the same paper Glouberman indicated that the accumulation of fat in the liver starts as soon as two days after the consumption of alcohol.

In an experiment with human subjects Lieber et al. (69) found that at an alcohol consumption level of 300 g/day fatty degeneration could be observed microscopically after 8 days; when 150 g (15 drinks) was consumed, this was not observed until after 21 days. Fatty degeneration was reversed upon termination of alcohol consumption.

According to Rao (110), who criticizes Lieber's experiments, nutritional factors are important for the development of fatty liver in non-alcoholics. With adequate diets 260 g alcohol did not produce fatty liver, whereas 130 g alcohol (46 en%) in combination with a diet of 36% fat, 2% carbohydrates and 16% protein resulted in fatty liver.

8.4.2. Hepatitis

Both the quantity of alcohol consumed per day and the duration of (excessive) alcohol consumption appear to be significant in the onset of hepatitis, but the individual differences found are large. However, frequently either only the period of alcohol use or the quantity of alcohol is reported. Brandt & Bronkhorst (13) stated a period of excessive alcohol consumption extending from one to five years. Glouberman (33) indicated that hepatitis was observed both among individuals who had been drinking for three months only and among persons who had done so for 36 years. In the same paper Glouberman mentioned that hepatitis rarely develops if daily alcohol consumption is less than 80 g.

Mezey & Santora (88) found that the five patients suffering from hepatitis (among 55 alcoholics they examined) all used to drink more than 200 g/day. However, also among alcoholics who reported to drink less than 80 g/day liver injury was found (fibrosis with fatty infiltration). Both in rodents and in primates an alcohol consumption corresponding with 36 en% resulted in fatty liver, but not in hepatitis or cirrhosis. At 50 en% hepatitis and cirrhosis developed in half of the primates, however (71–73). Although these results may not immediately apply to human beings, they prove to agree reasonably well with the results found for man (at a daily energy intake of 10 MJ, 36 and 50 en% correspond to 120 and 170 g alcohol per day respectively).

8.4.3. Cirrhosis

Piendl (106) indicated what quantities are regarded as injurious. Table 8.1 is a condensed representation of his survey. From these data it may be concluded that consumption of more than 100–160 g alcohol/day over a period of 10–20 years is necessary for the development of liver cirrhosis, but occasionally cirrhosis proves to be brought about by smaller quantities.

8.4.4. Sex differences in liver disease vulnerability

Many different groups have reported sex-related differences in alcohol-induced liver disease (27, 82, 90, 98, 149, 151). In general, female cirrhotic alcoholics drink less and for a shorter period of time than their male counterparts.

The results obtained by Péquignot and co-workers (104, 105) – in which they present threshold values for liver cirrhosis for men and women separately – are frequently quoted: a sharp increase in the risk of cirrhosis would occur when daily consumption exceeds 60 g for men and 20 g for women. In their survey Péquignot et al. compare the reported alcohol consumption of male and female cirrhotic patients with that of a random sample taken from the total population. In another investigation they found the slope of the increase in relative risk to be steeper for women than for men (147). In neither investigation they indicated whether the differences were statistically significant.

Norton (98) found in Australian women an increased and dose-dependent risk of alcohol-related cirrhosis from 41 g alcohol per day. Whereas 1% of the female Australian population drink 4 or more glasses per day, 90% of Australian cirrhotic women drink that much.

In most investigations alcohol consumption expressed in grammes per day has been examined. Chapter 2 of this review points out that women reach a higher BAC than men after the consumption of a given quantity of alcohol because of their smaller average body mass and their lower average percentage of body water. This may partly account for the differences in susceptibility observed.

Another explanation may be the distortion of actual consumption caused by the usual method of inquiry among the individuals tested. From several surveys the notion has emerged that excessive users tend to underreport their consumption (14, 21, 118), especially women (26). It is quite likely that women have a greater tendency to underreport since, according to generally received norms, a woman should not drink much (129).

Besides these possible explanations, a biologically determined susceptibility may exist. Sex hormones influence ADH and MEOS activity, and hence acetaldehyde production and lipid peroxidation. Possibly this plays a role in the increased susceptibility of women to alcoholic liver disease. Hormonal differences and differences in immune reactivity to alcohol have been suggested (27, 34, 126). Further research in this field is desirable.

8.4.5. Differences between beverages

In some publications it has been suggested that there are differences in cirrhogenic activity between the various types of beverages (70, 102, 107, 143, 153). A higher cirrhogenic activity has been attributed to wine in particular. In these studies attention has been drawn to the fact that there is a closer relationship between the consumption of wine and cirrhosis than between the consumption of spirits or beer and cirrhosis.

In four investigations cirrhosis mortality and per capita alcohol consumption, differentiated according to type of beverage, are compared for a number of countries. When cirrhosis mortality is related to the consumption of alcohol, both taken together and split up into beer, wine and spirits, a strong positive correlation between total alcohol consumption and cirrhosis mortality and between wine consumption and cirrhosis mortality is found, whereas no correlation is found between cirrhosis mortality and the consumption of spirits or beer. These results appear to indicate that cirrhosis mortality is largely caused by wine consumption.

However, a critical examination of these associations shows clearly that the results do not really contradict the hypothesis that all three types of beverages may cause cirrhosis to (almost) the same extent. Differences between countries in actual intake of alcohol prove to be determined mainly by variation in wine consumption and less so by variation in consumption of beer and spirits. In other words, from these investigations it can neither be concluded that all alcoholic beverages are (almost) equally cirrhogenic nor that wine is more cirrhogenic.

An important point is the fact that in these investigations it is not the alcohol consumption (quantity and type of beverage) by those who had died of cirrhosis that was examined. Instead, the average of the total population over 15 was taken as a measure.

Another methodological drawback of this type of investigation arises from the fact that the diagnosis of alcoholic liver cirrhosis may vary from one country to another. Moreover, countries (or states) have not only different drinking habits, but also different cultures and different eating habits and ways of life. These may also account for the observed variances in cirrhosis mortality. A high correlation coefficient does not necessarily point at causality.

Péquignot, too, pointed out that more wine is drunk in France and cirrhosis mortality in France is higher than in other countries (102), thus advancing the same statistically invalid argument. The same applies to another argument used by Péquignot, namely that 80–90% of the alcohol consumed by cirrhotic patients in France is drunk as wine: at the time of his investigation (1955) ca. 80% of all alcohol consumed in France was drunk as wine (3).

The relation between cirrhosis and alcoholic beverages has also been investigated by Lelbach (60). Of the 526 alcoholics he examined with respect to liver injury and alcohol consumption, the beer drinkers proved to have indeed less liver injury than the spirit drinkers, but the beer drinkers drank less excessively and for shorter periods. As Lelbach rightly points out, this does not justify the conclusion that spirits are more injurious to the liver than beer.

Di Luzio (83) and Mezey & Santora (88) did not find differences among the various types of alcoholic beverages as regards liver injury.

All in all, there is no conclusive evidence to justify the statement that a particular type of beverage would be more cirrhogenic than another type.

8.5. Epidemiology

Most epidemiological data concerning liver injury refer to cirrhosis mortality. There are regional as well as temporal relations to be observed between mortality due to cirrhosis and per capita alcohol consumption. These data are based on either the entire population or the population of 'drinking age', i.e. from the ages of 14, 15 or 20 years on.

Since most cases of cirrhosis occur among alcoholics, Jellinek has developed a method to estimate the number of alcoholics of a population on the basis of cirrhosis mortality (in addition to some other data): the 'Jellinek estimation formula'.

8.5.1. Temporal relations between cirrhosis mortality and per capita alcohol consumption

A positive temporal correlation between cirrhosis mortality and per capita alcohol consumption has been observed in various studies (25, 29, 30, 58, 124), whereas others have not found such a correlation (138).

Because cirrhosis tends to develop only after as many as 15 to 20 years of excessive alcohol consumption, it could be expected that brief fluctuations in alcohol consumption will not be reflected in cirrhosis mortality and that a permanent change in consumption will not be followed immediately by a change in cirrhosis mortality (time lag) (136). Therefore, it is remarkable that Sales (124) found changes in per capita consumption immediately reflected in changed cirrhosis mortality in the UK.

Longer periods of decreased alcohol consumption are usually followed by a decrease in cirrhosis mortality. Prohibition in the USA from 1916 to 1932 (109), the period between both World Wars in the Netherlands (28) and the scarcity of wine in Paris during the second World War (102) are examples of this. However, a decrease in cirrhosis mortality rate during a war cannot be attributed solely to a decrease of alcohol consumption, considering the change in total and specific mortality: someone suffering from cirrhosis who dies of wartime activities will not be included in the cirrhosis mortality rate!

The relation between disorders frequently attributed to alcohol consumption and national statistics on alcohol consumption proves less strong than has often been assumed. The fact that alcohol consumption in the Netherlands has quadrupled

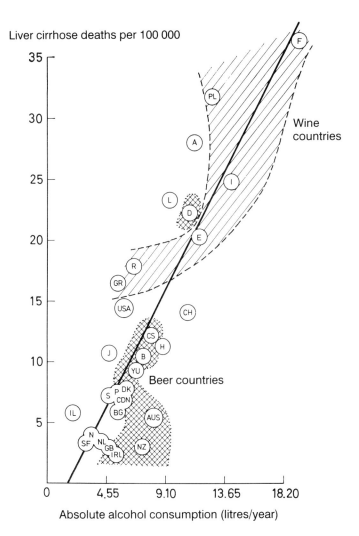

Fig. 8.3. Relation between per capita alcohol consumption and mortality from liver cirrhosis for 29 countries in 1967. A, Austria; AUS, Australia; B, Belgium; BG, Bulgaria; CDN, Canada; CH, Switzerland; CS, Czechoslovakia; D, Germany; DK, Denmark; E, Spain; F, France; GB, Great Britain; GR, Greece; H, Hungary; I, Italy; IL, Israel; IRL, Ireland; J, Japan; L, Luxembourg; N, Norway; NL, Netherlands; NZ, New Zealand; P, Poland; PL, Portugal; R, Rumania; S, Sweden; SF, Finland; USA, United States; YU, Yugoslavia. After: Piendl (106, 107) and Smith et al. (139).

between 1950 and 1975 has as yet not been followed by a comparable increase in mortality from illnesses (partly) caused by alcohol (cancer of the oesophagus, cirrhosis) (81).

8.5.2. Regional variation in the relation between cirrhosis mortality and per capita alcohol consumption

Several correlation studies have revealed a clear statistical relation between per capita alcohol consumption and cirrhosis mortality comparing the data for various countries or regions (16, 37, 107, 153). Piendl (106, 107) compared per capita alcohol consumption and liver cirrhosis mortality in 29 countries, on the basis of 1967 statistics (cf. Fig. 8.3).

Hoogendoorn (37) has done the same for 21 countries using alcohol consumption statistics over the period 1970–1972 and cirrhosis mortality statistics for 1974 (cf. Table 8.2). France had both the highest cirrhosis mortality and the highest per capita alcohol consumption. Other countries with a high cirrhosis mortality and a high alcohol consumption were Austria, Italy, Portugal, West Germany and Spain. In

Table 8.2. Per capita alcohol consumption (average of the period 1970–1972) and cirrhosis mortality (1974 data) in European and non-European countries. After: Hoogendoorn (37).

Country	Annual per capita alcohol consumption (l)	Liver cirrhosis mortality per 100 000
Europe		
France	16.0	32.8
Italy	13.7	31.9
Portugal	13.0	31.3
Austria	12.2	32.7
Spain	11.0	22.5
West Germany	10.9	26.9
Switzerland	10.7	14.8
Belgium	9.1	14.4
Denmark	7.3	10.4
England and Wales	6.7	3.6
Netherlands	5.9	4.5
Sweden	5.7	10.5
Ireland	5.7	3.7
Finland	4.7	5.5
Norway	3.7	4.1
Other regions		
Australia	8.3	8.3
New Zealand	7.9	5.4
Canada	7.1	11.6
USA	6.5	15.8
Japan	5.4	13.4
Israel	1.7	6.2

the Netherlands cirrhosis mortality per 100 000 inhabitants was less than one seventh of that in France. In Sweden a remarkably high cirrhosis mortality is found in spite of a low alcohol consumption. This could reflect a discrepancy between official and actual consumption data, owing to illigal alcohol production.

8.5.3. Alcoholics

Several surveys among groups of alcoholics have shown that, on average, alcoholics die some 12 years earlier than non-alcoholics. Total mortality is 2 to 4 times higher (14, 80) and the alcoholics' risk of dying of liver cirrhosis is 15–20 times higher than among non-alcoholics (81). Cirrhosis account for 5–15% of total mortality among alcoholics (14).

8.5.4. Cirrhosis mortality as an indicator of the prevalence of alcoholism

Jellinek has developed a formula which may be used to estimate the prevalence of alcoholism in a population on the basis of cirrhosis mortality – the 'Jellinek estimation formula'. Other methods to estimate the prevalence of alcoholism are the extrapolation of survey results, the Ledermann formula (on the basis of per capita consumption), and estimates founded on total mortality or mortality due to suicide. The Jellinek estimation formula and the Ledermann formula are the best known and most extensively used methods.

For the application of the Jellinek formula the following data are needed in addition to the total annual number of deaths from cirrhosis.
- The ratio between alcohol-induced cirrhosis mortality and total cirrhosis mortality. Initially, this ratio has been estimated from the change in cirrhosis mortality brought about by alcohol prohibition in the USA. Later, this estimation has been adapted several times. The ratio is presumably not constant, but depends on the mortality due to alcohol-induced cirrhosis and on the national health system of a country as a whole.
- The mortality due to liver cirrhosis among alcoholics showing complications.
- The ratio between alcoholics showing complications and all alcoholics. The value assigned to this relation depends on which definition of alcoholism is used.

There is so much insecurity in these variables that estimations based on the Jellinek formula must be interpreted with reserve.

8.6. Biochemical identification of alcohol abuse and alcoholism

Alcohol consumption, especially chronic alcohol abuse, affects many metabolic processes. Its major influence is exerted on processes in the liver. By this mechanism the enzyme contents of the liver change, which in turn is reflected in changed plasma levels of liver enzymes. Plasma levels of liver enzymes, as well as other blood

parameters, are often used as markers for alcohol abuse. The most commonly used parameters are:
- γ-glutamyltransferase (GGT)
- alkaline phosphatase (AP)
- erythrocyte mean corpuscular volume (MCV)
- alanine aminotransferase, or glutamic-pyruvic transaminase (ALAT, GPT)
- aspartate aminotransferase, or glutamic-oxaloacetic transaminase (ASAT, GOT)
- carbohydrate-deficient transferrin (CDT).

Other parameters have been used as well, such as urate, lactate, triglycerides, high-density lipoprotein (HDL) cholesterol, platelet count (91), δ-aminolaevulinic acid (ALA) dehydratase (58), α-amino-N-butyric acid/leucine ratio (132), and even chest radiography (112). Repeatedly new markers are proposed. Recent markers are urinary dolichol (115, 116), methanol (117), acetate (57), β-hexosaminidase (52), and reduced erythrocyte aldorase activity (35).

On a population level significant differences in the values of these parameters are usually obtained between users and non-users as controls. Compared to controls, the values for alcoholics commonly lie outside the normal ranges. Plasma GGT level is regarded to reflect alcoholism most reliably (8, 119) although the new marker CDT has been claimed to be more sensitive (11, 31).

Yet, the lack of sensitivity and specificity of these parameters hampers their use for individual cases. Sensitivity is defined as the percentage of alcoholics with a positive test and specificity is described as the percentage non-alcoholics with a negative test. Increased levels of GGT are found in approximately 50% of alcoholics (18, 23) as well as in most non-alcoholics with liver cell damage.

A combination of results of several tests may improve sensitivity and specificity (17, 18, 120, 121). Ryback et al. (120) identified correctly 94–100% of alcoholics (inpatients and outpatients) and 100% of non-alcoholics by using a combination of 25 tests.

The usefulness of distinctive tests depends on the way they are influenced by alcohol. Some tests react readily to a single alcohol dosage (MCV), while other ones only respond after hepatocellular damage or alcohol addiction. GGT seems to react to both liver injury (whether or not alcohol-related) and regular alcohol consumption (97, 101, 144). GGT is correlated only moderately with the extent of alcohol use (94). Probably a minimum amount of alcohol is necessary to increase the GGT level.

Aspartate aminotransferase (ASAT) and alanine aminotransferase (ALAT) levels are not affected by alcohol consumption, but are a reflection of organ damage. An increased ASAT/ALAT ratio is indicative of alcoholic liver injury (122).

MCV increases rapidly after alcohol intake and drops upon withdrawal (154). The simple and cheap MCV test provides an indicator for recent alcohol consumption. Even moderate alcohol drinkers can have an increased MCV level (148).

ALA dehydratase is also regarded as an indicator for recent alcohol consumption. In non-alcoholics ALA dehydratase declines to normal values only when BAC has been decreased to zero (92), whereas in alcoholics it takes several days before the value reaches the normal range (58).

Opinions on the role of α-amino-N-butyric acid diverge. It is regarded as a non-sensitive index for hepatocellular dysfunction by Morgan et al. (93) and as an indicator for alcoholism by Shaw et al. (131).

CDT is a promising new marker (11, 31). It has proved to be a more sensitive and specific marker than the eight markers commonly used including GGT. In people drinking 3 glasses or less per day no elevated levels were found, whereas from 4 glasses per day an elevated level was demonstrated in 68% of the subjects. In alcoholics 81% had an elevated CDT level. The relationship with alcohol consumption seems to be valid irrespective of alcoholic liver disease. It has been suggested that acetaldehyde mediates inhibition of enzymes involved in the synthesis of the trisaccharide component of transferrin, thus resulting in carbohydrate-deficient transferrin (141).

The choice of parameters depends on the aim which is pursued (e.g. screening, establishing alcoholism in suspected persons, controlling drop-outs in an alcohol treatment programme) and on the simplicity of testing (some tests are easily automated while others are time-consuming and expensive).

Summary

Prolonged consumption of large quantities of alcohol may cause fatty degeneration of the liver, hepatitis and liver cirrhosis. If alcohol is present in the liver the latter prefers alcohol to fat as a fuel. Because of the decreased oxidation of fat, and also because of an increased fat supply and synthesis, a reversible accumulation of fat in the liver will occur.

When normal liver metabolism is disturbed over a longer period of time liver cells may be destroyed and inflammation may occur (hepatitis). As liver injury becomes progressive, the fibrous tissue will begin to proliferate resulting in cirrhosis.

Susceptibility to the development of alcoholic liver disease varies widely. Only 8–30% of alcohol abusers actually develop structural liver damage. The variable response to alcohol makes sense if alcoholic liver injury is seen as a result of a complex of factors, part of which are alcohol-mediated, and part of which are genetically determined. A combination of some of the following processes presumably is involved in the onset of hepatitis and cirrhosis: hypoxia, increased peroxidation, mitochondrial injury, abnormal cellular immunological activity, changes in collagen production, disturbance of lymphocyte activity or antitubular activity of alcohol.

Hypoxia (an inadequate supply of oxygen to the cells) seems to be an important factor in hepatic damage, because necrosis (cell death), and later fibrosis (formation of fibrous tissue as a replacement of the damaged tissue), is observed first in the perivenular zones of the cells.

Evidence is accumulating for a major role for free radicals in the pathogenesis of alcohol-induced liver damage. The susceptibility of a tissue to peroxidation is a function of the balance between pro-oxidant and antioxidant systems. The pro-oxidant free radicals can originate from lipids, acetaldehyde and even alcohol itself. Iron is a powerful catalyst of free radical mechanisms. Glutathione (GSH), ascorbic acid, selenium and α-tocopherol have antioxidant properties. Alcohol seems to influence both pro-oxidant and antioxidant systems at various sites.

It has been suggested that acetaldehyde is capable of inducing liver damage not only by free radical production, but also by affecting immunity. However, its role in alcoholic liver disease remains speculative. Acetaldehyde binds to free amino groups, forming protein-acetaldehyde adducts with altered immunogenicity. It has been suggested that genetically determined immunological response to alcohol-induced antigens are partly responsible for the difference in susceptibility to alcoholic liver injury.

An increase in collagen content plays a direct role in the pathophysiology of alcoholic liver disease. It is not clear whether increased collagen in fibrotic cells should be attributed to increased synthesis or to decreased breakdown. Genetic factors may also play a role in enhanced collagen production.

The opinion is taking foot that a complexity of both environmental and genetic factors modify the association between alcohol consumption and alcoholic liver disease. So, threshold limits should be interpreted with caution. Life-style and eating and drinking habits could be seen as external, environmental modifying factors. After a period in which malnutrition was seen as the most important aetiological factor in alcoholic liver injury, followed by a period in which dietary factors were neglected, the diet is now assumed to have a modifying effect. There are no differences between beer, wine and spirits with respect to the risk of liver injury.

The quantity of alcohol and the duration of excessive consumption appear to influence the onset of fatty degeneration of the liver. Even short-term excessive alcohol consumption may cause a reversible accumulation of fat in the liver. Fatty liver does not develop below a daily alcohol consumption of ca. 70 g. Hepatitis appears to develop only rarely when less than 80 g alcohol is consumed daily over a period not exceeding 1-5 years is consumed. When daily alcohol consumption of over 100–160 g/day is prolonged for 10–20 years there will be an increased risk of developing cirrhosis. Occasionally, however, cirrhosis proves to develop when smaller quantities are consumed. It must be strongly emphasized that the values mentioned above are no more than rough indications.

Women appear to be more prone to developing liver cirrhosis than men. Their lower average body weight and smaller average body water volume can only partly explain this. Possibly hormonal effects play a role as well.

Both temporally (in the course of time) and regionally (between countries), per capita alcohol consumption appears to be associated with cirrhosis mortality.

Plasma levels of liver enzymes as well as other parameters are often used to determine the extent of alcohol (ab)use. On a population level significant differences are usually obtained. As yet, the lack of sensitivity and specificity of the parameters hamper their applicability to individual cases. However, a combination of various tests may improve sensitivity and specificity drastically.

References

1. Aaseth J, Smith-Kielland A, Thomassen Y. Selenium, alcohol and liver diseases. Ann Clin Res 1986;18: 43-7.
2. Achord JL. Nutrition, alcohol, and the liver. Am J Gastroenterol 1988;83: 244-8.
3. Anonymous. Hoeveel alcoholhoudende dranken worden er in de wereld gedronken?, 9th ed. Schiedam: Produktschap voor Gedestilleerde Dranken, 1970.
4. Anonymous. Alcohol and health: new knowledge. Washington DC: National Institute on Alcohol Abuse and Alcoholism, US-GPO, 1974. (DHEW Publication (ADM) 74-124.)
5. Anonymous. Informationen des Bundesministeriums für Jugend, Familie und Gesundheit, 'Alkohol- und Drogenmissbrauch und Kriminalität von Kindern und Jugendlichen. Bonn-Bad Godesberg: Bundesministerium für Jugend, Familie und Gesundheit, 1977; 79 pp.
6. Anonymous. Immunological abnormalities in alcoholic liver disease. Lancet 1983;1: 605-6.
7. Avenarius HJ, Eger W. Alkoholfettleber und Sekundärschaden. Mat Med Nordmark 1967;19: 283.
8. Baglin MC et al. Efficacité comparé du volume globulaire moyen (VGM) et de la gamma-gutamyl-transferase (γ-GT) serique comme tests de triage des buveurs excessifs d'alcool. Clin Chim Acta 1976;68: 321-6.
9. Baraona E., et al. Zonal redox changes as a cause of selective perivenular hepatotoxicity of alcohol. Pharmacol Biochem Behav 1983;18(suppl 1): 449-54.
10. Barry RE. Role of acetaldehyde in the pathogenesis of alcoholic liver disease. Br J Addict 1988;83: 1381-6.
11. Behrens UJ, Worner TM, Braly LF, Schaffner F, Lieber CS. Carbohydrate-deficient transferrin, a marker for chronic alcohol consumption in different ethnic populations. Alcoholism Clin Exp Res 1988;12: 427-32.
12. Bradford BU, Marotto M, Lemasters JJ, Thurman RG. New, simple models to evaluate zone-specific damage due to hypoxia in the perfused rat liver: time course and effect of nutritional state. J Pharmacol Exp Ther 1986;236: 263-8.
13. Brandt KH, Bronkhorst FB. Alcohol, lever en pancreas. Ned Tijdschr Geneesk 1979; 123: 1248-55.
14. Brody JA, Mills GS. On considering alcohol as a risk factor in specific diseases. Am J Epidemiol 1978;107: 462-7.
15. Bunout D, Gattás V, Iturriaga H, Pérez C, Pereda T, Ugarte G. Nutritional status of alcoholic patients: its possible relationship to alcoholic liver damage. Am J Clin Nutr 1983;38: 469-73.
16. Capocaccia R, Farchi G. Mortality from liver cirrhosis in Italy: proportion associated with consumption of alcohol. J Clin Epidemiol 1988;41: 347-57.
17. Chalmers DM et al. Biochemical and haematological indicators of excessive alcohol consumption. Gut 1981;22: 992-6.
18. Cushman P, Jacobson G, Barboriak JJ, Anderson AJ. Biochemical markers for alcoholism: sensitivity problems. Alcoholism Clin Exp Res 1984;8: 253-7.
19. Dicker E, Cederbaum AI. Increased oxygen radical-dependent inactivation of metabolic enzymes by liver microsomes after chronic ethanol consumption. FASEB J 1988;2: 2901-6.
20. Eddleston ALWF, Vento S. Relevance of immune mediated mechanisms in progressive alcoholic liver injury. In: Thomson AD, ed. Mechanisms of alcoholic liver damage. Oxford: Pergamon Press, 1988. Mol Asp Med 1988;10: 169-77.
21. Edwards G. Epidemiology applied to alcoholism. Q J Stud Alcohol 1973;34: 28-56.
22. Eghöje KN, Juhl E. Factors determining liver damage in chronic alcoholics. Scand J Gastroenterol 1973;8: 505-12.
23. Evans JR, Ogston S, Guthrie A, Johnston B, MacKechnie L. The relationship between liver function tests and alcohol intake in patients admitted to an alcoholism unit. Ann Clin Biochem 1984;21: 261-7.

24. French SW. Biochemical basis for alcohol-induced liver injury. Clin Biochem 1989;22: 41-9.
25. Gallagher RP, Elwood JM. Increases in alcohol related mortality in Canada, 1965–77. Lancet 1980;i: 775-6.
26. Garretsen HFL. Probleemdrinken: preventiebepaling, beïnvloedende factoren en preventiemogelijkheden. Lisse: Swets & Zeitlinger, 1983.
27. Gavaler JS. Sex-related differences in ethanol-induced liver disease: artificial or real? Alcoholism Clin Exp Res 1982;6: 186-96.
28. Gips CH. Recente gegevens over het alcoholgebruik en over de sterfte aan levercirrose in Nederland. Tijdschr Alcohol Drugs 1976;2: 131-3.
29. Gips CH. Alcohol, diseases of alcoholics and alcoholic liver disease. Neth J Med 1978;21: 83-90.
30. Gips CH. Alcohol, ziekten van alcoholisten en alcoholische leverziekten. Tijdschr Alcohol Drugs 1978;4: 175-6.
31. Gjerde H, Johnsen J, Bjorneboe A, Bjorneboe GEA, Morland J. A comparison of serum carbohydrate-deficient transferrin with other biological markers of excessive drinking. Scand J Clin Lab Invest 1988;48: 1-6.
32. Glen I, Skinner F, Glen E, MacDonell L. The role of essential fatty acids in alcohol dependence and tissue damage. Alcoholism Clin Exp Res 1987;11: 37-41.
33. Glouberman S. Alcohol and the liver. Ariz Med 1979;36: 186-8.
34. Gluud C. Testosterone and alcoholic cirrhosis: epidemiologic, pathophysiologic and therapeutic studies in men. Dan Med Bull 1988;35: 564-74.
35. Harada S. Reduced erythrocyte aldorase activity as a new marker of alcoholism and its mechanism. In: Kuriyama K, Takada A, Ishii H, eds. Biomedical and social aspects of alcohol and alcoholism: proceedings of the Fourth Congress of the International Society for Biomedical Research in Alcoholism (ISBRA), Kyoto, Japan, 26 June–2 July 1988. Amsterdam: Elsevier Science Publishers, 1988: 553-6.
36. Hasumura Y, Izumi N, Sakai Y, Takeuchi J. Lymphocyte infiltration in the liver of patients with alcoholic hepatitis. In: Kuriyama K, Takada A, Ishii H, eds. Biomedical and social aspects of alcohol and alcoholism: proceedings of the Fourth Congress of the International Society for Biomedical Research in Alcoholism (ISBRA), Kyoto, Japan, 26 June–2 July 1988. Amsterdam: Elsevier Science Publishers, 1988: 779-82.
37. Hoogendoorn D. Het toenemende gebruik van alcohol en de stijgende frequentie van enkele (mede) door alcohol veroorzaakte ziekten. Ned Tijdschr Geneesk 1978;122: 1275-80.
38. Horrobin DF. A biochemical basis for alcoholism and alcohol-induced damage including the fetal alcohol syndrome and cirrhosis: interference with essential fatty acid and prostaglandin metabolism. Med Hypotheses 1980;6: 929-42.
39. Horrobin DF. Essential fatty acids, prostaglandins, and alcoholism: an overview. Alcoholism Clin Exp Res 1987;11: 2-9.
40. Ingelman-Sundberg M, Ingelman-Sundberg M, Ekstrom G, Tindberg N. Lipid peroxidation dependent on ethanol-inducible cytochrome P-450 from rat liver. In: Nordman R, Ribiere C, Rouach H, eds. Alcohol toxicity and free radical mechanisms. Proceedings of the first Congress of the European Society for Biomedical Research on Alcoholism (ESBRA), Paris, 18–19 September 1987. Oxford: Pergamon Press, 1988: 43-8. (Advances in the biosciences vol 71).
41. Insunza I, Iturriaga H, Ugarte G, Altschiller H. Clinical and histological liver abnormalities in alcoholics. Acta Hepato-splenol 1971;31: 460-70.
42. Israel Y, et al. Hypermetabolic state, oxygen availability, and alcohol-induced liver damage. In: Majchrowicz E, Noble EP, eds. Biochemistry and pharmacology of ethanol, vol 1. New York: Plenum Press, 1979: 433-44.
43. Israel Y, et al. Studies on metabolic tolerance to alcohol, hepatomegaly and alcoholic liver disease. Drug Alcohol Depend 1979;4: 109-18.
44. Israel Y, Britton RS, Orrego H. Liver cell enlargement induced by chronic alcohol consumption; studies on its causes and consequences. Clin Biochem 1982;15: 189-92.
45. Israel Y, Orrego H, Colman JC, Britton RS. Alcohol-induced hepatomegaly: pathogenesis and role in the production of portal hypertension. Fedn Proc 1982;41: 2472-7.
46. Israel Y, Orrego H. On the characteristics of alcohol-induced liver enlargement and its possible hemodynamic consequences. Pharmacol Biochem Behav 1982;18(Suppl 1): 433-7.

47. Israel Y, Hurwitz E, Niemelä O, Arnon E. Monoclonal and polyclonal antibodies against acetaldehyde-containing epitopes in acetaldehyde-protein adducts. Proc Natl Acad Sci USA 1986;83: 7923-7.
48. Isselbacher KJ. Metabolic and hepatic effects of alcohol. N Engl J Med 1977;296: 612-6.
49. James R. Nutritional support in alcoholic liver disease: a review. J Hum Nutr Diet 1989;2: 315-22.
50. Jauhonen P, Baraona E, Miyakawa H, Lieber CS. Mechanism for selective perivenular hepatotoxicity of ethanol. Alcoholism Clin Exp Res 1982;6: 350-7.
51. Kamegaya K, Okazaki, Maruyama K. Ethanol, acetaldehyde and lactic acid stimulate collagen synthesis of cultured Chang liver cells. In: Kuriyama K, Takada A, Ishii H, eds. Biomedical and social aspects of alcohol and alcoholism: proceedings of the Fourth Congress of the International Society for Biomedical Research in Alcoholism (ISBRA), Kyoto, Japan, 26 June–2 July 1988. Elsevier Science Publishers, 1988: 685-8.
52. Kärkkäinen P, Poikolainen K, Salaspuro M. Serum beta-hexosaminidase as a marker of heavy drinking. Alcoholism Clin Exp Res 1990;14: 187-90.
53. Kato S, Kawase T, Alderman J, Lieber CS. The role of xanthine oxidase in ethanol-induced lipid peroxidation. In: Kuriyama K, Takada A, Ishii H, eds. Biomedical and social aspects of alcohol and alcoholism: proceedings of the Fourth Congress of the International Society for Biomedical Research in Alcoholism (ISBRA), Kyoto, Japan, 26 June–2 July 1988. Elsevier Science Publishers, 1988: 671-4.
54. Kawase T, Kato S, Lieber CS. Lipid peroxidation and antioxidant defense systems in rat liver after chronic ethanol feeding. Hepatology 1989;10: 815-21.
55. Knecht KT, Bradford BU, Mason RP, Thurman RG. In vivo formation of a free radical metabolite of ethanol. Mol Pharmacol 1990;38: 26-30.
56. Koga S, Hirayama C. Disturbed release of lipoprotein from ethanol-induced fatty liver. Experientia 1968;24: 438-9.
57. Korri UM, Salaspuro M. Characteristics of metabolic tolerance to alcohol as reflected in blood acetate levels during alcohol oxidation. In: Kuriyama K, Takada A, Ishii H, eds. Biomedical and social aspects of alcohol and alcoholism: proceedings of the Fourth Congress of the International Society for Biomedical Research in Alcoholism (ISBRA), Kyoto, Japan, 26 June 26–2 July 1988. Amsterdam: Elsevier Science Publishers, 1988: 511-5.
58. Krasner N, et al. Depression of erythrocyte δ-aminolaevulinic acid dehydratase activity in alcoholics. Clin Sci Mol Med 1974;46: 415-8.
59. Lauterburg BH, Velez ME. Glutathione deficiency in alcoholics: risk factor for paracetamol hepatotoxicity. Gut 1988;29: 1153-7.
60. Lelbach WK. Zur leberbeschädigenden Wirkung verschiedener Alkoholika. Dte Med Wochenschr 1967;92: 233-8.
61. Lelbach WK. Leberschäden bei chronischem Alkoholismus. Ergebnisse einer klinischen, klinisch-chemischen und bioptisch-histologischen Untersuchung an 526 Alkoholkranken während der Entziehungskur in einer offenen Trinkerheilstätte. Teil I: Klinische Ergebnisse. Acta Hepato-splenologica 1966;13: 321-34.
62. Lelbach WK. Leberschäden bei chronischem Alkoholismus. Ergebnisse einer klinischen, klinisch-chemischen und bioptisch-histologischen Untersuchung an 526 Alkoholkranken während der Entziehungskur in einer offenen Trinkerheilstätte. Teil II: Klinisch-chemische Ergebnisse. Acta Hepato-splenologica 1966;13: 334-49.
63. Lelbach WK. Leberschäden bei chronischem Alkoholismus. Ergebnisse einer klinischen, klinisch-chemischen und bioptisch-histologischen Untersuchung an 526 Alkoholkranken während der Entziehungskur in einer offenen Trinkerheilstätte. Teil III: Bioptisch-histologische Ergebnisse. Acta Hepato-splenologica 1967;14: 9-26.
64. Lelbach WK. Leberschäden bei chronischem Alkoholismus. Ergebnisse einer klinischen, klinisch-chemischen und bioptisch-histologischen Untersuchung an 526 Alkoholkranken während der Entziehungskur in einer offenen Trinkerheilstätte. Teil IV: Diskussion und Schlussfolgerungen. Acta Hepato-splenologica 1967;14: 27-39.
65. Lelbach WK. Leberzirrhose und Alkohol. Med Tribune (Int Wochenztg, Ausg Deutschland) 1970;5(48a): 19.

66. Lelbach WK. Dosis-Wirkungs-Beziehung bei Alkohol-Leberschäden. Dte Med Wochenschr 1972;97: 1435-6.
67. Lelbach WK. Alkohol und Leberzirrhose—ein quantitatives Problem? Therapiewoche 1974;24: 6025-6032.
68. Lelbach WK. Epidemiology of alcoholic liver disease: Continental Europe. In: Hall P, ed. Alcoholic liver disease. London: Edward Arnold, 1985: 130-66.
69. Lieber CS, Jones DP, DeCarli LM. Effects of prolonged ethanol intake: production of fatty liver despite adequate diets. J Clin Invest 1965;44: 1009-21.
70. Lieber CS. Liver adaptation and injury in alcoholism. N Engl J Med 1973;288: 356-62.
71. Lieber CS, et al. Effect of chronic alcohol consumption on ethanol and acetaldehyde metabolism. Adv Exp Med Biol 1975: 185-227.
72. Lieber CS, et al. Differences in hepatic and metabolic changes after acute and chronic alcohol consumption. Fedn Proc 1975;34: 2060-74.
73. Lieber CS. Alcohol, protein nutrition, and liver injury. Gastroenterology 1980;79: 373-90.
74. Lieber CS. Medical disorders of alcoholism; pathogenesis and treatment. Philadelphia, PA: Saunders, 1982. (Major problems in internal medicine 22.)
75. Lieber CS. Alcohol, protein nutrition, and liver injury. Curr Concepts Nutr 1983;12: 49-71.
76. Lieber CS. Alcohol and the liver. In: Arias IM, Frenkel M, Wilson JHP, eds. Liver annual 3: A series of critical surveys of the international literature. Amsterdam: Elsevier, 1983: 106-48.
77. Lieber CS. Precursor lesions of cirrhosis. Alcohol Alcoholism 1983;18: 5-20.
78. Lieber CS. Alcohol and the liver: metabolism of ethanol, metabolic effects and pathogenesis of injury. Acta Med Scand 1985; 11-55.
79. Lin RC, Lumeng L. Formation of a protein-acetaldehyde adduct in liver in vivo during chronic alcohol ingestion. In: Kuriyama K, Takada A, Ishii H, eds. Biomedical and social aspects of alcohol and alcoholism: proceedings of the Fourth Congress of the International Society for Biomedical Research in Alcoholism (ISBRA), Kyoto, Japan, 26 June—2 July 1988. Amsterdam: Elsevier Science Publishers, 1988; 325-8.
80. de Lint J. Current trends in the prevalence of excessive alcohol use and alcohol-related health damage. Br J Addict 1975;70: 3-14.
81. de Lint J. Liver cirrhosis mortality as an indicator of the prevalence of heavy alcohol use: A brief comment on Skog. Br J Addict 1980;75: 361-6.
82. Loft S, Olesen KL, Dossing M. Increased susceptibility to liver disease in relation to alcohol consumption in woman. Scand J Gastroenterol 1987;22: 1251-6.
83. di Luzio NR. The importance of ethanol metabolism to the development of the ethanol-induced fatty liver. In: Roach MK et al., eds. Biological aspects of alcohol. Austin, TX: University of Texas Press, 1971: 3-26.
84. Maclure KM, Hayes KC, Colditz GA, Stampfer MJ, Speizer FE, Willett WC. Weight, diet, and the risk of symptomatic gallstones in middle-aged women. N Engl J Med 1989;321: 563-9.
85. Maier KP, Seitzer D, Haag G, et al. Verlaufetormen alkoholischer Lebererkrankungen. Klin Wochenschr 1979;57: 311-7.
86. Marbet UA, Stalder GA, Thiel G, Bianchi L. The influence of HLA antigens on progression of alcoholic liver disease. Hepato-gastroenterology 1988;35: 65-8.
87. May SJ, Kuller LH, Perper JA. The relationship of alcohol to sudden natural death. T Stud Alcohol 1980;41: 693-701.
88. Mezey E, Santora PB. Liver abnormalities in alcoholism: Alcohol consumption and nutrition. In: Gastineau CF, Darby WJ, Turner TB, eds. Fermented food beverages in nutrition. New York: Academic Press, 1979: 303-16.
89. Mezey E. Alcoholic liver disease: roles of alcohol and malnutrition. Am J Clin Nutr 1980;33: 2709-18.
90. Mezey E, Kolman CJ, Diehl AM, Mitchell MC, Herlong HF. Alcohol and dietary intake in the development of chronic pancreatitis and liver disease in alcoholism. Am J Clin Nutr 1988;48: 148-51.
91. Mistilis SP, Barr GD. Alcohol and the liver III: Laboratory aids in the detection of alcohol abuse. Med J Aust 1981;ii: 538-40.
92. Moore MR, et al. Depression of δ-aminolaevulic acid dehydrase activity by ethanol in man and rat. Clin Sci 1971;40: 81-8.

93. Morgan MY, et al. Ratio of plasma alpha-amino-n-butyric acid to leucine as an empirical marker of alcoholism: diagnostic value. Science 1977;197: 1183-5.
94. Morgan MY, Colman JC, Sherlock S. The use of a combination of peripheral markers for diagnosing alcoholism and monitoring for continued abuse. Br J Alcohol Alcoholism 1981;16: 167-77.
95. Nakamura T, Nakamura S, Aikawa T, Karoji N, Suzuki O, Onodera A. Hepatic changes in heavy drinkers among in- and out-patients. Tohoku J Exp Med 1967;93: 191-7.
96. Niemelä O, Klajner F, Orrego H, Vidins E, Blendis L, Israel Y. Antibodies against acetaldehyde-modified protein epitopes in human alcoholics. Hepatology 1987;7: 1210-4.
97. Nishimura M, Teschke R. Alcohol and gamma-glutamyltransferase. Klin Wochenschr 1983;61: 265-75.
98. Norton R, Batey R, Dwyer T, MacMahon S. Alcohol consumption and the risk of alcohol related cirrhosis in women. Br Med J 1987;295: 80-2.
99. Okazaki I, Maruyama K, Kashiwazaki K, Kamegaya K, Ishii H, Tsuchiya, M. Does alcohol stimulate collagen synthesis or inhibit its degradation in the liver? In: Kuriyama K, Takada A, Ishii H, eds. Biomedical and social aspects of alcohol and alcoholism: proceedings of the Fourth Congress of the International Society for Biomedical Research in Alcoholism (ISBRA), Kyoto, Japan, 26 June–2 July 1988. Amsterdam: Elsevier Science Publishers, 1988: 667-70.
100. Olson RE. Absorption, metabolism, and excretion of ethanol including effects on water balance and nutritional status. In: Gastineau CF, Darby WJ, Turner TB, eds. Fermented food beverages in nutrition. New York: Academic Press, 1979: 197-212.
101. Penn R, Worthington DJ. Is serum gamma-glutamyltransferase a misleading test? Br Med J 1983;286: 531-5.
102. Péquignot G. Die Rolle des Alkohols bei der Ätiologie von Leberzirrhosen in Frankreich. Münch Med Wochenschr 1962;103:1464-8.
103. Péquignot G, Cyrulnik F. Chronic diseases due to overconsumption of alcoholic drinks (Excepting neuropsychiatric pathology). In: International encyclopedia of pharmacology and therapeutics, vol 2, Section 20: Alcohols and derivatives. Oxford: Pergamon, 1970: 375-412.
104. Péquignot G, et al. Augmentation du risque de cirrhose en fonction de la ration d'alcohol. Rev Alcool 1974;20: 191-202.
105. Péquignot G, Tuyns AJ, Berta JL. Ascitic cirrhosis in relation to alcohol consumption. Int J Eipdemiol 1978;7: 113-20.
106. Piendl A. Über die Grenze des massvollen Verzehrs und des Überkonsums an Alkohol I. Brauwelt 1979;119: 192-6.
107. Piendl A. Über die Grenze des massvollen Verzehrs und des Überkonsums an Alkohol II. Brauwelt 1979;119: 266-72.
108. Pimstone NR, French SW. Alcoholic liver disease. In: Dowgun K, Geokas MC, eds. Symposium on ethyl alcohol and disease. Philadelphia, PA: Saunders. Med Clin North Am 1984; 39-56.
109. Popham RE. The Jellinek alcoholism estimation formula and its application to Canadian data. Q J Stud Alcohol 1956;17: 559-93.
110. Rao GA, Larkin EC, Porta EA. Is alcohol itself hepatotoxic independent of nutritional factors in nonalcoholic humans? Biochem Arch 1989; 5:1-9.
111. Rappaport AM. The microcirculatory acinar concept of normal and pathological hepatic structure. Beitr Path 1976;157: 215-43.
112. Redmond AD. Chest radiography as a marker of alcoholism. Br Med J 1982;285: 810.
113. Reinke LA, Lai EK, MacCay PB. Administration of ethanol or acetaldehyde to rats results in the generation of free radicals in liver, heart, and other organs. In: Kuriyama K, Takada A, Ishii H, eds. Biomedical and social aspects of alcohol and alcoholism: proceedings of the Fourth Congress of the International Society for Biomedical Research in Alcoholism (ISBRA), Kyoto, Japan, 26 June–2 July 1988. Amsterdam: Elsevier Science Publishers, 1988: 663-6.
114. Renger F. Alkohol und Leber. Z Ges Inn Med 1981;36: 560-6.
115. Roine RP, Turpeinen U, Ylikhari R, Salaspuro M. Urinary dolichol: a new marker of alcoholism. Alcoholism Clin Exp Res 1987;11: 525-7.
116. Roine, R.P. Effects of moderate drinking and alcohol abstinence on urinary dolichol levels. Alcohol: Int Biomed J 1988;5: 229-31.

117. Roine RP, Eriksson CJP, Ylikahri R, Penttilä A, Salaspuro M. Methanol as a marker of alcohol abuse. Alcoholism Clin Exp Res 1989;13: 172-5.
118. Room R. Validity of alcohol expenditure data in consumer expenditure surveys. Drinking Drug Pract Surveyor 1971;4: 8.
119. Rosalki SB, Rau D. Serum γ-glutamyltranspeptidase activity in alcoholism. Clin Chim Acta 1972;39: 41-7.
120. Ryback RS, et al. Biochemical and hematological correlates of alcoholism. Res Commun Chem Path Pharmacol 1980;27: 533-50.
121. Ryback RS, Eckhardt MJ, Felsher B, Rawlings RR. Biochemical and hematological correlates of alcoholism and liver disease. JAMA 1982;248: 2261-5.
122. Salaspuro M. Characteristics of laboratory markers in alcohol-related organ damage. Scand J Gastroenterol 1989;24: 769-80.
123. Salem N. Alcohol, fatty acids, and diet. Alcohol Health Res World 1989;13: 211-8.
124. Sales J, Duffy J, Plant M, Peck D. Alcohol consumption, cigarette sales and mortality in the United Kingdom: an analysis of the period 1970–1985. Drug Alcohol Depend 1989;24: 155-60.
125. Sato N, Kawano S, Matsumura T, et al. Abnormal heterogenic distribution of hepatic sinusoidal blood flow and oxygenation after actue ethanol consumption in rats: direct evidence for pericentral hypoxemia. In: Kuriyama K, Takada A, Ishii H, eds. Biomedical and social aspects of alcohol and alcoholism: proceedings of the Fourth Congress of the International Society for Biomedical Research in Alcoholism (ISBRA), Kyoto, Japan, 26 June–2 July 1988. Amsterdam: Elsevier Science Publishers, 1988: 775-8.
126. Saunders J, et al. Do women develop alcoholic liver disease more readily than men? Br Med J 1981;i: 1140-3.
127. Saunders JB, Wodak AD, Haines A, et al. Accelerated development of alcoholic cirrhosis in patients with HLA B8. Lancet 1982;i: 1381-4.
128. Saunders JB. Treatment of alcoholic liver disease. Baillières Clin Gastroenterol 1989;3: 39-65.
129. Schippers GM. Constructie van de alcohol opvattingen vragenlijst. Internal report. Nijmegen: Vakgroep Klinische Psychologie, 1979.
130. Schwartzman K. In vino veritas?: alcoholics and liver transplantation. Can Med Assoc J 89;141: 1262-5.
131. Shaw S, et al. Plasma alpha-amino-n-butyric acid to leucine ratio: An empirical biochemical marker of alcoholism. Science 1976;194: 1057-8.
132. Shaw S, et al. Plasma alpha-amino-n-butyric acid/leucine: a biochemical marker for alcohol consumption. Application for the detection and assessment of alcoholism. In: Seixas FA, ed. Currents in alcoholism: biological, biochemical and clinical studies, vol 1. New York: Grune & Stratton, 1977: 17-31.
133. Shaw S. Lipid peroxidation, iron mobilization and radical generation induced by alcohol. Free Radical Biol Med 1989;7: 541-7.
134. Sherlock S. Current problems in alcoholic liver disease. Alcohol Alcoholism 1983;18: 99-118.
135. Simko V, Connell AM, Banks B. Nutritional status in alcoholics with and without liver disease. Am J Clin Nutr 1982;35: 197-203.
136. Skog OJ. Liver cirrhosis mortality as an indicator of prevalence of heavy alcohol use: some methodological problems. Oslo: National Institute for Alcohol Research, 1979.
137. Slater TF. Free radical mechanisms in tissue injury with special reference to the cytotoxic effects of ethanol and related alcohols. In: Nordmann R, Ribière C, Rouach H, eds. Alcohol toxicity and free radical mechanisms: proceedings of the first Congress of the European Society for Biomedical Research on Alcoholism (ESBRA), Paris, 18–19 September 1987. Oxford: Pergamon, 1988: 1-9.
138. Smart RG. Recent international reductions and increases in liver cirrhosis deaths. Alcoholism Clin Exp Res 1988;12: 239-42.
139. Smith LJ, Boudreau A, Chappel CI, Gillespie WI, Quastel DMJ. Beer, wine and spirits: beverage differences and public policy in Canda. Report of the Alcoholic Beverage Study Committee, Brewers Association of Canada, 1973; 164 pp.
140. Stabenau JR. Basic research on heredity and alcohol: implications for clinical application. Social Biol 1985;32: 297-321.

141. Stibler H, Borg S. The value of carbohydrate-deficient transferrin as a marker of high alcohol consumption. In: Kuriyama K, Takada A, Ishii H, eds. Biomedical and social aspects of alcohol and alcoholism: proceedings of the Fourth Congress of the International Society for Biomedical Research in Alcoholism (ISBRA), Kyoto, Japan, 26 June–2 July 1988. Amsterdam: Elsevier Science Publishers, 1988: 503-6.
142. Tanner AR, Bantock I, Hinks L, Lloyd B, Turner NR, Wright R. Depressed selenium and vitamin E levels in an alcoholic population: possible relationship to hepatic injury through increased lipid peroxidation. Dig Dis Sci 1986;31: 1307-12.
143. Terris M. Epidemiology of cirrhosis of the liver: national mortality data. Am J Public Health 1967;57: 2076-88.
144. Teschke R, Neuefeind M, Nishimura M, Strohmeyer G. Hepatic gamma-glutamyltransferase activity in alcoholic fatty liver: comparison with other liver enzymes in man and rats. Gut 1983;24: 625-630.
145. Thurman RG, Ji S, Matsumura T, Lemasters JJ. Is hypoxia involved in the mechanism of alcohol-induced liver injury? Fund Appl Texicol 1984;4: 125-33.
146. Tsuchimoto K, Ishii H, Tsuchiya M. Detection of autoantibodies to liver cell membrane and antibodies to ethanol altered liver cell determinants in alcoholic liver disease (ALD). In: Kuriyama K, Takada A, Ishii H, eds. Biomedical and social aspects of alcohol and alcoholism: proceedings of the Fourth Congress of the International Society for Biomedical Research in Alcoholism (ISBRA), Kyoto, Japan, 26 June–2 July 1988. Amsterdam: Elsevier Science Publishers, 1988: 783-787.
147. Tuyns AJ, Péquignot G. Greater risk of ascitic cirrhosis in females in relation to alcohol consumption. Int J Epidemiol 1984;13: 53-7.
148. Unger KW, Johnson D. Red blood cell mean corpuscular volume: A potential indicator of alcohol usage in a working population. Am J Med Sci 1974;267: 281-9.
149. Van Thiel DH, Gavaler JS, Rosenblum E, Tarter RE. Ethanol, its metabolism and hepatotoxicity as well as its gonadal effects: effects of sex. Pharmacol Ther 1989;41: 27-48.
150. Van Thiel DH, Gavaler JS, Tarter RE, et al. Liver transplantation for alcoholic liver disease: a consideration of reasons for and against. Alcoholism Clin Exp Res 1989;13: 181-4.
151. Van Thiel DH, Tarter RE, Rosenblum E, Gavaler JS. Ethanol, its metabolism and gonadal effects: does sex make a difference? In: Gordis E, Tabakoff B, Linnoila M, Stimmel B, eds. Alcohol research from bench to bedside. New York: The Haworth Press, 1989: 131-69.
152. Weiner FR, Eskreis DS, Compton KV, Orrego H, Zern MA. Haplotype analysis of a type I collagen gene and its association with alcoholic cirrhosis in man. In: Thomson AD, ed. Mechanisms of alcoholic liver damage. Oxford: Pergamon Press, 1988. Mol Asp Med 1988;10: 159-68.
153. Whitlock FA. Liver cirrhosis, alcoholism and alcohol consumption. Q J Stud Alcohol 1974;35: 586-605.
154. Wu A, et al. Macrocytosis of chronic alcoholism. Lancet 1974;i: 829-31.
155. Younes M, Strubelt O. Enhancement of ethanol-induced hepatotoxicity by hypoxia. In: Nordman R, Ribiere C, Rouach H. Alcohol toxicity and free radical mechanisms: Proceedings of the first Congress of the European Society for Biomedical Research on Alcoholism (ESBRA), Paris, 18–19 September 1987. Oxford: Pergamon Press, 1988: 165-9. (Advances in the biosciences vol 71.)
156. Younes M, Wagner H, Strubelt O. Enhancement of acute ethanol hepatotoxicity under conditions of low oxygen supply and ischemia/reperfusion: the role of oxygen radicals. Biochem Pharmacol 1989;38: 3573-81.

CHAPTER 9

Effects of alcohol on the brain and the nervous system

W.G. Vrij-Standhardt

Alcohol is widely known to affect functioning of the brain and of the nervous system. The changes of mood and behaviour brought about by the consumption of a few drinks are the most common and the least drastic. On the other hand, irreversibele mental disorders are observed in some alcoholics. A distinction can be made between the effects of acute and chronic alcohol consumption.

In this chapter the effects of acute and chronic alcohol consumption and possible mechanisms of the action of alcohol on the brain and the nervous system will be discussed. Attention is paid to the aetiology and mechanisms of tolerance, physical dependence and withdrawal.

9.1. Effects of acute alcohol consumption

The effect of alcohol is greatest during the increasing stage of the blood alcohol concentration (BAC). In broad outline, there is a relation between the height of the BAC and the changes observed in mood, behaviour and physical condition (tipsiness, intoxication) (23, 34). When the BAC has returned to (almost) normal values, hangover symptoms may appear.

9.1.1. Effects of increasing BAC

When someone's BAC has reached a value of 0.2 g/l, he will feel relaxed, pleasantly mellow and talkative. In these quantities alcohol has a 'lubricant' effect on social contacts. At 0.3–0.5 g/l an euphoric condition comes about (an increased sense of well-being, accompanied with unfounded optimism). Powers of concentration decrease and reaction time increases. Subjects whose BAC ranges between 0.3 and 0.8 g/l perform concentration and reaction tests, and also driving tests, less successfully than in a sober state, whereas most of them are convinced of the opposite (27, 49, 51). Also, at a BAC of ca. 0.3 g/l sport achievements prove to be diminished, mainly as a result of poor co-ordination and reflex control (9). A BAC of ca. 0.5 g/l is the safety limit for participating in traffic in the Netherlands. In other countries this limit has been set at 0.8 g/l (France, Great Britain) or 1.0 g/l (USA).

At ca. 1 g/l intoxication will be distinctly observed, with its attendant symptoms such as slurred speech, unsteady gait and antisocial, possibly aggressive, behaviour.

A BAC of 2–3 g/l may cause unconsciousness or a black-out (loss of memory without loss of consciousness). During a black-out, which may last as long as 48 hours, one is unable to add any perception to his long-term memory (92). The short-term memory is not affected, neither is the ability to recall from the long-term memory. So, the individual in question does not behave differently during blackout drinking as compared to drinking without amnesia.

A BAC of 5 g/l may be considered as the LD_{50} dose. Death is usually caused by respiratory embarrassment (80). From post-mortem investigations the average lethal BAC, calculated retrospectively, has been estimated at 4–4.5 g/l (76). It ranges between 1.5 and 6.7 g/l, but there are people who have survived BACs of up to 11 g/l (68, 76, 80). In deaths resulting from a combination of alcohol and drugs lower BACs have been found (average ca. 3 g/l, range 1.4–6.2 g/l).

There are large intra- and inter-individual differences not only in lethal dose, but in all reactions to any BAC (23, 34). The character of a particular person, the mood in which he begins to drink, his rate of drinking and his surroundings all have influence on the way he will react to alcohol. Alcoholics who have developed tolerance to alcohol show less changes in the central nervous system induced by high BACs than others (21). Differences in cognitive and mood changes between men and women upon alcohol consumption have been observed (38, 55, 82). The phase of the menstrual cycle may also influence the female reaction to alcohol (58, 91). Moreover, caffein, which in itself does not affect reaction time, proves to reinforce the effect of alcohol on reaction time (72).

9.1.2. Mechanisms of action

The blood circulating in the brain is separated from the remaining blood by several functional and structural barriers. A number of substances cannot reach the brain because these barriers are impermeable to them. Ethanol, as well as other fat-soluble small-molecular substances, crosses these barriers by diffusion, however (16). The alcohol concentration in the brain depends on the plasma concentration. It is uncertain whether alcohol can be metabolized in the brain. Some alcohol dehydrogenase (ADH) activity (27) and possibly also some katalase activity (17) have been demonstrated. Presumably acetaldehyde, which is generated from ethanol in the liver and which may pass the brain barrier, can be metabolized in the brain, considering the presence of aldehyde dehydrogenase (ALDH) in (rat) brain (89).

There is substantial evidence that at least the acute effects of alcohol on the brain are caused by alcohol per se (56). The specific mechanisms of acute alcohol action are still uncertain, albeit several mechanisms seem to be involved. Alcohol increases the fluidity of neuronal membranes by disturbing the composition of the bilipid layer. This results in a diminished ability to elevate intraneuronal calcium ion concentrations, which could explain the acute intoxicational effects of alcohol (22). Inhibition of glutamate (a neurotransmitter) receptors in the hippocampus by alcohol has been observed (64). It has recently been suggested that impairment of memory during alcohol intoxication is mediated by deficiency of the neurotransmitter serotonin, and can be reversed by dietary tryptophan (a precursor of serotonin)(103).

9.1.3 Hangover

A hangover – the aftereffects of alcohol use – can appear at about the time when the ethanol concentration in the blood approaches zero and often lasts for several hours. A hangover is associated with unpleasant mood changes, drowsiness, thirst, headache, sleep disturbances and sweating. The occurrence of a hangover is often related to the quantity of fusel alcohols consumed (75, 83). Since the higher alcohols are absorbed and metabolized more slowly than ethanol, the former are longer detectable in the blood (78). These pharmacologically active higher alcohols, such as n-propanol, n-butanol, and iso-amyl alcohol, are present in larger quantities in beverages like whisky, rum and brandy than in, for instance, vodka. Murphree (68, 69) found bourbon (American whisky) to have more effect on the EEG (electro-encephalogram) than vodka and also to induce a state of greater stupefaction. Another survey (75) reports for various alcoholic beverages a relation between hangover symptoms and the congener content of the beverage after the consumption of 15 ml alcohol per kg body weight.

Yet a hangover is not caused by congeners only. Pure alcohol may also cause a hangover. A decrease of the blood sugar concentration and an increase in lactate, triglyceride, ketone bodies, and free fatty acid concentrations caused by alcohol consumption may all play a part in this, although no clear relations with the extent of hangover symptoms have been found (106). A disturbance of the water balance, caused by the influence of alcohol on hormones regulating the water balance, may also (partly) account for hangover symptoms (57).

Infusion of prostaglandins may produce symptoms that have much in common with hangover (24). Alcohol may increase the formation of prostaglandins (18), whereas inhibitors of the prostaglandin synthesis, such as aspirin and tolfenamic acid, alleviate some hangover symptoms (45). Some consider a hangover to be an early withdrawal symptom, also because taking an alcoholic drink during a hangover may alleviate its symptoms (25).

The occurrence and the extent of a hangover do not only depend on the quantity of alcohol and congeners consumed, but are also influenced by personality, behaviour, social and environmental factors, smoking and tiredness. Hangover symptoms are difficult to reproduce in laboratory experiments (46).

9.2. Effects of chronic alcohol consumption

9.2.1. Brain damage

Research on the effect of chronic alcoholism on the brain can be classified into psychometric research related to brain function, and research into structural changes in the brain. Since computerized tomography (CT scanning, a non-invasive bra... scanning X-ray technique) came up, our knowledge of the structural changes in the alcoholic brain has substantially increased. The abundance of data was intially confusing, but has gained structure by now, and theories and hypotheses have emerged (60, 105).

More than half of the abstinent alcoholics in treatment programmes have some detectable cognitive dysfunction. The disorders are usually mild and would not be detected in routine clinical examinations. Most patients improve after weeks to years of abstinence (30, 105). This common mild form of cerebral dysfunction in alcoholics is called intermediate brain syndrome (IBS). Deficits revealed by tests evaluating visuospatial capacity, memory, abstracting ability, learning, problem-solving, perceptual-motor functions, analogic reasoning, and planning have been reported frequently (30, 42, 65, 94, 95). Occasionally disturbed motor activity is observed (23). Both the left and the right hemisphere seem to be affected (11). Some test results suggest that the picture is worse in alcoholics with a longer drinking history (11).

Although some have found a relation between alcohol consumption and psychometric tests, only a very small part of the variation in test results can be explained by alcohol consumption. Bowden (13) concludes in his review that there is no concern for brain impairment in social drinkers.

Cognitive impairment caused by alcohol abuse may be compared with the pattern going with ageing. The association between alcohol and ageing is not clear. There are three hypotheses: the premature ageing hypothesis (alcohol speeds up the ageing process), the age sensitivity hypothesis (elderly people are more sensitive to the brain-toxicating effects of alcohol) and the independent-decrements hypothesis. The results of investigations are controversial (15, 74, 81, 85).

The most pronouncing – but relatively rare – effect of alcoholism on the brain is Korsakoff's psychosis, with impairment of both long-term and short-term memory. Korsakoff's psychosis, also called alcohol amnestic disorder (AAD) or Wernicke-Korsakoff's syndrome, develops mostly in survivors of Wernicke's encephalopathy, which is caused by thiamin deficiency. AAD without a clear history of Wernicke's encephalopathy has been suggested to result from subclinical episodes of Wernicke's encephalopathy (60, 105). Wernicke's encephalopathy can be effectively treated with thiamin, whereas Korsakoff's psychosis is irreversible. The memory is located in the deeper, subcortical parts of the brain. In Korsakoff patients both cortical and subcortical lesions are found (60, 94). There is some evidence that memory impairment is due to repeated periods of alcohol withdrawal, rather than to continuous alcohol consumption (102).

About 4% of all dementia patients have a history of alcoholism. This type of dementia is called alcoholic dementia and has probably another aetiology than the primary progressive dementia (Alzheimer's disease)(59). Willenbring (105) suggests that alcoholic dementia is a combination of ISB and AAD.

Alcoholism affects both the cortical and the subcortical regions of the brain. At autopsy, but also with CT scanning, brain shrinkage is found (36, 60). Cortical atrophy, sulcal widening and ventricular enlargement is seen (33, 60, 65, 105). The cortex of alcoholics contains less neurons than the cortex of controls, especially the frontal cortex. A selective loss of the larger neurons – which are also more vulnerable in both Alzheimer's disease and the normal ageing process – has been demonstrated (35). The dendritic arborization in the frontal cortex has decreased (36).

Also at an alcohol consumption of less than 120 g/day cerebral atrophy is observed, which proves to have decreased after six months of abstinence (14). Even in heavy alcoholics brain damage is not entirely irreversible. Abstinence is followed by improvement (6, 36, 84, 98). There is no clear relation between CT-scan results and severity and length of drinking history (60).

Several investigations have revealed a moderate but significant relation between brain atrophy and impairment of cognitive abilities (33, 65, 104), but other investigations have not (60). The weak correlation between morphological brain damage and cognitive impairment is understandable if cognitive impairment is regarded as resulting from dysfunctioning of nerve cells. Not until much later this dysfunctioning is reflected in detectable morphological changes.

It is not known exactly how brain damage caused by chronic alcohol consumption is brought about, but both the direct toxicity of alcohol – or of its much more injurious metabolite acetaldehyde – and thiamin depletion seem to be involved. One of the effects of alcohol could be cerebral oedema, caused by an inappropriate secretion of vasopressin in alcoholics (50). The synthesis of arginine vasopressin (AVP), a fragment of vasopressin, is altered by chronic alcohol ingestion (40). AVP controls brain functions and modulates memory processes (39, 100). The toxic substances which accumulate abnormally in hepatic failure may also cause brain damage (hepatic encephalopathy)(86). The alcoholics examined by Lee (52) showed a greater incidence of brain degeneration than of liver injury, so liver injury per se is not the only cause of brain damage.

There seem to be different aetiologies underlying different types of brain damage among alcoholics. A number of these disorders, Wernicke's encephalopathy in particular, result from malnutrition and vitamin deficiency. Glucose utilization is thiamin-dependent and is likely to be a major factor in thiamin-deficient brain damage (96). Other disorders appear to be caused by the toxicity of alcohol (41) or of acetaldehyde (27).

Lishman (60) hypothesizes that alcohol neurotoxicity affects both the cortex and subcortical structures, whereas thiamin depletion has a marked effect on subcortical regions only. The personal biochemical vulnerability to neurotoxicity and/or thiamin depletion determines the type and severity of brain damage in alcoholics.

9.2.2. Tolerance, dependence and withdrawal

Chronic alcohol consumption leads to habituation; more and more alcohol is needed to bring about a specific mental effect, due to an acquired resistance to the physiological and behavioural effects of alcohol. This phenomenon is called tolerance, a distinction being made between metabolic and functional tolerance (34).

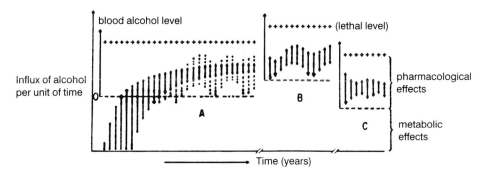

Fig. 9.1. Changes in tolerance during prolonged alcohol consumption. A, range of normal rate of metabolism; B, increased rate due to adaptation; C, decreased rate due to malnutrition or other factors. After: Lelbach (53).

The term 'metabolic' or 'pharmacokinetic' tolerance refers to the fact that a certain quantity of alcohol consumed leads to a lower BAC owing to an increased metabolic rate in the liver, partly due to MEOS induction (see Chapter 2).

Metabolic tolerance disappears within three weeks of abstinence (67). When liver injury and/or malnutrition occurs in case of continuing alcohol abuse metabolic tolerance decreases again. In this case alcohol metabolizes more slowly than before tolerance developed because of a disturbance of alcohol metabolism (see Fig. 9.1). In case of functional (also called cellular or pharmacodynamic) tolerance the organism has changed in such a way that it can adapt to the presence of alcohol. A particular BAC thus has less effect on the organism than before (34). Alcoholics with a BAC of 3 g/l are mostly sleepy or less affected (21).

It has been shown that the development of alcohol tolerance can be influenced genetically and environmentally, so individuals may be particularly vulnerable to the reinforcing effects of alcohol. Tabakoff (93) distinguishes an environment-independent and an environment-dependent form of tolerance. The latter is a kind of Pavlovian conditioning model of tolerance, an adaptive response of the central nervous system to stimuli, produced by the ritual of drinking.

A different phenomenon is acute tolerance; the effects of any given BAC are less during the decreasing stage than during the increasing stage (56, 77).

Functional tolerance is presumably a prerequisite of the development of physical dependence (61, 62). Initially the body is capable of adapting rapidly again to the absence of alcohol. Later on this adaptive capacity decreases and physical dependence on alcohol will be observed. Adaptation to alcohol has then developed in the body (presumably in the central nervous system) necessitating the presence of alcohol to maintain homeostasis (34). During chronic alcohol consumption some underlying biologic necessity of alcohol develops in the psychologically dependent individual, resulting in a need, craving, for alcohol. In the absence of alcohol the body cannot function 'normally' any longer and abstinence or withdrawal symptoms will occur, such as disturbances of consciousness, nausea, hallucinations, depressions, tremor and anxiety (34).

One of the most critical withdrawal symptoms is delirium tremens (101). Symptoms are mental clouding (hallucinations, misperceptions, disorientation), tachycardia, hypertension, termors and psychomotor agitation (19, 26). Delirium tremens is best managed, and possibly can be prevented, before it has become fully manifest. In most patients the deliria have subsided after some days or a week. Treatment of delirium tremens includes general measures (replacement of fluids, glucose, vitamins, magnesium or phosphates) as well as specific therapy for alcohol withdrawal (benzodiazepines, chlormethiazole or other CNS drugs (19, 37). The symptoms appear to be related to magnesium deficiency in the blood. The administration of magnesium proves an effective therapy in a number of cases (83). What cause these decreased magnesium values to occur just during the withdrawal phase is unknown, albeit there are specific indicators for higher mortality (spirit drinking, overweight, old age, treatment with phenothiazines). In the widely used Diagnostic and Statistical Manual of Mental Disorders (DSMIII), the term 'delirium tremens' has been replaced with 'alcohol withdrawal delirium'.

9.2.3. Mechanisms of action

There is consensus with regard to the notion that a change in neurotransmission is an important factor in the development of functional alcohol tolerance and physical alcohol dependence. The biochemical mechanism by which chronic alcohol consumption causes this adaption has not been elucidated sufficiently yet. There are several research lines. It has been suggested that alcohol dependence is actually acetaldehyde dependence. Mice can be made dependent on acetaldehyde; abstinence causes the same withdrawal symptoms as for alcohol (62). In the brain acetaldehyde can form condensation products with biogenic amines, i.e. tetrahydroisoquinolines (TIQs) and tetrahydrobetacarbolines (THBC). Both classes of aldehyde adducts can exert specific actions on neurons in different regions of the brain (71), acting as false neurotransmitters. It has been hypothesized that the metabolite of methanol, formaldehyde, has a stronger affinity to neurotransmitter amines than acetaldehyde to form TIQs and THBCs, and that methanol contributes partly to the effects of alcoholism. All alcoholic beverages contain methanol (5–5000 mg/l)(90).

So, the changes in neurotransmission can be caused directly by alcohol, by acetaldehyde, or by condensation products of acetaldehyde and formaldehyde. These substances can affect neurotransmission on various sites. One of the sites of action is the synaptic membrane, which is affected by alcohol. Other research is directed at the influence of alcohol or acetaldehyde on opioid peptides (endorphins, enkephalins) and other neuropeptides (arginine vasopressin).

Permanent changes in neurotransmission due to chronic alcohol consumption have been described by Hunt (41). The activity of pre- and postsynaptic receptors – selective binding sites – influence neurotransmission. Effects of alcohol on the functions of the neurotransmitters noradrenaline (norepinephrine)(58), γ-aminobutyric acid (GABA)(97), serotonin (87) and dopamine (3, 5) have been found (73). On a molecular level Linnoila (58) found that chronic alcohol consumption results in a down-regulation of one of the presynaptic receptors of the sympathetic (noradrenergic) nervous system. These alterations may account for the mood changes which follow upon alcohol consumption and, later on, during withdrawal. Part by part the neurobiological mechanisms have been elucidated, which leads to the development of drugs that may eventually be used to treat alcohol intoxication and withdrawal symptoms (58, 87).

Alcohol has been found to increase neuronal membrane fluidity, thus altering almost all membrane functions, i.e. neurotransmitter release, function of membrane-bound enzymes (ATPases), ion conductance channels, and calcium-binding sites (1, 8, 32, 56, 61, 62, 88). This can result in inhibition of transmembrane signalling (i.e. the translation of extracellular receptor activation to intracellular response)(2). It has

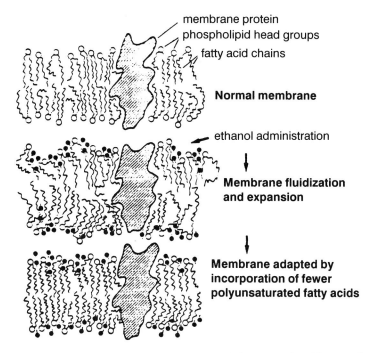

Fig. 9.2. Membrane changes induced by alcohol consumption. The postulated mechanism for development of tolerance to ethanol at the level of the synaptic membrane is shown. When ethanol enters the phospholipid bilayer it causes some disruption of the lipids as the head groups are separated and causes slight expansion of the membrane. The cell responds by increasing the proportion of saturated fatty acids in the membrane phospholipids rendering ethanol less effective in this respect. After: Locke et al. (63).

been suggested that prostaglandins, which are formed when membranes are disturbed, mediate the effects of alcohol on noradrenaline- and calcium-depending processes (7). Functional alcohol tolerance may consist in incorporating more saturated fatty acids and cholesterol in order to compensate for the increased membrane fluidity (cf. Fig. 9.2).

Like other addictions and obesitas, alcoholism is now associated with the endogenous opiate system. There are three known families of endogenous opioid peptides: endorphins, enkephalins and dynorphins. The existence of different types of opioid peptides is associated with the existence of multiple types of opiate-binding sites, which are classified depending on their affinity to specific opiate types. These binding sites – receptors – can be found in the synapses of specific opioid neurons. An important function of endogenous opioids is to regulate the responses to physical and physiological stress. In non-physiological doses endogenous opioids have an analgesic effect, just like morphine. After repeated administration tolerance and physical dependence develop as is the case for morphine (100).

Both acute and chronic alcohol treatment influence the activity of the endogenous opioid system, but there is considerable conflict as to the precise effects (31). Alcohol may interact with the endogenous opiate system by producing ethanol metabolites – TIQs – which bind to opiate receptors (20), by altering the binding properties of opiate receptors (4, 54), and by altering the synthesis and release of endogenous opioid peptides (31).

In relation to alcohol, β-endorphin is the most intensively studied endogenous opiate. Beta-endorphin and ACTH share the same precursor and are co-released from the pituitary gland under a number of physiological conditions such as stress. Alcohol is known to stimulate the release of ACTH, but the effects on β-endorphin release is not clear yet.

Naloxone is a morphine antagonist which ousts endorphin (and morphine) from the receptors and thus counteracts the effects of (non-physiological doses of) endorphin. Injection of naloxone proves to decrease the effects of alcohol (decreased reactivity, nystagmus) slightly (42, 43, 66). The endorphin concentration in cerebrospinal fluid is affected during intoxication and withdrawal (12). An individual's plasma endorphin concentration and his mood appear to be related, but the plasma endorphin concentration does not correlate with the BAC (47, 48).

Evidence is accumulating that vasopressin, oxytocin and some of their fragments, called neuropeptides, modulate neurotransmission. They decrease under certain conditions addictive behaviour of experimental animals and man and seem to be involved in reward processes, tolerance and dependence (39, 40, 99).

Summary

The consumption of one or more glasses of alcoholic beverage brings about a temporary, reversible change of mood and behaviour. The specific mechanisms of acute alcohol action are still uncertain. They may consist in increasing the fluidity of neuronal membranes or another disturbance of neurotransmission.

A hangover can occur when there is hardly any alcohol left in the blood. It is partly caused by the effects of congeners, partly by the metabolic, hormonal and central nervous system changes due to alcohol itself.

During chronic alcohol consumption tolerance develops. Two different effects may be distinguished in this process, namely metabolic and functional (cellular) tolerance. Metabolic tolerance implies an increased rate of alcohol metabolism, resulting from MEOS induction. In functional tolerance the body – particularly the central nervous system – may become adapted to the presence of alcohol in the organism, thus causing a specific BAC to have less effect than it had before. Physical dependence on alcohol is established if the presence of alcohol has become necessary for maintaining homeostasis of the body, and abstinence is followed by withdrawal symptoms.

Recent research has shed considerable light on the intracellular and molecular processes that play a role in cellular tolerance and physical dependence. Neurotransmission in and near the synapses is affected owing to alcohol, acetaldehyde or condensation products from aldehydes and neurotransmitter amines.

Chronic alcohol consumption causes diffuse cortical and cerebellar atrophy. Local degenerative changes also occur in the central brain structures. Cortical atrophy consists in death of neurons and/or degeneration of the dendritic arborization. Psychometric testing of chronic alcoholics have displayed a variety of forms of impairment, ranging from hardly measurable reversible changes to severe irreversible mental damage (i.e. Wernicke-Korsakoff's psychosis). The various types of brain damage are partially caused by alcohol itself, and partially by thiamin depletion.

References

1. Alling C. Alcohol effects on cell membranes. Subst Alcohol Actions Misuse 1983;4: 67-72.
2. Änggård E. Ethanol, phosphoinositides, and transmembrane signalling: towards a unifying mechanism of action. Br Assoc Psychopharmacol Monogr 1988;10: 50-9.
3. Annunziato L, Amoroso S, Renzo G di et al. Increased GH responsiveness to dopamine receptor stimulation in alcohol addicts during the late withdrawal syndrome. Life Sci 1983;33: 2651-5.
4. Anokhina IP, Brusov OS, Nechaev NV et al. Effect of acute and chronic ethanol exposure on the rat brain opiate receptor function. Alcohol Alcoholism 1983;18: 21-6.
5. Anokhina IP, Kogan BM, Drozdov AZ. Disturbances in regulation of catecholamine neuromediation in alcoholism. Alcohol Alcoholism 1988;23: 343-50.
6. Anonymous. Alcohol-induced brain damage and its reversibility. Nutr Rev 1980;38: 11-2.
7. Anton RF, Randall CL. Central nervous system prostaglandins and ethanol. Alcohol Clin Exp Res 1987;11: 10-8.
8. Beaugé F, Fleuret-Balter C, Barin F et al. Brain membrane sensitivity to ethanol during development of functional tolerance to ethanol in rats. Alcohol Clin Exp Res 1984;8: 167-71.
9. Berendsen F et al. Alcohol in sport (scriptie). 's-Gravenhage: Nederlandse Sport Federatie, 1980.

10. Blum K et al. Alcohol and opiates: neurochemical behavioral mechanisms. New York: Academic Press, 1977.
11. Bolter JF, Hannon R. Lateralized cerebral dysfunction in early and late stage alcoholics. J Stud Alcohol 1986;47: 213-8.
12. Borg S, Kvande H, Rijdberg U et al. Endorphin levels in human cerebrospinal fluid during alcohol intoxication and withdrawal. Psychopharmacology 1982;78: 101-3.
13. Bowden SC. Brain impairment in social drinkers? No cause for concern. Alcohol Clin Exp Res 1987;11: 407-10.
14. Cala LA, Joens B, Burns P et al. Results of compterized tomography, psychometric testing, and dietary studies in social drinkers, with emphasis on reversibility after abstinence. Med J Aust 1983;141: 264-9.
15. Cermak LS, Peck E. Continuum versus premature aging theories of chronic alcoholism. Alcohol Clin Exp Res 1982;6: 89-95.
16. Chin JH. Ethanol and the blood-brain barrier. In: Majchrowicz E, Noble EP, eds. Biochemistry and pharmacology of ethanol. New York: Plenum Press, 1979;2: 101-18.
17. Cohen G et al. Ethanol oxidation by rat brain in vivo. Alcohol Clin Exp Res 1980;4: 366-70.
18. Collier HOJ, MacDonald-Gibson WJ, Saeede SA. Stimulation of prostaglandin biosynthesis by capsaicin, ethanol and tyramine. Lancet 1975;i: 702.
19. Cushman P. Delirium tremens: update on an old disorder. Postgrad Med 1987;82: 117-22.
20. Davies VG, Walsh MJ. Alcohol, amines and alkaloids: a possible basis for alcohol addiction. Science 1970;167: 1005-7.
21. Davis AR, Lipson AH. Central nervous system tolerance to high blood alcohol levels. Med J Aust 1986;144: 9-12.
22. Deutsch SI, Huntzinger JA, Rosse RB, Kaushik M, Mastropaolo J. The role of excitatory amino acids and intraneuronal calcium in the acute intoxicational effects of ethanol. Clin Neuropharmacol 1989;12: 483-9.
23. Dreyfus PM. Effects of alcohol on the nervous system. In: Gastineau CF, Darby WJ, Turner TB, eds. Fermented food beverages in nutrition. New York: Academic Press, 1979: 342-58.
24. Eklund B, Carlson LA. Central and peripheral circulatory effects and metabolic effects of different prostaglandins given i.v. to man. Prostaglandins 1980;20: 333-7.
25. Esser PH. Hangover symptoms. Br J Addict 1963;59: 65-73.
26. Feuerlein W, Reiser E. Parameters affecting the course and results of delirium tremens treatment. Acta Psychiatr Scand 1986;73: 120-3.
27. Fischer G, Kampmann H. Untersuchungen des Acetaldehydestoffwechsels im Blut: Zum Problem der Kausalität alkoholbedingter Organschäden. Blutalkohol 1982;19: 53-74.
28. Flanagan NG, Strike PW, Rigby CJ. The effects of low doses of alcohol on driving performance. Med Sci Law 1983;23: 203-8.
29. Freund G. The interaction of chronic alcohol consumption and aging on brain structure and function. Alcohol Clin Exp Res 1982;6: 13-21.
30. Gardner MK, Clark E, Bowman MAC, Miller PJ. Analogical reasoning abilities of recovering alcoholics. Alcohol Clin Exp Res 1989;13: 508-11.
31. Gianoulakis C. The effect of ethanol on the biosynthesis and regulation of opioid peptides. Experientia 1989;45: 428-35.
32. Goldstein DB. Alcohol and biological membranes. In: Goedde HW, Agarwal DP. Alcoholism: biomedical and genetic aspects. New York: Pergamon Press, 1989: 87-98.
33. Graff-Radford NR, Heaton RK, Earnest MP et al. Brain atrophy and neurophysiologic impairment in young alcoholics. J Stud Alcohol 1982;43: 859-68.
34. Gross MM. Psychobiological contributions to the alcohol dependence syndrome : a selective review of recent research. In: Edwards G et al, eds. Alcohol-related disabilities. Geneva: WHO, 1977: 107-31. (WHO offset publication 32.)
35. Harper C, Krill J. Patterns of neuronal loss in the cerebral cortex in chronic alcoholic patients. J Neurol Sci 1989; 92: 81-9.
36. Harper CG, Kril JJ. Neuropathology of alcoholism. Alcohol Alcoholism 1990;25: 207-16.
37. Hartmann H. Alkoholeinwirkung: Pharmakodynamik. In: Mallach HJ, Hartmann H, Schmidt V, eds. Alkoholwirkung beim Menschen: Pathophysiologie, Nachweis, Intoxication, Wechselwirkungen. Stuttgart: Thieme Verlag, 1987: 70-99.

38. Haut JS, Beckwith BE, Petros TV, Russell S. Gender differences in retrieval from long-term memory following acute intoxication with ethanol. Physiol Behav 1989;45: 1161-5.
39. Hoffman PL. Structural requirements for neurohypophyseal peptide maintenance of ethanol tolerance. Pharmacol Biochem Behav 1982;17: 685-90.
40. Hoffman PL, Tabakoff B. Mechanisms of alcohol tolerance. In: Thirteenth annual symposium, The North Carolina Alcoholism Research Authority. Alcohol Alcoholism 1989;24: 251-2.
41. Hunt WA, Majchrowicz E. Alterations in neurotransmitter furnction after acute and chronic treatment with ethanol. In: Majchrowicz E, Noble EP, eds. Biochemistry and pharmacology of ethanol. New York: Plenum Press, 1979;2: 167-86.
42. Jeffcoate WJ. Prevention of effects of alcohol intoxication by naloxone. Lancet 1979;ii: 1157-9.
43. Jeffcoate WJ, Hastings AG, Cullen MH. Naloxone and ethanol antagonism. Lancet 1981;I: 1052.
44. Jones AW, Sato A, Forsander OA. Liquid/air partition coefficients of acetaldehyde : values and limitations in estimating blood concentrations from analysis of breath. Alcohol Clin Exp Res 1985;9: 461-4.
45. Kaivola S, Parantainen J, Osterman T, Timonen H. Hangover headache and prostaglandins : prophylactic treatment with tolfenamic acid. Cephalalgia 1983;3: 31-6.
46. Keller M. A lexicon of disablements related to alcohol consumption. In: Edwards G et al, eds. Alcohol-related disabilities. Geneva: WHO, 1977:23-60. (WHO offset publication 32.)
47. Kimball CD, Huang SM, Torget CE et al. Plasma ethanol, endorphin, and glucose experiment. Lancet 1980;ii: 418-9.
48. Kimball CD, Houck JC. Plasma ethanol, endorphin, and glucose experiment. Lancet 1981;i: 212.
49. Lamb MR, Robertson LC. Effect of acute alcohol on attention and the processing of hierarchical patterns. Alcohol Clin Exp Res 1987;11: 243-8.
50. Lambie DG. Alcoholic brain damage and neurological symptoms of alcohol withdrawal : manifestations of overhydration. Med Hypotheses 1985;16: 377-88.
51. Landauer AA. Low and moderate alcohol doses, psychomotor performance and perceived drowsiness. Ergonomics 1983;26: 647-57.
52. Lee K, Hardt F, Møller L et al. Alcohol-induced brain damage and liver damage in young males. Lancet 1979;ii: 759-61.
53. Lelbach WK. Cirrhosis in the alcoholic and its relation to the volume of alcohol abuse. Ann N Y Acad Sci 1975;252: 85-105
54. Levine AS, Hess S, Morley JE. Alcohol and the opiate receptor. Alcohol Clin Exp Res 1983;7: 83-4.
55. Lex BW, Greenwald NE, Lukas SE, Slater JP, Mendelson JH. Blood ethanol levels, self-rated ethanol effects and cognitive-perceptual tasks. Pharmacol Biochem Behav 1988;29: 509-15.
56. Lieber CS. Medical disorders of alcoholism: pathogenesis and treatment. Philadelphia: Saunders, 1982.
57. Linkola J, Fykrquist F, Ylikahri RH. Renin, aldosterone and cortisol during ethanol intoxication and hangover. Acta Physiol Scand 1979;106: 75-82.
58. Linnoila M. Alcohol withdrawal syndrome and sympathetic nervous system function. Alcohol Health Res World 1989;13: 355-7.
59. Lishman WA. Alcoholic dementia : a hypothesis. Lancet 1986;i: 1184-6.
60. Lishman WA. Alcohol and the brain. Br J Psychiatry 1990;156: 635-44.
61. Littleton JM. The biological basis of alcoholism: some recent experimental evidence. In: Edwards G, Grant M eds. Alcoholism: new knowledge and new responses. London: Croom-Helm, 1977: 107-16.
62. Littleton JM. Neuropharmacological aspects of ethanol tolerance and dependence. In: Mendlewicz J, Van Praag HM, eds. Alcoholism: a multidisciplinary approach. Basel: Karger, 1979;3: 75-87.
63. Locke S, Cohen G, Dembiec D. Uptake and accumulation of 3H-6,7-dihydroxytetrahydroisoquinoline by peripheral sympathetic nerves in vivo. J Pharmacol Exp Ther 1973;187: 56-67.
64. Lovinger DM, White G, Weight FF. Ethanol inhibits N MDA-activated ion current in hippocampal neurons. Science 1989;243: 1721-4.

65. Lusins J et al. Alcoholism and cerebral atropy: a study of 50 patients with CT scan and psychologic testing. Alcohol Clin Exp Res 1980;4: 406-11.
66. Mattila MJ, Nuotto E, Seppälä T. Naloxone is not an effective antagonist of ethanol. Lancet 1981;i: 775-6.
67. Mendelson JH. Ethanol-1-C-14 metabolism in alcoholics and nonalcoholics. Science 1968;159: 319-20.
68. Minion GE, Slovis CM, Boutiette L. Severe alcohol intoxication: a study of 204 consecutive patients. J Toxicol Clin Toxicol 1989;27: 375-84.
69. Murphree HB, Price LM. Computer time-series analysis of the EEG effects of alcoholic beverages. Fedn Proc 1966;25: 503.
70. Murphree HB, Greenberg LA, Carroll RB. Neuropharmacological effects of substances other than ethanol in alcoholic beverages. Fedn Proc 1967;26: 1468-73.
71. Myers RD. Isoquinolines, beta-carbolines and alcohol drinking: involvement of opioid and dopaminergic mechanisms. Experientia 1989;45: 436-43.
72. Oborne DJ, Rogers Y. Interactions of alcohol and caffeine on human reaction time. Aviat Space Environ Med 1983;54: 528-34.
73. Ollat H, Parvez H, Parvez S, Tipton KF. Alcohol and central neurotransmission. Neurochem Int 1988;13: 275-300.
74. Page RD, Cleveland MF. Cognitive dysfunction and aging among male alcoholics and social drinkers. Alcohol Clin Exp Res 1987;11: 376-84.
75. Pawan GLS. Alcoholic drinks and hangover effects. Proc Nutr Soc 1973;32: 15a.
76. Poikolainen K. Estimated lethal ethanol concentrations in relation to age, aspiration, and drugs. Alcohol Clin Exp Res 1984;8: 223-5.
77. Portans I, White JM, Staiger PK. Acute tolerance to alcohol: changes in subjective effects among social drinkers. Psychopharmacology 1989;97: 365-9.
78. Prokop L, Machata G. Höhere Alkohole und Äthanolwirkung beim Menschen. Blutalkohol 1974;11: 80-7.
79. Rahwan RG. Toxic effects of ethanol: possible role of acetaldehyde, tetrahydroisoquinolines, and tetrahydro-beta-carbolines. Toxicol Appl Pharmacol 1975;34: 3-27.
80. Redmond AD. Blood alcohol concentration and conscious level. Alcohol Alcoholism 1983;18: 89-91.
81. Riege WH, Tomaszewski R, Lanto A et al. Age and alcoholism: independent memory decrements. Alcohol Clin Exp Res 1984;8: 42-7.
82. Robbins BJ, Brotherton PL. Mood change with alcohol intoxication. Br J Soc Clin Psychol 1980;19: 149-55.
83. Roe DA. Alcohol and the diet. Westport: AVI, 1979.
84. Ron MA. The alcoholic brain. Psychol Med 1982;82: 1-33.
85. Ryan C, Butters N. Learning and memory impairments in young and old alcoholics: evidence for the premature-aging hypothesis. Alcohol Clin Exp Res 1980;4: 288-93.
86. Schafer DF. Hepatocerebral Interactions. Alcohol and the brain: chronic effects. New York: Plenum, 1985: 121-30.
87. Sellers EM, Naranjo CA, Lawrin MO. New drugs to decrease alcohol use. In: Kuriyama K, Takada A, Ishii H, eds. Biomedical and social aspects of alcohol and alcoholism: proceedings of the Fourth Congress of the International Society for Biomedical Research in Alcoholism (ISBRA), Kyoto, Japan, 26 June-2 July 1988. Amsterdam: Elsevier Science Publishers (Biomedical Division), 1988: 609-12.
88. Shaw GK. Alcohol dependence and withdrawal. Br Med Bull 1982;38: 99-102.
89. Sloviter HA. Ethanol, acetaldehyde, and the metabolism of perfused brain. In: Majchrowicz E, Noble EP, eds. Biochemistry and pharmacology of ethanol. New York: Plenum, 1979;1: 505-20.
90. Sprung R, Bonte W, Lesch OM. Methanol: ein bisher verkannter Bastandteil aller alkoholischen Getränke: eine neue biochemische Annäherung an das Problem des chronischen Alkoholismus. Wien Klin Wochenschr 1988;100: 282-5.
91. Sutker PB, Goist KC, King AR. Acute alcohol intoxication in women: relationship to dose and menstrual cycle phase. Alcohol Clin Exp Res 1987;11: 74-9.
92. Sweeney DF. Alcohol versus Mnemosyne: blackouts. J Subst Abuse Treatment 1989;6: 159-62.

93. Tabakoff B, Hoffman PL. Adaptive responses to ethanol in the central nervous system. In: Goedde HW, Agarwal DP, eds. Alcoholism: biomedical and genetic aspects. New York: Pergamon, 1989: 99-112.
94. Tarter RE, Arria AM, Van Thiel DH. Neurobehavioral disorders associated with chronic alcohol abuse. In: Goedde HW, Agarwal DP, eds. Alcoholism: biomedical and genetic aspects. New York: Pergamon, 1989: 113-29.
95. Taylor TH. Psychometric testing and alcohol-induced brain damage. Lancet 1979;ii: 1019.
96. Thomson AD, Ryle PR, Shaw GK. Ethanol, thiamine and brain damage. Alcohol Alcoholism 1983;18: 27-43.
97. Ticku MK. Ethanol and the benzodiazepine-GABA receptor-ionophore complex. Experientia 1989;45: 413-8.
98. US Department of Health and Human Services. Alcohol and health. Fourth special report to the US Congress. Washington: US Department of Health and Human Services, 1981.
99. Van Ree JM. Endorfinen, lichaamseigen morfine-achtige stoffen. Tijdschr Alcohol Drugs 1978;4: 59-65.
100. Van Ree JM. Role of pituitary and related neuropeptides in alcoholism and pharmacodependence. Prog Neuropsychopharmacol Biol Psychiatry 1986; 10: 219-28.
101. Victor M. Part II: withdrawal, neurological syndrome, and EEG. Ann N Y Acad Sci 1973;215: 210-3.
102. Wagner Glenn S, Wagner Glenn S, Parsons OA, Sinha R, Stevens L. The effects of repeated withdrawals from alcohol on the memory of male and female alcoholics. Br J Addict 1988;23: 337-42.
103. Westrick ER, Shapiro AP, Nathan PE, Brick J. Dietary tryptophan reverses alcohol-induced impairment of facial recognition but not verbal recall. Alcohol Clin Exp Res 1988;12: 531-3.
104. Wilkinson DA. Examination of alcoholics by computed tomographic (CT) scans: a critical review. Alcohol Clin Exp Res 1982;6: 1.
105. Willenbring ML. Organic Mental Disorders associated with heavy drinking and alcohol dependence. Clin Geriatr Med 1988;4: 869-87.
106. Ylikahri RH et al. Metabolic studies on the pathogenesis of hangover. Eur J Clin Invest 1974;4: 93-100.

CHAPTER 10

Effects of alcohol on the cardiovascular system

J. Veenstra

Most of the early publications on the effects of alcohol on the cardiovascular system emphasize the detrimental effects of alcohol abuse on the heart, causing such diseases as cardiomyopathy and cardiac beriberi. Since the late 1970s, however, an overwhelming majority of epidemiological studies indicate that, although excessive alcohol consumption is associated with increased risk of cardiovascular diseases, moderate alcohol consumption is associated with a reduced risk as compared to non-drinkers, especially of coronary heart disease. These epidemiological studies have initiated a vast number of epidemiological as well as experimental studies into the possible mechanism of this protective effect. In this chapter the effects of both moderate and excessive alcohol consumption will be reviewed.

10.1. Blood lipids and lipoproteins

In epidemiological studies the levels of blood lipids and lipoproteins are clearly associated with the risk of cardiovascular diseases (CVD) of which the most common form is coronary heart disease (CHD). The levels of total cholesterol, low-density lipoprotein (LDL)-cholesterol and apolipoprotein-B (Apo-B) are found to be positively associated with atherosclerosis and CHD, whereas the levels of high-density lipoprotein (HDL)-cholesterol and the apolipoproteins Apo-A1 and Apo-A2 show negative associations. Successive experimental studies have shown that several dietary factors may influence blood lipid and lipoprotein concentrations and consequently probably also influence the risk of CVD.

    A large number of epidemiological studies (3, 5, 16, 19, 21, 28, 33, 34, 44, 53, 54, 61, 75, 83, 103, 104, 108, 112, 134, 135, 185, 196, 209–211) all over the world have consistently shown a positive association between alcohol consumption and HDL-cholesterol levels. Fig. 10.1 shows some results of the Tromsø Heart Study (28). Significant positive associations were found for all three types of alcoholic beverage. Even low alcohol consumption was associated with increased HDL-cholesterol levels. HDL, in which Apo-A1 and Apo-A2 are the major proteins, is considered to be an essential factor in the body's defence against accumulation of cholesterol in tissues, in a process called 'reverse cholesterol transport'. The initial excitement evoked by this apparent explanation for the observed lower risk of CHD in moderate alcohol consumers was tempered when it was found that it was mainly

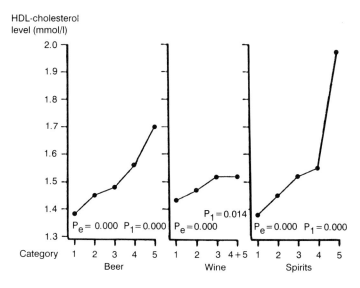

Fig. 10.1. Relationship between HDL-cholesterol and alcohol consumption. P_e, P value for test of equal group means; P_1, P value for test of similar trend. Values adjusted for time since the last meal.

the HDL3 subfraction of HDL-cholesterol that was associated with alcohol consumption, whereas it was the HDL2 subfraction that showed the strongest association with CHD (151). In more recent studies, however, it was found that HDL3-cholesterol was also associated with lower risk of CHD (180, 188). In addition, Miller et al. (150), in the first large-scale study (4860 men) on the effects of alcohol consumption on HDL-cholesterol subfraction levels, found comparable positive associations for HDL2- and HDL3-cholesterol.

In a relatively small number of epidemiological studies attention has been paid to the relationship between alcohol consumption and the apolipoprotein concentrations. The levels of Apo-A1 and Apo-A2 may be even better predictors of a low CHD risk than HDL-cholesterol. Phillips et al. (161) have found a positive association between alcohol use and Apo-A1 in a group of 289 Californian men and women, but have not measured Apo-A2. Haffner et al. (86) and Williams et al. (211) report positive correlations for both Apo-A1 and Apo-A2 in smaller groups (77 men and 50 men and women, respectively).

The relationships between alcohol consumption and HDL-cholesterol and the apolipoproteins A1 and A2, observed in epidemiological studies, have been confirmed in a number of experimental studies. The experimental studies, however, varied widely in study design with respect to the length of the experimental period, the beverage type used, the amount of alcohol consumed and the variables measured. Table 10.1 gives an overview of the variation in study design.

In most of the long-term studies (22, 30, 39, 42, 45, 69, 90–92, 146, 162, 197) a rise in HDL- or HDL-cholesterol was observed within a few weeks. In only two

Table 10.1. Survey of studies into the effects of alcohol consumption on blood lipids and lipoproteins.

Alcohol dose (g/day)	Type of beverage	Duration of study (weeks)	Reference
28	beer	3	90
39	white wine	6	197
31 (average)	optional	6	92
31	red wine	5	42
18.4 (average)	optional	4	30
24	beer	6	146
23 or 46	red wine	5	162
75	alcohol or beer	5	22
75	red wine	5	39
90	diluted alcohol	4	45
34 or 53	wodka	2	73
12.6	beer	8	152
40	beer	6	69
12.5 or 25	beer	3	91

studies (42,43) no effects were found. In one of these studies (42), however, the experimental period was rather short, and in the other one (152) the dose of alcohol was very low. In the latter study, despite the absence of an effect on HDL, a significant increase in Apo-A1 was observed after consumption of only one glass of beer (12.6 g of alcohol) per day for 8 weeks.

In the long-term studies blood lipids have been examined in the fasting state, usually after an overnight fast. In a few studies (66, 74, 204) the acute effects of alcohol on blood lipids in the postprandial state have been investigated. Goldberg et al. (74) and Franceschini et al. (66) have studied the effects of a single dose of 120 ml whisky (40 g alcohol), using comparable experimental designs. The alcohol was drunk at 09.00 after a 14-hour fast. In the study of Franceschini et al. whisky was consumed mixed with a fluid meal of 100 g mayonnaise and 25 g bread. In both studies it was assumed that the alcohol had been fully metabolized after 4 hours, and measurements were performed 4, 6, 8, 10 and 12 hours after consumption. Goldberg et al. found significantly higher levels of HDL-cholesterol after alcohol consumption as compared to the control treatment. In addition, a 67% decrease was observed for activity of hepatic lipase, an important enzyme in the reverse cholesterol transport. By contrast, Franceschini et al. found a lower HDL-cholesterol level after consumption of alcohol, mayonnaise and bread relative to the treatment without alcohol.

Only one study has been reported (204) in which the effects of alcohol on blood lipids were studied shortly after consumption, when the alcohol is still in the circulation. In that study, Veenstra et al. tried to approximate reality as closely as possible in a carefully controlled experiment. Sixteen male volunteers (8 aged 20–30 and 8 aged 45–55) consumed 30 g of alcohol as port and wine during a normal evening dinner. For comparison, they drank similar amounts of mineral water instead on another day. One hour after dinner, when the blood alcohol levels reached a maximum (mean 0.175 g/L, SD 0.055), significantly higher levels of HDL-cholesterol, Apo-A2 and triglycerides were found for the alcohol treatment than for the mineral water treatment. The HDL-cholesterol-enhancing effect was more prominent in the middle-aged men and was most evident for the HDL2-cholesterol fraction.

In a number of studies (23, 47, 48, 50, 109, 195, 200) HDL-cholesterol levels have been studied in alcoholics during a period of abstention. Most alcoholics have increased HDL-cholesterol levels, which rapidly decline as soon as alcohol consumption is stopped. The changes upon abstention in alcoholics are mainly found in the HDL2 subfraction, whereas most epidemiological studies suggest a relationship between alcohol and the HDL3 subfraction. This controversy prompted Välimäki et al. (56) to compare the effects of moderate (30 g/day) and heavy (60 g/day) alcohol consumption. The consumption levels differed in their effect on the HDL-subfraction. Heavy alcohol consumption increased HDL2 levels within two days and gradually increased HDL3 levels later on, whereas moderate alcohol consumption affected HDL3 levels only.

The epidemiological and experimental studies mentioned above clearly show a relationship between alcohol consumption and levels of blood lipids which is the most frequently cited explanation for a beneficial effect of moderate alcohol consumption on risk of CHD. However, whether these changes are beneficial indeed depends strongly on the exact mechanism causing the increase in HDL-cholesterol, which may be indicative of an increased reverse cholesterol transport, but could just as well be due to a decreased clearance of HDL-cholesterol. More detailed studies into the precise mechanism are needed. In this respect, the possibility that the observed changes originate from acute effects of alcohol on blood lipids as observed by Veenstra et al. as well as the differences between moderate and heavy alcohol consumption observed by Välimäki et al. deserve further attention.

10.2. Haemostasis

Haemostasis is the balance between the formation of blood clots, in which blood platelets (thrombocytes) are involved, and fibrinolysis, the dissolution of such clots. In most cases of myocardial infarction a coronary artery is blocked by a clot primarily consisting of thrombocytes (51). In normal physiological conditions clotting of thrombocytes, called platelet aggregation, most probably takes place constantly to some extent. The fibrinolytic system, on the other hand, plays a part in continually dissolving the newly formed clots. An effect of alcohol on this complex and delicate haemostatic balance would, of course, also have an effect on the risk of CVD.

Measurement of platelet aggregation is complicated and labour-intensive; So far, only one epidemiological study has been devoted to the relationship between alcohol consumption and platelet aggregation. Meade et al. (148) found an inverse association in a population of 685 British men and 273 British women (Northwick

Park Heart Study). In the same study Meade et al. (147) found the level of alcohol consumption to be associated with a decreased plasma fibrinogen level and an increased fibrinolytic activity. These latter results were confirmed in several other studies (13, 27, 174, 217). However, no correlation between alcohol consumption and fibrinogen was observed by Balleisen et al. (14) in 2880 men and 1306 women in the Munster Arteriosclerosis Study.

The experimental studies into the effects of alcohol consumption on platelet aggregation and other platelet function tests (52, 55, 60, 63, 71, 95, 96, 99, 113, 131, 136, 149, 154, 162, 206) are characterized by a large diversity in experimental design (Table 10.2).

First, the amount of alcohol consumed varied widely between 15 g and ad libitum consumption. In six studies, the amount of alcohol consumed in a single binge was 100 g or more (1 drink contains approximately 10 g). Furthermore, the studies varied considerably in type of alcoholic beverage. In almost all studies the alcohol was consumed on an empty stomach, mostly in the morning. The results are conflicting: no effects (55, 71, 99, 154, 206) inhibitory effects (60, 63, 149) as well as stimulating effects (95, 96) on platelet aggregation have been observed. The data on thromboxane production are also conflicting, with some studies showing an increase (95, 96) and other ones showing no effect (71, 99, 154), whereas one study showed a decrease (149) in platelet thromboxane production. Consequently, no definite conclusions can be drawn from the experimental studies. In the only study reported on the longer-term effects of moderate alcohol consumption on platelet

Table 10.2. Survey of studies into the effects of alcohol use on platelet function.

Alcohol dose[1] (g)	Term	Number of volunteers	Platelet function[2]	Year of publication	Reference
Ad libitum	acute	20	0	1981	55
50	acute	12	0	1982	52
120	acute	12	−	1982	113
120	acute	8	−	1982	131
64	acute	7	−	1983	149
51	acute	10	−	1984	60
32	acute	6	−	1984	136
100	acute	15	−	1984	63
32	acute	4	0	1984	71
120	acute	8	+	1985	96
120	acute	10	+	1985	95
88	acute	12	0	1987	99
20	acute	6	0	1987	154
23 and 46	5 weeks	12	−	1987	162
40	acute	8	0	1990	206

[1] The alcohol dose, irrespective of the volunteer's body weight, has been converted for a hypothetical 80-kg subject for comparison sake.
[2] Platelet function: + increased; 0 no effect; − decreased.

aggregation (162), a decrease in platelet aggregation was observed in agreement with the epidemiological findings of Meade et al. (148). In the only study published on the acute effects (206), in which the alcohol was consumed at a customary time of the day and under habitual conditions (during dinner instead of early in the morning and on an empty stomach) no acute effects of alcohol on platelet aggregation could be found.

Table 10.3. Survey of studies into the acute effects of alcohol on fibrinolysis.

Parameter	Amount of alcohol	Type of beverage	Effect[1]	Year of publ.	Reference
Blood clot lysis time	20–40 g	beer/cider/white wine	+	1960	62
Blood clot lysis time	20–40 g	gin/whisky/alcohol	0	1960	62
Fibrinolytic acticity	25–100 g	beer/wine	–	1961	155
Fibrinolytic activity	25–100 g	alcohol	0	1961	155
Euglobulin lysis time	90 g	cider	+	1983	4
Fibrinolytic activity	1.5 g/kg	alcohol	–	1983	97, 98
Euglobulin lysis time	1.5 g/kg	alcohol	+	1983	97, 98
Fibrin degr. products	1.5 g/kg	alcohol	0	1983	97, 98
Factor VIII: C activity	1.5 g/kg	alcohol	+	1983	97, 98
Factor VIII: antigen	1.5 g/kg	alcohol	+	1983	97, 98
Factor VIII: R cofactor	1.5 g/kg	alcohol	+	1983	97, 98
Antithrombin III	1.5 g/kg	alcohol	0	1983	97, 98
Factor VIII	0.8 g/kg	whisky	0	1984	59
Fibrin degr. products	0.8 g/kg	whisky	0	1984	59
Plasminogen	0.8 g/kg	whisky	0	1984	59
Fibrinogen	0.8 g/kg	whisky	0	1984	59
Fibrinolytic activity	0.8 g/kg	whisky	0	1984	59
Euglobulin lysis time	0.8 g/kg	whisky	0	1984	59
Antithrombin III	0.8 g/kg	whisky	0	1984	59
Blood clot lysis time	0.5–1.0 g/l	ethanol	+	1987	156
Euglobulin lysis time	24–47 g	shochu/sake/beer	–	1988	193
Fibrin degr. products	24–47 g	shochu/sake/beer	0	1988	193
Fibrinolytic activity	24–47 g	shochu/sake/beer	+	1988	193
t-PA activity	30 g	red wine/port	–	1990	203
PAI activity	30 g	red wine/port	+	1990	203
PAI activity	40 g	red wine/gin	+	1990	202
PAI antigen	40 g	red wine/gin	0	1990	202
t-PA activity	40 g	red wine/gin	–	1990	202
t-PA antigen	40 g	red wine/gin	+	1990	202
Total UK-PA antigen	40 g	red wine/gin	0	1990	202
Pro-UK-PA antigen	40 g	red wine/gin	0	1990	202
UK-PA inhibitor complex	40 g	red wine/gin	0	1990	202

[1] Effect: + increased; 0 no effect; – decreased.

Besides platelet aggregation, Pikaar et al. investigated in the same study (162) the long-term effects of moderate alcohol consumption on a number of fibrinolytic factors. No effects on fibrinogen could be observed after 5 weeks of moderate alcohol consumption (23 or 46 g/day) or after 5 weeks of binge drinking (14 glasses of wine per week consumed on the three weekend days). The levels of plasminogen and tissue-plasminogen activator (t-PA) activity, however, appeared to be clearly affected by wine consumption. Plasminogen levels moderately increased with an increasing dose of wine. The t-PA activity, on the other hand, showed a large and dose-dependent decrease. There was no effect on bleeding time. In additional analyses (128), the levels of plasminogen activator inhibitor 1 (PAI-1) were measured, and significantly higher levels were observed after 5 weeks of moderate alcohol consumption than in the control treatment. Since high PAI-1 levels inhibit fibrinolytic activity these results are in contrast with the epidemiological findings of Meade et al. (148).

In a number of studies (4, 59, 62, 97, 98, 155, 156, 193, 202, 203) over a period of 30 years, the acute effects of alcohol on fibrinolytic parameters have been investigated (Table 10.3).

In the earlier studies, overall fibrinolytic activity of plasma or whole blood was measured in global assays. These measurements are very aspecific because fibrinolytic activity is influenced by many variables and the results obtained hence strongly depend on the experimental conditions used. This probably explains some of the contradictions shown in Table 10.3. In recent years, however, new analytical techniques have made possible the measurement of more specific fibrinolytic factors. A strong and consistent decrease in t-PA activity and an increase in PAI activity have been observed in two studies (202, 203), in agreement with the long-term study of Pikaar et al. (162). The levels of t-PA antigen, on the other hand, increased whereas no effects on the urokinase-plasminogen activator (UK-PA) system could be observed.

In conclusion, the results of studies on alcohol and fibrinolysis are contradictory. Alcohol consumption is epidemiologically associated with increased fibrinolytic activity, whereas decreases in fibrinolytic activity and t-PA activity and increases in PAI activity and t-PA antigen have been observed in experimental studies. Further research is needed to explain this contradiction and to investigate the mechanisms and consequences of these effects of alcohol. Some recent epidemiological studies suggest a relationship between lipid metabolism and fibrinolysis. This possible relationship also needs further attention.

10.3. Blood pressure

There is strong epidemiological evidence that regular use of large quantities of alcohol is related to higher blood pressure and a greater prevalence of hypertension (10, 12, 31, 41, 43, 56, 58, 65, 70, 77, 85, 88, 89, 116, 123, 138, 153, 159, 163, 190, 198, 207, 213), even after correction for confounding factors such as obesity and excessive salt intake. In the Kaiser Permanente study, the prevalence of hypertension was twice as high among people who had 6 or more drinks per day as among people who had 2 or less drinks per day (123). It seems prudent for health practitioners to advice hypertensive patients who are used to take three or more drinks a day to reduce their alcohol use. Potter & Beevers (166) and Malhotra et al. (143) have reported significant favourable responses to a reduction of alcohol intake within a few days (4 and 5 days, respectively). Puddy et al. (169), in a study among 44 hypertensive men with an average alcohol consumption of 6 to 7 glasses per day, demonstrated a significant reduction in blood pressure after a switch from regular beer (5% v/v alcohol) to low-alcohol beer (0.9% v/v alcohol). In the study of Puddy et al. not only blood pressure, but also body weight fell significantly in the period of low-alcohol beer consumption. This body weight reduction has undoubtedly contributed to a lowering of blood pressure. Unfortunately, no studies among hypertensives have been conducted yet in which the independent effect of alcohol, i.e. the energy from alcohol versus non-alcohol energy, has been investigated.

There is much less consensus with regard to the effects of moderate alcohol consumption. Some studies have found a continuous increase in blood pressure from abstainers to heavy alcohol users (10, 31, 41, 198). Other studies (43, 123, 213) have established a threshold above which alcohol was found to be positively associated with blood pressure. In still other studies (85, 88, 89, 116) a J-shaped relationship between alcohol consumption and blood pressure was found.

In an epidemiological study, Periti et al. (159) classified 1190 Italian men and women into two groups. In one group the people were asked to abstain for 3 days from alcohol before their blood pressure was measured, the others continued their normal alcohol consumption habits until the day of blood pressure measurement. In the men, but not the women, of the latter group, systolic blood pressure increased significantly with increasing alcohol use, regardless of age, body weight, smoking or coffee consumption. The systolic blood pressure increased by 4.6 mmHg per 100 g of alcohol per day. No significant relationship between alcohol consumption and diastolic blood pressure could be observed. In the group that was asked to abstain from alcohol for 3 days, no relation between alcohol consumption and blood pressure was found. These results are consistent with earlier findings showing an association of increased blood pressure with excessive alcohol consumption and also indicate that this increase disappears after a few days of abstention.

The effects of a single dose of alcohol have been investigated in a number of studies (105, 106, 168, 191, J Veenstra & E. te Wierik, in preparation). Potter et al. (168) have studied the effect of a single dose of alcohol on blood pressure in 16 young male students. Blood pressure was measured over a 5-hour period after the ingestion of either 600 ml non-alcoholic beer or 600 ml non-alcoholic beer spiked with alcohol (0.75 g per kg body weight). All volunteers were subjected to both treatments in a cross-over design. The period between the treatments was at least one week. The blood pressure course roughly paralleled the blood alcohol level: after a rapid increase by ca. 7 mmHg within one hour, blood pressure gradually decreased. Similar acute effects have been observed by Ireland et al. (105, 106).

In a recent study by J. Veenstra & E. te Wierik (in preparation) the effects of a moderate dose of beer, wine and spirits (40 g alcohol) and mineral water were studied in a randomized cross-over design. The drinks were consumed during a normal evening dinner, the volunteers being 8 healthy middle-aged men aged 45–55 years. In contrast to the studies mentioned above, a lower blood pressure was observed after alcohol consumption as compared to the mineral water treatment. There were no significant differences among the alcoholic beverages, albeit the effect was not as strong for beer as for wine or spirits, possibly due to the larger volume of the beer. The contradiction with the earlier studies can possibly be explained by the lower alcohol dose, the combination with a meal, the age of the volunteers or the time of the day. In the study of Veenstra & te Wierik the aim was to study the effects of alcohol under conditions mimicking normal life as closely as possible. Therefore, normal alcoholic beverages were used in combination with a normal but standardized evening dinner in a non-laboratory environment.

The acute studies mentioned so far have the drawback that the treatments are not isocaloric, i.e. the caloric value of the alcohol is not compensated for in the control treatment. This energy effect was taken into account in a study by Stott et al. (191). In that study, the design of which was comparable with that of Potter et al., an isocaloric amount of glucose was added to the non-alcoholic drink. In agreement with the results of Potter et al., a slight increase of blood pressure was observed one hour after alcohol ingestion, followed by a decrease. However, a similar increase and subsequent decrease in blood pressure was observed after the consumption of the glucose-enriched non-alcoholic drink. From this study it was concluded that the increase of blood pressure after the consumption of a single moderate dose of alcohol does not differ from the increase after the ingestion of an isocaloric amount of glucose.

Kelbaek et al. (117, 118) recently studied the effect of a single alcohol dose (0.9 g per kg body weight) on the heart function of 20 male CHD and cardiopathy patients. Ten controls with a similar clinical picture received an equal amount of an isocaloric non-alcoholic drink. A slight but significant decrease (–6%) of the systemic arterial blood pressure was found after alcohol ingestion. Alcohol was found to have no effect on central venous pressure, pulmonary artery pressure, cardiac output, stroke volume or global peripheral resistance. Kelbaek et al. concluded that ingestion of a moderate dose of alcohol by CHD patients is unlikely to evoke disease symptoms.

In some studies (101, 105, 167, 170) an increase in plasma adrenaline and noradrenaline after alcohol consumption has been connected with a rise in blood pressure. In the carefully controlled study of Stott et al., however, no effects of alcohol consumption on plasma catecholamine levels were observed. An alternative explanation for the relation between alcohol consumption and blood pressure is that alcohol induces changes in electrolyte excretion (36). J Veenstra & E. te Wierik (in preparation) observed large changes in mineral excretion in the urine and a strong diuretic effect of alcohol, which could possibly explain the acute decrease of blood pressure observed in their study. Finally, the results of a study by Finch et al. (64) suggest that an increase in peripheral blood flow and subsequently in skin temperature may be responsible for a decrease in blood pressure after a moderate dose of alcohol.

10.4. Atherosclerosis

In several studies the relationship between level of alcohol consumption and degree of atherosclerosis has been investigated. The degree of atherosclerosis has been established by autopsy (72, 100, 139) or in vivo by means of arteriography (15, 17, 18, 20, 67, 82, 87, 158, 183).

As early as 1904, it was concluded from an autopsy study that atherosclerosis is uncommon in alcoholics. However, judging by present standards, this study was poorly designed, lacking controls. Later, in 1965, Hirst et al. (100) observed no differences in degree of occlusion between non-cirrhotic alcoholics and controls. In cirrhotic alcoholics, however, he found fewer occlusions, which made him conclude that cirrhosis may have a protective effect against atherosclerosis, possibly by a change in oestrogen metabolism or blood coagulation. Some studies have confirmed this association between cirrhosis and atherosclerosis, others did not find any effect (11, 72).

In a few studies among non-alcoholics the extent of occlusion of coronary arteries in CHD cases has been related to the level of alcohol consumption (17, 18, 67, 82, 158). Barboriak et al. (17) have studied the effects of smoking and of alcohol consumption on coronary artery occlusion in a group of 2989 men. Fig. 10.2 summarizes the results of that study. Smoking was found to have a dose-dependent

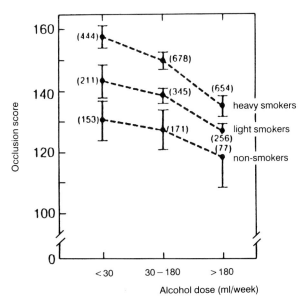

Fig. 10.2. Interrelationships between the extent of coronary occlusion, alcohol intake and smoking (17). Number of patients in parenthesis. Vertical bars: SEM.

occlusive effect, whereas the effect of alcohol was associated with lower occlusion scores in non-smokers, light smokers as well as heavy smokers.

Within the same study framework Gruchow et al. (82) investigated the effect of alcohol consumption patterns by classifying the subjects into non-drinkers, occasional drinkers, regular drinkers adhering to a more or less fixed dose, and regular drinkers varying the amounts consumed. Regular users of moderate doses of alcohol were found to be at significantly lower risk of occlusion of coronary arteries. Irregular drinkers were at higher risk irrespective of the level of alcohol consumption.

In a very recent study by Handa et al. (87), the relation between alcohol consumption and severity of coronary atherosclerosis was examined in 212 men undergoing coronary angiography. In addition, a blood sample was taken to measure blood lipids. The men were categorized as follows: non-drinkers, light drinkers (1–100 ml alcohol weekly), moderate drinkers (101–300 ml weekly) and heavy drinkers (>300 ml weekly). HDL-cholesterol levels increased and total cholesterol levels decreased with increasing alcohol consumption. After adjustment for age, smoking, hypertension, HDL-cholesterol, total cholesterol and triglycerides the relative risk of coronary stenosis for moderate drinkers was only 0.29 (95% confidence interval 0.13–0.63). The relative risk of light and of heavy drinkers did not differ significantly from that of non-drinkers. Also the severity of coronary atherosclerosis was significantly less in the moderate drinkers category (about 8–24 glasses weekly).

Fried et al. (67) studied in a group of 31 men with coronary arteries of normal diameter the effects of smoking and of alcohol consumption on the diameter of three main coronary arteries. Smoking and alcohol use appeared to affect these diameters highly significantly and independently. Smoking was found to have a vasoconstrictive effect, whereas alcohol consumption promoted vasodilatation.

In contrast with the studies mentioned above, two earlier studies, an autopsy study by Gent et al. (72) and an arteriographic study by Sirtori et al. (183), have reported a positive correlation, i.e. an increase in alcohol consumption associated with a larger number or more extensive atherosclerotic lesions.

10.5. Coronary heart disease and mortality

The relationship between level of alcohol consumption and risk of CHD has been investigated in many epidemiological studies. In ecological studies the per capita alcohol consumption in various countries has been correlated with the mortality rate for CHD (29, 93, 119, 137, 181, 184, 208). St. Leger et al. (184), for example, compared these values for 18 countries and found a clear inverse relation between average alcohol consumption and CHD. The observed effect of alcohol was not attributable to differences in known CHD risk factors such as smoking, cholesterol intake, fat intake or total energy intake. The strongest association was observed for wine consumption and CHD (Fig. 10.3).

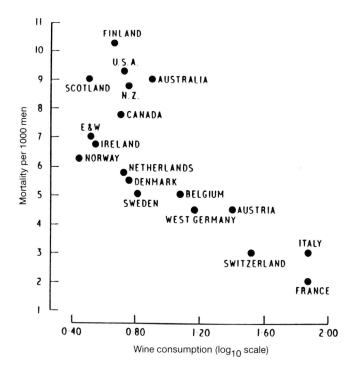

Fig. 10.3. Relationship between CHD mortality rate in men aged 55–64 and wine consumption (143).

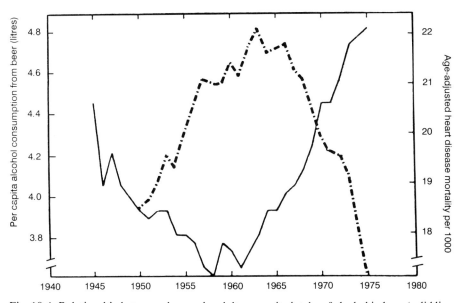

Fig. 10.4. Relationship between changes in adult per capita intake of alcohol in beer (solid line, 1945–1975) and changes in total age-adjusted heart disease death rates in the USA (broken line, 1950–1975) (137).

Comparable results have been reported by others (29, 93, 119, 137, 181, 208). Schmidt & Popham (181) and Werth (208) also reported significant negative correlations between total consumption of alcohol, wine and spirits and CHD mortality in the 50 states of the USA. LaPorte et al. (137) have also paid attention to trends in alcohol use and CHD incidence rates between 1945 and 1975 and found significant correlations between these, the strongest for beer. Taking into account a 5-year lag period, the correlation coefficient between beer consumption and CHD mortality was –0.943 (Fig. 10.4).

Ecological studies are no real proof of a favourable effect of alcohol consumption on CHD development. The results just indicate a relationship on a population level, and differences between countries with regard to culture and living and working conditions are not taken into account. A second drawback of these studies is that the alcohol consumption data are derived from alcohol sales data, uncorrected for home production, for sales to visitors from abroad or for consumption abroad. Finally, the hypothetical U-shaped curve between alcohol consumption and CHD poses an insurmountable problem for ecological studies since the use of sales statistics does not differentiate between non-drinkers, moderate drinkers and heavy drinkers.

In several case-control studies the alcohol consumption of CHD patients has been compared with a control group (94, 115, 126, 160, 171, 175, 177, 189). The results of these studies fairly consistently point at an inverse relation between moderate alcohol consumption and CHD for both men and women. Only Kaufman et al. (115) found no effect of moderate alcohol consumption.

Hennekens et al. (94) classified their study population into three categories, with an average daily alcohol consumption of 0, 0–3.5 and >3.5 glasses respectively. The relative risk was found to be significantly lower for moderate drinkers than for non-drinkers (relative risk (RR) = 0.4). The relative risk for drinkers of more than 3.5 glasses per day, on the other hand, was not significantly different from that of non-drinkers (RR = 0.7). In contrast, Klatsky et al. (126) found the relative risk to decrease with increasing consumption level, with relative risks of 1.0, 0.7 and 0.4 for categories with a daily alcohol consumption of <3, 3–5 and >5 glasses respectively.

Case-control studies, just as ecological studies, have a number of drawbacks. Probably the most serious one is that alcohol consumption is asked about after the disease has been diagnosed. This may result in specific underreporting of alcohol consumption by the patients. Moreover, the patients may have gradually reduced their alcohol use in the preclinical phase of the disease.

The strongest support for a protective effect of moderate alcohol consumption on CHD risk comes from cohort studies. In cohort studies, also called prospective or longitudinal studies, large populations are classified into categories varying in alcohol consumption level. After a number of years the association between alcohol consumption and CHD incidence is studied. In a remarkably large number of these studies performed in various parts of the world, a reduced risk of CHD in moderate drinkers has been found (7–9, 25, 26, 32, 38, 46, 57, 68, 76, 78, 79, 81, 114, 120, 122, 124, 129, 130, 133, 145, 179, 182, 186, 192, 214–216). In two studies – one in Ireland (81), another in California (32) – no effects of alcohol use were found. A recent study in Finland showed, in contrast to all other studies, a positive association between consumption of alcohol, in particular of spirits, and CHD incidence (192).

It is also from cohort studies in particular that the idea of a U-shaped curve has arisen, i.e. moderate drinkers would be at lower risk of CHD than both teetotallers and heavy drinkers. Dyer et al. (57), in 1980, were the first to suggest such a U-shaped curve on the basis of a study in which they followed 1899 employees of the Western Electric Company from 1957 to 1974. A decrease in CHD incidence was found with increasing alcohol consumption up to a daily dose of 5 glasses. At higher doses CHD incidence was found to increase again. However these results lost their significance after correction for smoking habits, serum cholesterol and blood pressure. In 1981, the existence of a U-shaped relationship was confirmed in two independent studies, one by Klatsky et al. (124), the other by Marmot et al. (145). Later support came from a study of Friedman & Kimball (68) among 5209 men and women in Framingham during a 24-year follow-up period.

Although the vast majority of all studies have reported a decreased incidence of CHD among moderate drinkers as compared to non-drinkers, not all studies have found a U-shaped relationship. There is, however, no doubt that heavy alcohol consumption, besides increasing other health risks, damages the cardiovascular system (see also Sections 10.3 and 10.6–10.8). Possibly alcoholics and excessive alcohol users are less likely to be included in the cohort studies. This is, for example, the case in the Nurses Health Cohort Study by Stampfer et al. (186), in which the alcohol consumption of the cohort was relatively low.

A serious drawback of many cohort studies is that the category of non-drinkers considered in these studies may comprise former heavy drinkers as well as people who have given up alcohol on medical advice owing to, for example, heart complaints. This potential bias has recently been studied by Shaper et al. (182) in a cohort of 7735 men living in 24 different towns in England, Scotland and Wales. They found a U-shaped relationship between alcohol consumption and total mortality and an inverse relationship between alcohol consumption and cardiovascular disease mortality, in fair agreement with the results of earlier studies. At the start of the follow-up study, when the average age of the subjects was 50 years, 24.2% of the subjects already had heart complaints. The associations between alcohol use and total mortality and between alcohol use and CVD mortality were also studied for both sub-cohorts separately (men suffering from and men free from heart disorders). In the category initially free from heart disorders a U-shaped or inverse relationship could no longer be found. Shaper et al. concluded that the U-shaped curve could largely be explained from the fact that people suffering from CVD disorders tend to reduce or even give up alcohol consumption so that the curvature cannot be attributed to any favourable effects of moderate alcohol use. An alternative explanation, however, could be that by excluding 24.2% from the cohort, those people are removed that are most susceptable to CVD and therefore could benefit most from a potential protective effect of moderate alcohol consumption. The Lancet paper of Shaper et al. (182) has evoked a lively discussion in that journal as to whether moderate alcohol use protects against CHD (7–9).

In a number of studies the non-drinkers group has been clearly defined. In the Honolulu Heart Study (216), in which the non-drinkers were classified into lifelong teetotallers and former drinkers, a higher CHD incidence as compared to drinkers was found for both categories. This finding is supported by the results of a case-control study among 513 cases and 918 controls (175) and other studies (122, 129, 130). In the Honolulu Heart Study (184) and in a study by Klatsky et al. (122) no significant differences in CHD risk were found between lifelong teetotallers and former drinkers. Finally, Stampfer et al. (186), in a cohort of as many as 87 526 nurses, has recently searched for any changes in drinking habits over the past ten years. Subjects who had given up drinking on medical advice could be excluded from the study this way. The CHD risk for the non-drinkers group was found to be up to 2.5 times as high as for the drinkers categories.

Interestingly, Stampfer et al., in the Nurses Health Cohort Study, also describe a decreased risk of ischaemic stroke in light and moderate alcohol consumers, which indicates a possible common physiological mechanism. The risk of haemorrhagic stroke, on the other hand, showed a dose-dependent increase with increasing alcohol consumption. These results were recently confirmed for both men and women in a cohort study by Klatsky et al. (121) comprising over 100 000 people. The increased risk of haemorrhagic stroke could partly be explained by the effects of alcohol on blood pressure.

An important question with respect to the proposed beneficial effects of moderate alcohol consumption on risk of CHD is whether total mortality rates are also favourably influenced. Obviously, it is important to show that, when the risk of CHD in moderate drinkers decreases, the risk of other death causes does not simultaneously increase. Large-scale epidemiological studies, such as the Chicago Western Electric Company Study (57), the Honolulu Heart Study (25), the London Civil Servant Study (145), the Kaiser Permanente Study (124), the Framingham Study (78) and – very recently – the large American Cancer Society Prospective Study (26), however, have shown that not only CHD risks are lower in moderate alcohol consumers, but that there is also a U- or J-shaped relationship between alcohol consumption and risk of total mortality.

Finally, the effects of alcohol consumption on blood lipids, haemostasis, blood pressure and atherosclerosis (see sections 10.1–10.4) all have been put forward to explain the observed beneficial effects of moderate alcohol consumption in epidemiological studies. An alternative explanation, however, might be that moderate alcohol consumption is associated with a healthier life-style. In a few studies the relationship between level of alcohol consumption and dietary patterns has been investigated (49, 84, 107, 110, 205). With respect to the intake levels for various nutrients the results are not very consistent and may vary due to cultural differences in dietary habits between the populations studied. It was fairly consistently shown, however, that total energy intake increases with increasing alcohol consumption. In other words, the energy consumed in the form of alcohol is not compensated for by omitting calories from other dietary sources. Furthermore, it was shown that alcohol consumption, even in moderate amounts, is strongly associated with smoking, which is in agreement with findings in many other epidemiological studies (125, 144, 187). These findings do not provide an alternative explanation for the observed decreased risk of CHD in moderate drinkers. Therefore, the physiological mechanisms described in sections 10.1–10.4 still form the most likely explanation.

10.6. Angina pectoris

According to Orlando et al. (157), angina pectoris (pain in the chest during exercise, caused by ischaemia of the coronary arteries), occurs more frequently after alcohol consumption. In earlier studies alcohol was proved either to have no effect (40) or to attenuate angina pectoris (178). Hrubec et al. (102) found a statistical relationship between angina pectoris and the number of times individuals are intoxicated, but found no clear relation with the quantity of alcohol consumed weekly.

Takizawa et al. (194) concluded that alcohol consumption may provoke attacks of variant angina, a type of angina pectoris which may occur spontaneously in rest.

In view of the conflicting evidence no definite relationship between alcohol consumption and angina pectoris, either positive or negative, can be said to exist.

10.7. Cardiomyopathy

A single large dose of alcohol has a deleterious effect on cardiac performance such as myocardial contractility and depression of the left ventricular function. Van Zwieten (218) reported consumption of 80 g alcohol in 2 hours to be associated with this phenomenon. Regan et al. (172) reported this change to occur at a blood alcohol level of 0.75 g/l. Others mention with regard to this syndrome blood alcohol levels of 0.1 to 2.85 g/l (1, 35). The consumption of a single large dose of alcohol may also cause arrhythmia of the heart muscle such as atrial fibrillation (132, 140). In a study by Lowenstein et al. (140), 35% of the cases admitted to the hospital with new-onset atrial fibrillation could be linked with excessive alcohol consumption. In a large prospective study by Cohen et al. (37) it was shown that arrhythmia in general was 2.3 times more common in heavy drinkers (more than 6 glasses per day) than in light drinkers (less than 1 glass per day). The so-called 'holiday heart syndrome', which is characterized by arrhythmia, is also associated with acute excessive alcohol consumption (80, 111).

Alcoholism may cause myocardial injury (173, 218). This effect of excessive alcohol consumption is not specific for the heart muscle, but may affect other muscles as well (see Chapter 11). Recently, Urbano-Marquez et al. (199) showed strong correlations between the total lifetime amount of alcohol consumed and the performance of both skeletal and cardiac muscle, in a group of severe alcoholics. There is now sufficient evidence (see Table 10.4) to support the causal role of alcohol and/or acetaldehyde in injury of the heart muscle (cardiomyopathy) in

Table 10.4. Evidence in support of the syndrome called 'alcoholic cardiomyopathy' (from 127).

1. Association of drinking and heart muscle disease noted by numerous authorities.
2. High proportion of chronic users of large amounts of alcohol among patients with congestive cardiomyopathy.
3. Cases that show convincing evidence of heart muscle dysfunction in relation to episodic drinking.
4. Acute impairment of heart muscle contractility due to alcohol in man and animals.
5. Acute rhythm disturbances related to alcohol in man ('holiday heart syndrome').
6. Impaired heart function in alcoholics without acute alcohol load.
7. Heart muscle metabolic dysfunction in animals related to acute alcohol load.
8. Alcohol-produced heart muscle cellular abnormalities in animals.
9. Autopsy evidence of heart muscle damage in alcoholics without a history of clinical heart disease.
10. Well-documented acute and chronic skeletal muscle syndromes owing to alcohol.

alcoholics (127, 172, 218), also in the absence of the nutritional deficiencies often observed in alcoholics (176, 199).

Among the possible causes mentioned are: changes in ion permeability of muscle cells (164), inhibition of protein synthesis, inhibition of actin and myosin binding in the heart muscle (6), and mitochondrial abnormalities (1).

Cardiomyopathy may also be caused by intoxication due to trace elements, cobalt in particular (1). Alexander (2) found myocardial injury in 27 heavy drinkers who used to drink one particular brand of beer to which 1–1.5 mg/l cobalt had been added to stabilize the froth.

Alcohol abuse must have continued over a long period of time in order to cause severe heart muscle injury. Wink (212) mentioned an alcohol limit of 250 g/day over 10 years as the borderline above which definite cardiac risk will occur. A limit of 10 years of excessive alcohol intake is also indicated elsewhere (1, 6). Damage to other organs in the body is usually observed prior to cardiomyopathy (6, 218).

10.8. Cardiac beriberi

Beriberi is a nutritional disorder caused by deficiency of thiamin (vitamin B-1). The disease has been widespread in poor communities in East Asia, with rice as the staple food and an otherwise very poor diet. However, in rich countries thiamin deficiency may also occur in people whose main source of energy is alcohol. Thiamin deficiency may lead to the Wernicke-Korsakoff syndrome (see Chapter 9) in some alcoholics, or cause cardiac failure with generalized oedema, pulmonary congestion and breathing difficulties in others (28, 142, 165). These diseases can occur together in one patient, but this is uncommon. There is no explanation as to why in some patients the heart is affected by the thiamin deficiency and in others the brain. In beriberi among alcoholics lactacidosis is regularly observed (141), which is not the case for beriberi in non-alcoholics. In its early stages the disease can also be treated with thiamin, but recovery of severe cases is slow and often incomplete (127). Supplementation with thiamin is a very successful and cost-effective way of preventing the disease. Doctors and social workers should therefore pay attention to the alcoholic's diet and be aware of the risks of thiamin deficiency.

10.9. Final considerations

In this chapter the effect of alcohol on the cardiovascular system is described. Excessive alcohol use has obviously detrimental effects on the vascular system, by increasing blood pressure, and on the heart muscle itself. Moderate alcohol use, on the other hand, has been shown to have a protective effect on CHD. The exact mechanism of this effect is not yet fully known, but a number of plausible mechanisms have been put forward. These results underline the sense of prudent drinking limits. On the one hand, they should not be used to promote alcohol consumption among non-drinkers. It should be realized that the lower risk of CHD is mainly observed in population studies and is not necessarily applicable to each single individual. On the other hand, these results may be used to persuade excessive drinkers to moderate their alcohol use.

Summary

Consumption of alcohol has an influence on several processes that play an important role in the development of diseases of the cardiovascular system.
– Alcohol consumption is positively correlated with HDL-cholesterol in epidemiological studies. In experimental studies alcohol increases the HDL-cholesterol level within some weeks.
– Studies on the acute effects of alcohol on platelet aggregation have revealed conflicting results. Epidemiological and long-term experimental evidence shows that alcohol consumption probably reduces platelet aggregation.
– Alcohol consumption, even at moderate doses, strongly affects the components of the fibrinolytic system, acutely and in the long term. PAI activity and the level of t-PA antigen is increased, whereas the level of t-PA activity is strongly decreased.
– Excessive alcohol consumption is clearly associated with an increased blood pressure. The effects of moderate alcohol use are less evident. Some epidemiological studies have shown a continual dose-dependent increase, other ones have shown an increase above a certain level, and in some studies a U-shaped relationship between alcohol and blood pressure has been found.
– Studies among hypertensive excessive alcohol consumers have clearly shown a marked reduction in blood pressure within days after reducing alcohol intake.
– A number of studies have shown that moderate alcohol use is associated with lower scores for occlusion of the coronary arteries. In these studies, the drinking pattern was shown to be an important factor. Binge drinking was associated with a higher risk.

The relationship between alcohol consumption and cardiovascular diseases has been the subject of a large number of studies.
- In a number of ecological studies an inverse relationship between alcohol use and the incidence of CHD has been found.
- A protective effect of moderate alcohol consumption against the development of CHD has been confirmed in many patient-control and cohort studies. Total mortality was also shown to decrease among moderate drinkers as compared to non-drinkers.
- Since excessive alcohol use is associated with an increased risk of CHD, the relationship between alcohol consumption and CHD is said to be U-shaped.
- Moderate alcohol use has been found to be associated with a lower incidence of ischaemic stroke. The risk of haemorrhagic stroke, on the other hand, increases with alcohol consumption.
- In a number of studies moderate alcohol consumption was found not to be associated with a healthier life style as far as dietary patterns and smoking habits are concerned.
- Long-term as well as acute excessive alcohol consumption has been shown to increase the risk of cardiac arrhythmia.
- Long-term excessive alcohol consumption can cause cardiomyopathy, even in the absence of nutritional deficiencies. In alcoholics, the total lifelong consumption of alcohol has been shown to be inversely correlated with both cardiac performance and skeletal muscle strength.
- In alcoholics, thiamin deficiency may result in cardiac beriberi, which in an early stage can be prevented easily and cost-effectively by thiamin supplementation.

References

1. Alderman EL, Coltart DJ. Alcohol and the heart. Br Med Bull 1982;38: 77-80.
2. Alexander CS. Cobalt-beer cardiomyopathy. Circulation 1970;128(suppl III): 470.
3. Allen JK, Adena MA. The association between plasma cholesterol, high-density lipoprotein cholesterol, triglycerides and uric acid in ethanol consumers. Ann Clin Biochem 1985;22: 62-6.
4. Anderson JA, Gow LA, Ogston D. Influence on the fibrinolytic enzyme system. Acta Haemat 1983;69: 344-8.
5. Angelico F, Bucci A, Capocaccia R, Morisi G, Terzino M, Ricci G. Further considerations on alcohol intake and coronary risk factors in a Rome working population group: HDL-cholesterol. Ann Nutr Metab 1982;26: 73-6.
6. Anonymous. Alcoholic heart disease. Lancet 1980;i: 961-2.
7. Anonymous. Alcohol and the U-shaped curve. Lancet 1989;i: 105.
8. Anonymous. Alcohol and the U-shaped curve. Lancet 1989;i: 224-5.
9. Anonymous. Alcohol and the U-shaped curve. Lancet 1989;i: 336.
10. Arkwright PD, Beilin LJ, Rouse I, Armstrong BK, Vandongen R. Alcohol: effect on blood pressure and predisposition to hypertension. Clin Sci 1981;61: 373s-5s.
11. Ashley MJ, Rankin JG. Alcohol consumption and hypertension: the evidence from hazardous drinking and alcoholic populations. Aust N Z J Med 1979;9: 201-6.
12. Ashley MJ. Alcohol consumption, ischaemic heart disease and cerebrovascular disease: an epidemiological perspective. J Stud Alcohol 1982;43: 869-87.
13. Bain BJ. Some influences on the ESR and the fibrinogen level in healthy subjects. Clin Lab Haemat 1983;5: 45-54.

14. Balleisen L, Bailey J, Epping PH, Schulte H, van de Loo J. Epidemiological study on factor VII, factor VIII and fibrinogen in an industrial population. Thrombosis Haemostasis 1985;5: 475-9.
15. Barboriak JJ et al. Alcohol and coronary arteries. Alcoholism 1979;3: 29-32.
16. Barboriak JJ, Anderson AJ, Hoffmann RG. Interrelationship between coronary artery occlusion high-density lipoprotein cholesterol, and alcohol intake. J Lab Clin Med 1979;94: 348-53.
17. Barboriak JJ, Anderson AJ, Hoffmann RG. Smoking, alcohol and coronary artery occlusion. Atherosclerosis 1982;43: 277-82.
18. Barboriak JJ, Anderson AJ, Rimm AA, Tristani FE. Alcohol and coronary arteries. Alcoholism: Clin Exp Res 1979;3: 29-32.
19. Barboriak JJ, Gruchow HW, Anderson AJ. Alcohol consumption in the diet-heart controversy. Alcohol Clin Exp Res 1983;7: 31-4.
20. Barboriak JJ. Coronary artery occlusion and alcohol intake. Br Heart J 1977;39: 289-93.
21. Barrett-Connor E, Suarez L. A community study of alcohol and other factors associated with the distribution of high-density lipoprotein cholesterol in older vs. younger men. Am J Epidemiol 1982;115: 888-93.
22. Belfrage P, Berg B, Hagerstrand I, Nilsson-Ehle P, Tornquist H, Wiebe T. Alterations of lipid metabolism in healthy volunteers during long-term ethanol intake. Eur J Clin Invest 1977;7: 127-31.
23. Bell H, Strømme JH, Steensland H, Bache-Wiig JE. Plasma HDL cholesterol and estimated ethanol consumption in 104 patients with alcohol dependence syndrome. Alcohol Alcoholism 1985;20: 35-40.
24. Blacket RB, Palmer AJ. Haemodynamic studies in high output beriberi. Br Heart J 1960;22: 483-501.
25. Blackwelder WC, Yano K, Rhoads GG, Kagan A, Gordon T, Palesch Y. Alcohol and mortality: The Honolulu Heart Study. Am J Med 1980;68: 164-8.
26. Boffetta P, Garfinkel L. Alcohol drinking and mortality among men enrolled in an American Cancer Society Prospective Study. Epidemiology 1990;1: 342-8.
27. Boniton-Kopp C, Scarabin P, Bara L, Castanier M, Jacqueson A, Roger M. Relationship between sex hormones and haemostatic factors in healthy middle-aged men. Atherosclerosis 1988;71: 71-6.
28. Brenn T. The Tromsø Heart Study: alcoholic beverages and coronary risk factors. J Epidemiol Commun Health 1986;40: 249-56.
29. Brummer P. Coronary mortality and living standard, II. Coffee, tea, cacao, alcohol and tobacco. Acta Med Scand 1969;186: 61-3.
30. Burr ML, Fehily AM, Butland BK. Alcohol and high-density lipoprotein cholesterol: a randomized controlled trial. Br J Nutr 1986;56: 81-6.
31. Cairns V, Keil U, Kleinbaum D, Doering A, Stieber J. Alcohol consumption as a risk factor for high blood pressure. Hypertension 1984;6: 124-31.
32. Camacho TC, Kaplan GA, Richard D. Alcohol consumption and mortality in Almeda County. J Chronic Dis 1987;40: 229-36.
33. Castelli WP, Gordon T, Hjortland MC, Kagan A, Doyle JT, Hames CG, Hulley SB, Zukel WJ. Alcohol and blood lipids. The cooperative lipoprotein phenotyping study. Lancet 1977;ii: 153-55.
34. Chen H, Zhuang H, Han Q. Serum high-density lipoprotein cholesterol and factors influencing its levels in healthy Chinese. Atherosclerosis 1983;48: 71-9.
35. Cigarroa RG, Lange RA, Popma JJ, Yurow G, Sills MN, Firth BG, Hillis LD. Ethanol-induced coronary vasodilation in patients with and without coronary artery disease. Am Heart J 1990;119: 254-9.
36. Clark LT. Role of electrolytes in the etiology of alcohol-induced hypertension. Magnesium 1989;8: 124-31.
37. Cohen EJ, Klatsky AL, Armstrong MA. Alcohol use and supraventricular arrhytmia. Am J Cardiol 1988;62: 971-3.
38. Colditz GA, Branch LG, Lipnick RJ et al. Moderate alcohol and decreased cardiovascular mortality in an elderly cohort. Am Heart J 1985;109: 886-9.

39. Contaldo F, D'Arrigo E, Carandente V et al. Short-term effects of moderate alcohol consumption on lipid metabolism and energy balance in normal men. Metabolism 1989;38: 166-71.
40. Conway N. Haqemodynamic effects of ethyl alcohol in patients with coronary heart disease. Br Heart J. 1968;30: 638-44.
41. Cooke KM, Frost GW, Stokes GS. Blood pressure and its relationship to low levels of alcohol consumption. Clin Exp Pharmacol Physiol 1983;10: 229-33.
42. Couzigou P, Fleury B, Crockett R et al. High-density lipoprotein cholesterol and apoprotein A1 in healthy volunteers during long-term moderate alcohol intake. Ann Nutr Metab 1984;28: 377-84.
43. Criqui MH, Wallace RB, Mishkel M, et al. Alcohol consumption and blood pressure: The Lipid Research Clinics Prevalence Study. Hypertension 1981;3: 557-65.
44. Croft JB, Freedman DS, Cresanta JL, Srinivasan SR, Burke GL, Hunter SM, Webber LS, Smoak CG, Berenson GS. Adverse influences of alcohol, tobacco, and oral contraceptive use on cardiovascular risk factors during transition to adulthood. Am J Epidemiol 1987;126: 202-213.
45. Crouse JR, Grundy SM. Effects of alcohol on plasma lipoproteins and cholesterol and triglyceride metabolism in man. J Lipid Res 1984;25: 486-96.
46. Cullen K, Stenhouse NS, Wearne KL. Alcohol and mortality in the Busselton Study. Int J Epidemiol 1982;11: 67-70.
47. Cushman P, Barboriak J, Kalbfleisch J. Alcohol: High-density lipoproteins, apolipoproteins. Alcohol Clin Exp Res 1986;10: 154-7.
48. Danielsson B, Ekman R, Fex G et al. Changes in plasma high-density lipoproteins in chronic male alcoholics during and after abuse. Scand J Clin Lab Invest 1978;38: 113-9.
49. de Castro JM, Orozco S. Moderate alcohol intake and spontaneous eating patterns of humans: evidence of unregulated supplementation. Am J Clin Nutr 1990;52: 46-53.
50. Devenyi P, Robinson GM, Kapur BM, Roncari DAK. High-density lipoprotein cholesterol in male alcoholics with and without severe liver disease. Am J Med 1981;71: 589-94.
51. DeWood MA, Spores J, Notske R. Prevalance of total, coronary occlusion during the early hours of transmural, myocardial infarction. N Engl J Med 1980;303: 897-902.
52. Deykin D, Janson P, McMahon L. Ethanol potentiation of aspirin-induced prolongation of the bleeding time. N Engl J Med 1982; 306: 852-4.
53. Diehl AK, Fuller JH, Mattock MB, Salter AM, Gohari RE, Keen H. The relationship of high-density lipoprotein subfractions to alcohol consumption, other lifestyle factors and coronary heart disease. Atherosclerosis 1988;69: 145-53.
54. Donahue RP, Orchard TJ, Kuller LH, Drash AL. Lipids and lipoproteins in a young adult population. Am J Epidemiol 1985;122: 458-67.
55. Dunn EL, Cohen RG, Moore EE, Hamstra RD. Acute alcohol ingestion and platelet function. Arch Surg 1981;116: 1082-3.
56. Dyer AR, Stamier E, Oglesby P. Alcohol consumption, cardiovascular risk factors, and mortality in two Chicago epidemiologic studies. Circulation 1977;56: 1067-74.
57. Dyer AR, Stamler J, Paul O et al. Alcohol consumption and 17-year mortality in the Chicago Western Electric Company Study. Prev Med 1980;9: 78-90.
58. Elliott P, Rogers S, Scally G, et al. Sodium, potassium, body mass, alcohol and blood pressure in three United Kingdom centres (the INTERSALT Study). Eur J Clin Nutr 1990;44: 637-45.
59. Elmer O, Göransson G, Zoucas E. Impairment of primary hemostasis and platelet function after alcohol ingestion in man. Haemostasisis 1984;14: 223-8.
60. Elmér O, Göransson G, Zoucas E. Impairment of primary hemostasis and platelet function after alcohol ingestion in man. Haemostasis 1984;14: 223-8.
61. Ernst N, Fisher M, Smith W et al. The association of plasma high-density lipoprotein cholesterol with dietary intake and alcohol consumption. The Lipid Research Clinics program prevalence study. Circulation 1980;62 (suppl IV): 41-52.
62. Fearnly GR, Ferguson J, Chakrabarti R, Vincent CT. Effect of beer on blood fibrinolytic activity. Lancet 1960;i: 184-6.
63. Fenn CG, Littleton JM. Interactions between ethanol and dietary fat in determining human platelet function. Thrombosis Haemostasis 1984;51: 50-3.

64. Finch MB, Copeland S, Leahey WJ, Johnston GD. Short-term effects of alcohol on peripheral blood flow, platelet aggregation and noradrenaline output in normal man. Int J Tissue React 1988;10: 257-60.
65. Fortmann SP et al. The association of blood pressure and dietary alcohol: differences by age, sex, ans estrogen use. Am J. Epidemiol 1983;118: 497-507.
66. Franceschini G, Moreno Y, Apebe P et al. Alterations in high-density lipoprotein subfractions during postprandial lipidaemia induced by fat with and without ethanol. Clin Sci 1988;75: 135-42.
67. Fried LP, Moore RD, Pearson TA. Long-term effects of cigarette smoking and moderate alcohol consumption on coronary artery diameter. Am J Med 1986;80: 27-44.
68. Friedman LA, Kimball AW. Coronary heart disease mortality and alcohol consumption in Framingham. Am J Epidemiol 1986;124: 481-9.
69. Frimpong NA, Lapp JA. Effects of moderate alcohol intake in fixed or variable amounts on concentration of serum lipids and liver enzymes in healthy young men. Am J Clin Nutr 1989;50: 987-91.
70. Funke E, Schieffer B, Baikie M, et al. Intersalt in Newfoundland and Labrador. Ann Nutr Metab 1990;34: 253-8.
71. Galli C, Colli S, Gianfranceschi G et al. Acute effects of ethanol, caffeine, or both on platelet aggregation, thromboxane formation, and plasma-free fatty acids in human subjects. Drug Nutr Interact 1984;3: 61-7.
72. Gent M et al. Further exploration of the relationship between aortic atherosclorosis and the use of cigarettes and alcohol. Circulation 1970;(suppl 3): 156.
73. Glueck CJ, Hogg E, Allen C, Gartside PS. Effects of alcohol ingestion on lipids and lipoproteins in normal men: Isocaloric metabolic studies. Am J Clin Nutr 1980;33: 2287-93.
74. Goldberg CS, Tall AR, Krumholz S. Acute inhibition of hepatic lipase and increase in plasma lipoproteins after alcohol intake. J Lipid Res 1984;25: 714-20.
75. Gordon T, Doyle JT. Alcohol consumption and its relationship to smoking, weight, blood pressure, and blood lipids: The Albany study. Arch Intern Med 1986;146: 262-5.
76. Gordon T, Doyle JT. Drinking and coronary heart disease: The Albany Study. Am Heart J 1985;110: 331-4.
77. Gordon T, Kannel WB. Drinking and its relation to smoking, blood pressure, blood lipids, and uric acid. Arch Intern Med 1983;143: 1366-74.
78. Gordon T, Kannel WB. Drinking and mortality: the Farmingham Study. Am J Epidemiol 1984;120: 97-107.
79. Gordon T, Kannel WB. Drinking habits and cardiovascular disease: The Framingham Study. Am Heart J 1983;105: 667-73.
80. Greenspon AJ, Schaal SF. The 'Holiday Heart': electrophysiological studies of alcohol effects in alcoholics. Ann Int Med 1983;98: 135-9.
81. Greig M, Pemberton J, Hay I, MacKenzie G. A proptective study of the development of coronary heart disease in a group of 1202 middle-aged men. J Epidem Commun Health 1980;34: 23-30.
82. Gruchow HW, Hoffmann RG, Anderson AJ, Barboriak JJ. Effects of drinking patterns on the relationship between alcohol and coronary occlusion. Atherosclerosis 1982;43: 393-404.
83. Gruchow HW, Hoffmann RG, Anderson AJ, Barboriak JJ. Effects of drinking patterns on the relationship between alcohol and coronary occlusion. Atherosclerosis 1982;43: 393-404.
84. Gruchow HW, Sobocinski KA, Barboriak JJ, Scheller JG. Alcohol consumption, nutrient intake and relative body weight among US adults. Am J Clin Nutr 1985;42: 289-95.
85. Gyntelberg F, Meyer J. Relationship between blood pressure and physical fitness and alcohol consumption in Copenhagen males aged 40–59. Acta Med Scand 1974;195: 375-80.
86. Haffner SM, Applebaum-Bowden D, Wahl PW et al. Epidemiological correlates of high-density lipoprotein subfractions, apolipoproteins A-I, A-II, and D, and lecithin cholesterol acyltransferase. Effects of smoking, alcohol, and adiposity. Arteriosclerosis 1985;5: 169-77.
87. Handa K, Sasaki J, Saku K, Kono S, Arakawa K. Alcohol consumption, serum lipids and severity of angiographically determined coronary artery disease. Am J Cardiol 1990;65: 287-9.
88. Harburg E, Ozgoren F, Hawthorne VM, Schork MA. Community norms of alcohol usage and blood pressure: Tecumseh Michican. Am J Public Health 1980;70: 813-20.

89. Harlan WR, Hull AL, Schmouder RL, Landis JR, Thompson FE, Larkin FA. Blood pressure and nutrition in adults. Am J Epidemiol 1984;120: 17-28.
90. Hartung GH, Foreyt JP, Mitchell RE, Mitchell JG, Reeves RS, Gotto AM. Effect of alcohol intake on high-density lipoprotein cholesterol levels in runners and inactive men. J Am Med Assoc 1983;249: 747-50.
91. Hartung GH, Foreyt JP, Reeves RS, Krock LP, Patsch W, Patsch JR, Gotto AM. Effect of alcohol dose on plasma lipoprotein subfractions and lipolytic enzyme activity in active and inactive men. Metabolism 1990;39: 81-5.
92. Haskell WL, Camargo C, Williams PT et al. The effect of cessation and resumption of moderate alcohol intake on serum high-density lipoprotein subfractions. N Engl J Med 1984;310: 805-10.
93. Hegsted DM, Ausman LM. Diet, alcohol and coronary heart disease in men. J Nutr 1988;118: 1184-9.
94. Hennekens CH, Rosner B, Cole DS. Daily alcohol consumption and fatal coronary heart disease. Am J Epidemiol 1978;107: 196-200.
95. Hillbom M, Kangasaho M, Kaste M, Numminen H, Vapaatalo H. Acute ethanol ingestion increases platelet reactivity: Is there a relationship to stroke? Stroke 1985;16: 19-23.
96. Hillbom M, Kangasaho M, Löwbeer C, Kaste M, Muuronen A, Numminen H. Effects of ethanol on platelet function. Alcohol 1985;2: 429-32.
97. Hillbom M, Kaste M, Rasi V. Can ethanol intoxication affect hemocoagulation to increase the risk of brain infarction in young adults? Neurology 1983;33: 381-4.
98. Hillbom M, Kaste M, Rasi V. Effects of ethanol on the hemocoagulation in healthy male volunteers. Acta Neurol Scand 1982;65: 182-3.
99. Hillbom M, Muuronen A, Neiman J, Björk G, Egberg N, Kangasaho M. Effects of vitamin E therapy on ethanol-induced changes in platelet aggregation, thromboxane formation, factor VIII levels and serum lipids. Eur J Clin Invest 1987;17: 68-74.
100. Hirst AE, Hadly GG, Gore I. The effect of chronic alcoholism and cirrhosis of the liver on atherosclerosis. Am J. Med Sci 1965;249: 143-9.
101. Howes LG, Reid JL. Changes in plasma free 3,4-dihydroxyphenylethylene glycol and noradrenaline levels after acute alcohol administration. Clin Sci 1985;69: 423-8.
102. Hrubec Z, Cederlöf R, Friberg L. Background of angina pectoris: social and environmental factors in relation to smoking. Am J Epidemiol 1976;103: 16-29.
103. Hulley SB, Cohen R, Widdowson G. Plasma high-density lipoprotein cholesterol level. JAMA 1977;238: 2269-71.
104. Hulley SB, Gordon S. Alcohol and high-density lipoprotein cholesterol. Causal inference from diverse study designs. Circulation 1981;64(suppl III): 57-63.
105. Ireland M, Vandongen R, Davidson L, Beilin LJ, Rouse IL. Acute effects of moderate alcohol consumption on blood pressure and plasma catecholamines. Clin Sci 1984;66: 643-8.
106. Ireland M, Vandongen R, Davidson L, Beilin LJ, Rouse IL. Pressor effect of moderate alcohol consumption in man: A proposed mechanism. Clin Exp Pharmacol Physiol 1983;10: 375-9.
107. Jacques PF, Sulsky S, Hartz SC, Russell RM. Moderate alcohol intake and nutritional status in nonalcoholic elderly subjects. Am J Clin Nutr 1989;50: 875-83.
108. Jacqueson A, Richard JL, Ducimetiere P, Claude JR. High-density lipoprotein cholesterol and alcohol consumption in a French male population. Atherosclerosis 1983;48: 131-8.
109. Johansson BG, Medhus A. Increase in plasma alpha-lipoproteins in chronic alcoholics after acute abuse. Acta Med Scand 1974;195: 273-7.
110. Jones BR, Barrett-Connor E, Criqui MH, Holdbrook MJ. A community study of calorie and nutrient intake in drinkers and nondrinkers of alcohol. Am J Clin Nutr 1982;35: 135-9.
111. Jordaens L et al. Het 'hiliday heart'-syndroom. Ned Tijdschr Geneesk 1983;127: 1307-10.
112. Kagan A, Yano K, Rhoads GG, McGee DL. Alcohol and cardiovascular disease: The Hawaiian experience. Circulation 1981;64(suppl III): 27-31.
113. Kangasaho M, Hillbom M, Kaste M, Vapaatalo H. Effects of ethanol intoxication and hangover on plasma levels of thromboxane B_2, and 6-Keto-Prostaglandin F_{1a}, and on Thromboxane B_2, formation by platelets in man. Thrombosis Haemostasis 1982;48: 232-34.
114. Kannel WB, Castelli WP, McNamara PM. The coronary profile: 12-year follow-up in the Framingham Study. J Occup Med 1967;12: 611-19.

115. Kaufman DW, Rosenberg L, Hemrich SP, Shapiro S. Alcoholic beverages and myocardial infarction in young men. Am J Epidemiol 1985;121: 548-54.
116. Keil U, Chambless L, Remmers A. Alcohol and blood pressure: results from the Lübeck blood pressure study. Prevent Med 1989;18: 1-10.
117. Kelbaek H, Heslet L, Skagen K, Christensen NJ, Godtfredsen J, Munck O. Hemodynamic effects of alcohol at rest and during upright exercise in coronary artery disease. Am J Cardiol 1988;61: 61-4.
118. Kelbaek H, Heslet L, Skagen K, Munck O, Christensen NJ, Godtfredsen J. Cardiac function after alcohol ingestion in patients with ischemic heart disease and cardiomyopathy: A controlled study. Alcohol Alcoholism 1988;23: 17-21.
119. Keys A. Wine, garlic and CHD in seven countries. Lancet 1977;i: 145-6.
120. Kittner SJ, Garcia-Palmleri MR, Costas R, Cruz-Vidal M, Abbott RD, Havlik RJ. Alcohol and coronary heart disease in Puerto Rico. Am J Epidemiol 1983;117: 538-50.
121. Klatsky AL, Armstromg MA, Friedman GD. Alcohol use and subsequent cerebrovascular disease hospitalizations. Stroke 1989;20: 741-6.
122. Klatsky AL, Armstrong MA, Friedman GD. Relations of alcoholic beverage use to subsequent coronary artery disease hospitalization. Am J Cardiol 1986;58: 710-4.
123. Klatsky AL, Friedman GD, Siegelaub AB, Gerard MJ. Alcohol consumption and blood pressure. N Engl J Med 1977;296: 1194-200.
124. Klatsky AL, Friedman GD, Siegelaub AB. Alcohol and mortality. A ten-year Kaiser-Permanente Experience. Ann Intern Med 1981;95: 139-45.
125. Klatsky AL, Friedman GD, Siegelaub AB. Alcohol and mortality. A ten-year Kaiser-Permanente Experience. Annals of Internal Medicine 1981;95: 139-45.
126. Klatsky AL, Friedman GD, Siegelaub AB. Alcohol consumption before myocardial infarction. Results from the Kaiser-Permanente epidemiologic study of myocardial infarction. Ann Intern Med 1974;81: 294-301.
127. Klatsky AL. The relationship of alcohol and the cardiovascular system. In: Biochemical processes and consequences of alcohol use. Rockville, TN: NIAAA, 1982: 173-209. (Alcohol and Health Monograph 2.)
128. Kluft C, Veenstra J, Schaafsma G, Pikaar NA. Regular moderate wine consumption for five weeks increases plasma activity of the plasminogen activator inhibitor 1 (PAI-1) in healthy young volunteers. Fibrinolysis 1990;4(suppl 2): 69-70.
129. Kono S, Ikeda M, Ogata M, Tokudome S, Nishizumi M, Kuratsune M. The relationship between alcohol and mortality among Japanese physicians. Int J Epidemiol 1983;12: 437-41.
130. Kono S, Ikeda M, Tokudome S, Nishizumi M, Kuratsune M. Alcohol and mortality: A cohort of male Japanese physicians. Int J Epidemiol 1986;15: 527-32.
131. Kontula K, Viinikka L, Ylikorkala O, Ylikahri R. Effect of acute ethanol intake on thromboxane and prostacyclin in human. Life Sci 1982;31: 261-4.
132. Koskinen P, Kupari M, Leinonen H, Luomanmaki K. Alcohol and new onset atrial fibrillatio: a case-control study of a current series. Br Heart J 1987;57: 468-73.
133. Kozararevic D, Vojvodic N, Dawber T, McGee D, Racic Z, Gordon T. Frequency of alcohol consumption and morbidity and mortality: The Yugoslavia cardiovascular disease study. Lancet 1980;i: 613-6.
134. Kromhout D, Nissinen A, Menotti A, Bloemberg B, Pekkanen J, Giampaoli S. Total and HDL cholesterol and their correlates in elderly men in Finland, Italy, and the Netherlands. Am J Epidemiol 1990;131: 855-63.
135. Kuller LH, Hulley SB, LaPorte R, Neston J, Dai WS. Environmental determinants, liver function and high-density lipoprotein cholesterol levels. Am J Epidemiol 1983;117: 406-18.
136. Landolfi R, Steiner M. Ethanol raises prostacyclin in vivo and in vitro. Blood 1984;64: 679-82.
137. LaPorte RE, Cresanta JL, Kuller LH. The relationship of alcohol consumption to atherosclerotic heart disease. Prev Med 1980;9: 22-40.
138. Lian C. L'alcoholisme, cause d'hypertension artérielle. Bull Acad Natl Med Paris 1915;74: 525-8.
139. Lifsic AM. Alcohol consumption and atherosclerosis. Bull WHO 1976;53: 623-30.
140. Lowenstein SR, Gabow PA, Cramer J, Olivia PB, Ratner K. The role of alcohol in new-onset atrial fibrillation. Arch Intern Med 1983;143: 1882-5.

141. Majoor CLH, Hillen HFP. Cardiale beriberi met melkzuuracidose en cardiovasculaire collaps (sjosjin), een bij alcoholici niet zeldzaam ziektebeeld, dat gemakkelijk wordt miskend. Ned Tijdschr Geneesk 1982; 126: 749-57.
142. Majoor CLH. Alcoholism as a cause of beri-beri heart disease. J R Coll Physic Lond 1978;12: 142-52.
143. Malhotra H, Mehta SR, Mathur D, Khandelwal PD. Pressor effects of alcohol in normotensive and hypertensive subjects. Lancet 1985;ii: 584-6.
144. Marmot MG, Rose G, Shipley MJ, Thomas BJ. Alcohol and mortality: A U-shaped curve. Lancet 1981;i: 580-3.
145. Marmot MG, Shipley MJ, Rose G, Thomas BJ. Alcohol and mortality: A U-shaped curve. Lancet 1981;i: 580-3.
146. Masarei JRL, Puddey IB, Rouse IL, Lynch WJ, Vandongen R, Beilin LJ. Effects of alcohol consumption on serum lipoprotein lipid and apolipoprotein concentrations: Results from an intervention study in healthy subjects. Atherosclerosis 1986;60: 79-87.
147. Meade TW, Chakrabarti R, Haines AP, North WRS, Stirling Y. Characteristics affecting fibrinolytic activity and plasma fibrinogen concentrations. Br Med J 1979;1: 153-6.
148. Meade TW, Vickers MV, Thompson SG, Stirling Y, Haines AP, Miller GJ. Epidemiological characteristics of platelet aggregability. Br Med J 1985;290: 428-32.
149. Mikhailidis DP, Jeremy JY, Barradas MA, Green N, Dandona P. Effect of ethanol on vascular prostacyclin prostaglandin I_2 synthesis, platelet aggregation, and platelet thromboxane release. Br Med J 1983;287: 1495-98.
150. Miller NE, Bolton CH, Hayes TM et al. Associations of alcohol consumption with plasma high-density lipoprotein cholesterol and its major subfractions: The Caerphilly and Speedwell Collaborative Heart Disease Studies. J Epidemiol Community Health 1988;42: 220-5.
151. Moore RD, Pearson TA. Moderate alcohol consumption and coronary artery disease. A review. Medicine 1986;65: 242-67.
152. Moore RD, Smith CR, Kwiterovich PO, Pearson TA. Effect of low-dose alcohol use versus abstention on apolipoproteins A1 and B. Am J Med 1988;84: 884-90.
153. Myrhed M. Alcohol consumption in relation to factors associated with ischemic heart disease. Acta Med Scand 1974;195(S): 567.
154. Neiman J, Jones AW, Numminen H, Hillbom M. Combined effect of a small dose of ethanol and 36 hr fasting on blood-glucose response, breath-acetone profiles and platelet function in healthy men. Alcohol Alcoholism 1987;22: 265-70.
155. Nilsson IM, Björkman SE, v. Studnitz W, Hallen A. Antifibrinolytic activity of certain pectins. Thromb Diath Haemorrh 1962;6: 177-87.
156. Olsen H, Østerud B. Effects of ethanol on human blood fibrinolysis and coagulation. Alcohol Alcoholism 1987(suppl 1): 591-4.
157. Orlando J et al. Effect of ethanol on angina pectoris. Ann Int Med 1976;84: 652-5.
158. Pearson TA, Bulkley BH, Achuff SC, Kwiterovich PO, Gordis L. The association of low levels of HDL-cholesterol and arteriographically defined coronary artery disease. Am J Epidemiol 1979;109: 285-95.
159. Periti M, Salvaggio A, Quaglia G, di Marzio L. Alcohol consumption and blood pressure: an Italian Study. Eur J Epidemiol 1988;4: 477-81.
160. Petitti DB, Wingerd J, Pellegrin F, Ramcharan S. Risk of vascular disease in woman. Smoking, oral contraceptives, noncontraceptive estrogens, and other factors. JAMA 1979;242: 1150-54.
161. Phillips NR, Havel RJ, Kane JP. Serum apolipoprotein A-1 levels. Relationship to lipoprotein lipid levels ans selected demographic variables. Am J Epidemiol 1982;116: 302-13.
162. Pikaar NA, Wedel M, van der Beek E et al. Effects of moderate alcohol consumption on platelet aggregation, fibrinolysis, and blood lipids. Metabolism 1987;36: 538-47.
163. Pincherle G, Robinson D. Mean blood pressure and its relation to other factors determined at a routine executive health examination. J Chronic Dis 1974;27: 245-60.
164. Portal RW. Alcoholic heart disease. Br Med J 1981;283: 1202-3.
165. Portal RW. Alcoholic heart disease. Br Med J 1981;283: 1202-3.
166. Potter JF, Beevers DG. Pressor effects of alcohol in hypertension. Lancet 1984;i: 119-22.
167. Potter JF, Macdonald IA, Beevers DG. Alcohol raises blood pressure in hypertensive patients. J Hypertension 1986;4: 435-41.

168. Potter JF, Watson RDS, Skan W, Beevers DG. The pressor and metabolic effects of alcohol in normotensive subjects. Hypertension 1986;8: 625-31.
169. Puddey IB, Beilin LJ, Vandongen R. Regular alcohol use raises blood pressure in treated hypertensive subjects: A randomised controlled trial. Lancet 1987;i: 647-51.
170. Puddey IB, Vandongen R, Beilin LJ, Rouse IL. Alcohol stimulation of renin release in man: its relationship to the haemodynamic, electrolyte, and sympathoadrenal responses to drinking. J Clin Endocrinol Metab 1985;61: 37-42.
171. Ramsay LE. Alcohol and myocardial infarction in hypertensive men. Am Heart J 1979;98: 402-3.
172. Regan TJ et al. The role of ethanol in cardiac disease. Ann Rev Med 1977;28: 393-409.
173. Regan TJ et al. Ventricular function in nonacrdiacs with alcoholic fatty liver: role of ethanol in the production of cardiomyopathy. J Clin Invest 1969;48: 397-407.
174. Rogers S, Yarnell JWG, Fehily M. Nutritional determinants of haemostatic factors in the Caerphilly study. Eur J Clin Nutr 1988;42: 197-205.
175. Rosenberg L, Slone D, Shapiro S, Kaufman DW, Miettinen OS, Stolley PD. Alcoholic beverages and myocardial infarction in young woman. Am J Publ Health 1981;71: 82-5.
176. Ross MA, Alcohol and malnutrition in the pathogenesis of experimental alcoholic cardiomyopathy. J Path 1980;130; 105-16.
177. Ross RK, Mack TM, Paganini-Hill A, Arthur M, Henderson BE. Menopausal oestrogen therapy and protection from death from ischaemic heart disease. Lancet 1981;i: 858-60.
178. Russek HI, Naegele CF, Regan FD. Alcohol in the treatment of angina pectoris. JAMA 1950;143: 355-7.
179. Salonen JT, Puska P, Nissinen A. Intake of spirits and beer and risk of myocardial infarction and death. A longitudinal study in eastern Finland. J Chron Dis 1983;36: 533-43.
180. Salonen JT, Seppanen K, Rauramaa R. Serum high-density lipoprotein cholesterol subfractions and the risk of acute myocardial infarction: A population study in eastern Finland (Abstr). Circulation 1988;78 (suppl II): 11281.
181. Schmidt W, Popham RE. Alcohol consumption and ischemic heart disease: Some evidence from population studies. Br J Addict 1981;76: 407-17.
182. Shaper AG, Wannamethee G, Walker M. Alcohol and mortality in British men: explaining the U-shaped curve. Lancet 1988;ii: 1267-73.
183. Sirtori CR et al. Diet, lipids and lipoproteins in patients with peripheral vascular disease. Am J. Med Sci 1974;268: 325-32.
184. St.Leger AS, Cochrane AL, Moore F. Factors associated with cardiac mortality in developed countries with particular reference to the consumption of wine. Lancet 1979;i: 1017-20.
185. Stamford BA, Matter S, Fell RD, Sady S, Cresanta MK, Papanek P. Cigarette smoking, physical activity, and alcohol consumption: Relationship to blood lipids and lipoproteins in premenopausal females. Metabolism 1984;33: 585-90.
186. Stampfer MJ, Colditz GA, Willett WC, Speizer FE, Hennekens CH. A prospective study of moderate alcohol consumption and the risk of coronary disease and stroke in women. N Engl J Med 1988;319: 267-73.
187. Stampfer MJ, Colditz GA, Willett WC, Speizer FE, Hennekens CH. A prospective study of moderate alcohol consumption and the risk of coronary heart disease and stroke in women. New Engl J Med 1988;319: 267-73.
188. Stampfer MJ, Sacks FM, Hennekens CH. A prospective study of lipids and apoproteins and risk of myocardial infarction (Abstr). Circulation 1988;78(suppl II): 11281.
189. Stason WB, Neff RK, Miettinen OS, Jick H. Alcohol consumption and nonfatal myocardial infarction. Am J Epidemiol 1976;104: 603-8.
190. Stokes GS. Hypertension and alcohol: is there a link? J Chron Dis 1982;35: 759-62.
191. Stott DJ, Ball SG, Inglis GC et al. Effects of a single moderate dose of alcohol on blood pressure, heart rate and associated metabolic and endocrine changes. Clin Sci 1987;73: 411-6.
192. Suhonen O, Aromaa A, Reunanen A, Knekt P. Alcohol consumption and sudden coronary death in middle-aged Finnish men. Acta Med Scand 1987;221: 335-41.
193. Sumi H, Hamada H, Tsushima H, Mihara H. Urokinase-like plasminogen activator increased in plasma after alcohol drinking. Alcohol Alcoholism 1988;23: 33-43.
194. Takizawa A et al. Variant angina induced by alcohol ingestion. Am Heart J 1984;107: 25-7.

195. Taskinen MR, Välimäki M, Nikkilä EA, Kurisi T, Ehuholm C, Ylikahri R. High-density lipoprotein subfractions and postheparin plasma lipases in alcoholic men before and after alcohol withdrawal. Metabolism 1982;31: 1168-73.
196. Taylor KG, Carter TJ, Valente AJ, Wright AD, Smith HJ, Matthews KA. Sex differences in the relationship between obesity, alcohol consumption and cigarette smoking in serum lipid and apolipoprotein concentrations in a normal population. Atherosclerosis 1981;38: 11-18.
197. Thornton J, Symes C, Heaton K. Moderate alcohol intake reduces bile cholesterol saturation and raises HDL cholesterol. Lancet 1983;ii: 819-21.
198. Ueshima JH, Shimamoto Y, Lida M. Alcohol intake and hypertension among urban and rural Japanese populations. J Chron Dis 1984;37: 585-92.
199. Urbano-Marquez A, Estruch R, Navarro-Lopez F, et al. The effects of alcoholism on skeletal and cardiac muscle. N Engl J Med 1989;320: 409-15.
200. Välimäki M, Nikkilä EA, Taskinen MR, Ylikahri R. Rapid decrease in high-density lipoprotein subfractions and postheparin plasma lipase activities after cessation of chronic alcohol intake. Atherosclerosis 1986;59: 147-53.
201. Välimäki M, Taskinen MR, Ylikahri R, Roine R, Kuusi T, Nikkilä EA. Comparison of the effects of two different dosis of alcohol on serum lipoproteins, HDL-subfractions and apolipoproteins A-I and A-II: a controlled study. Eur J Clin Invest 1988;18: 472-480.
202. Veenstra J, Kluft C, van de Pol H, Dooijewaard G, Schaafsma G. Acute effects of moderate alcohol consumption on fibrinolytic factors in healthy middle-aged men. In preparation.
203. Veenstra J, Kluft C, van de Pol H, Wedel M, Schaafsma G. Effects of moderate alcohol consumption on platelet function, tissue-type plasminogen activator and plasminogen activator inhibitor. Thrombosis Haemostasis 1990;63: 345-8.
204. Veenstra J, Ockhuizen Th, van de Pol H, Wedel M, Schaafsma G. Effects of a moderate dose of alcohol on blood lipids and lipoproteins postprandially and in the fasting state. Alcohol Alcoholism 1990;25: 371-7.
205. Veenstra J, Schenkel JAA, van Erp-Baart AMJ, et al. Alcohol consumption in relation to food intake and smoking habits in the Dutch National Food Consumption Study. Eur J Clin Nutr, in preparation.
206. Veenstra J, van de Pol H, Schaafsma G. Moderate alcohol consumption and platelet aggregation in healthy middle-aged men. Alcohol 1990;7: 547-9.
207. Viamontes JA, Schwerdtfeger T. Hypertension and alcoholics. Curr Alcohol 1981;8: 155-8.
208. Werth J. A little wine for thy heart's sake. Lancet 1980;ii: 1141.
209. Willett W, Hennekens CH, Siegel AJ, Adner MM, Castelli WP. Alcohol consumption and high-density lipoprotein cholesterol in marathon runners. N Engl J Med 1980;303: 1159-61.
210. Williams P, Robinson D, Baily A. High-density lipoprotein and coronary risk factors in normal men. Lancet 1979;i: 72-75.
211. Williams PT, Kraus RM, Wood PD, Albers JJ, Dreon D, Ellsworth N. Associations of diet and alcohol intake with high-density lipoprotein subclasses. Metab Clin Exp 1985;34: 524-30.
212. Wink K. Diagnose und Differentialdiagnose der alkoholischen Myokardiopathie. Intern Praxis 1978;18: 369-76.
213. Witteman JCM, Willet WC, Stampfer MJ, Colditz GA, Kok, FJ, Sacks FM, Speizer FE, Rosner B, Hennekens CH. Relation of moderate alcohol consumption and risk of systemic hypertension in women. Am J Cardiol 1990;65: 633-7.
214. Wolf PA, Kannel WB, Verter J. Current status of stroke risk factors. Neurol Clin 1983;1: 317-343.
215. Yano K, Reed DM, McGee DL. Ten-year incidence of coronary heart disease in the Honolulu Heart Program. Am J Epidemiol 1984;119: 653-66.
216. Yano K, Rhoads GG, Kagan A. Coffee, alcohol and risk of coronary heart disease among Japanese men living in Hawaii. N Engl J Med 1977;297: 405-9.
217. Yarnell JWG, Fehily AM, Milbank J, Kubicki AJ, Eastham R, Hayes TM. Determinants of plasma lipoproteins and coagulation factors in men from Caerphilly, South Wales. J Epidemiol Community Health 1983;37: 137-40.
218. Zwieten PA van. Myocardaandoeningen door alcoholisme. Hart Bull 1980;11: 151-3.

CHAPTER 11

Effects of alcohol on various tissues and organs

E. te Wierik and W.G. Vrij-Standhardt

In other chapters the effects of alcohol consumption on the liver (Chapter 8), the cardiovascular system (Chapter 10), the brain and the nervous system (Chapter 9) have been discussed. In this chapter the effects of alcohol on some other organs will be briefly discussed, in particular the effects on the gastrointestinal tract, the pancreas, the eye and the lung. The effect of alcohol on muscles, blood and bone marrow will also be mentioned.

11.1. The gastrointestinal tract

Alcohol consumption affects the function and structure of the gastrointestinal tract in many direct and indirect ways (23, 57, 81). Alcohol itself affects the mucosa of mouth, oesophagus, stomach and intestines, and modifies the secretion of hormones and enzymes that are involved in digestion and absorption. Changes in the function of liver, pancreas and gall-bladder resulting from excessive alcohol consumption also affect digestion. The poor nutritional status of many alcoholics is a consequence of primary malnutrition (decreased or imbalanced dietary intake) on the one hand, and of secondary malnutrition on the other. The latter term refers to a decreased utilization of nutrients, caused by (among other factors) impaired digestion and absorption. The resulting vitamin deficiencies will in turn cause a further decrease of the function of the gastrointestinal tract, thus causing a vicious circle. In this section the influence of alcohol consumption on the gastrointestinal tract will be discussed, except for the role of alcohol in carcinogenesis which is discussed in Chapter 12.

11.1.1. Mouth and oesophagus

Alcohol affects the salivary glands. Hyperamylasaemia frequently occurs in alcoholics (14), possibly due to to an injurious effect of ethanol on salivary glands or to a decreased renal clearance of amylase (15). Further, chronic alcohol consumption can lead to a reduced salivary secretion rate (16, 46). Since saliva seems to be able to protect the mucosa against the local actions of chemical carcinogens, reduction of the salivary gland function might play a role in carcinogenesis of the upper digestive tract.

In the oesphagus chronic alcohol consumption may cause oesophagitis or oesophageal motility disturbances (37).

11.1.2. Influence of alcohol on the integrity of gastric mucosa

11.1.2.1. Acute effects
After once-only excessive alcohol consumption, both in man and in laboratory animals, distinct lesions of the gastric mucosa (hyperaemia, erosion) are detectable by means of gastroscopy. These lesions will heal after 1 to 3 weeks (58).

In most studies in which the harmful effects of acute alcohol consumption on gastric mucosa were demonstrated, concentrated ethanol solutions or spirits were used (17, 84). Diluted solutions (10%) have hardly any effect on the mucosa (17). According to Davenport (10) the deleterious effect of alcohol depends on the concentration of the alcohol that comes into contact with the gastric mucosa, rather than on the quantity of concentrated alcohol. Both the concentration and the quantity of alcohol have been found to play a role. Kawashima & Glass (34, 35) observed in mice that a 10% ethanol solution, irrespective of the dose, did not cause mucosal lesions, whereas the administration of 25–50% solutions instilled into the stomach caused mucosal erosion. The higher the concentration and (for concentrations of 25% or higher) the larger the quantity, the more extensive and the more severe the lesions observed. The effects of solutions containing 10–25% alcohol have not been reported in these papers. Further research in this intermediate range will be quite interesting in view of the broad variety of alcoholic beverages in this range (wine, sherry, port, long drinks).

11.1.2.2. Chronic effects
In about 66% of the 70 alcoholics studied by Dinoso et al. (13) chronic atrophic gastritis in the gastric antrum was found. Gastritis is an important cause of gastric haemorrhage in alcoholics (20). The mucosa generally heals upon abstinence (78).

Gastric lesions are presumably caused by the increased permeability of the mucosa induced by alcohol. According to Davenport & Arbor (9), administration of alcohol has a barrier-breaking effect: the junctions between the epithelial cells are loosened. Because of the increased permeability, the capacity of the epithelial layer of the gastric mucosa to protect the underlying tissue from the being entered by hydrogen ions is lost. The invading hydrogen ions stimulate other mucosal cells to secrete histamine, which stimulates acid secretion and causes vascular dilatation. In a study in rats an altered gastric mucosal permeability has been observed after the administration of a solution containing 25% alcohol. In this study the presence of alcohol was accompanied with an increased passage of larger agents across the gastric mucosa (73).

In dog experiments the barrier-breaking effect was found only for solutions with an alcohol content of 14% or more, and no effect at all for those containing 8% alcohol or less (8). Acetylsalicylic acid (aspirin) has the same barrier-breaking capacity (10). It is therefore advisable to refrain from taking aspirin in case of alcohol intoxication. However, animal studies have shown that bland irritating agents (e.g. alcohol solutions) have a protective effect towards later administration of strong irritating agents. In human studies it has been observed that the antral injury occurring after a single dose of aspirin is reduced when the administration of aspirin has been preceded by a cocktail of vodka and tomato juice. According to Cohen et al. (5) this protective effect may be mediated by an enhanced production of endogenous prostaglandin after the consumption of the mildly irritating agent.

It has been reported that free radicals may be involved in the pathogenesis of mucosal injury (61, 80), but the mechanisms of action are not known yet.

11.1.3. Influence of alcohol on gastric motility

Both animal experiments and human studies have indicated that acute alcohol administration decreases gastric motility and delays gastric emptying (20, 29, 32, 85). This may cause nausea and vomiting in cases of alcohol intoxication.

The mechanism underlying the influence of alcohol on gastric motility is not clear. It may function via its influence on one or more of the gastrointestinal hormones, such as gastrin or cholecystokinin (50), on smooth muscles or the nervous system, or on the release of neurotransmitters.

11.1.4. Effect of alcohol on gastric acid secretion

Investigations on a possible stimulating effect of alcohol on gastric acid secretion have not produced unambiguous results. In some studies no effect was found (6, 7), whereas in other studies effects were found after the consumption of low-alcohol solutions, beer or wine (11, 27, 38, 43, 69–71).

Recent investigations have shown that particularly red and white wines and beer cause gastrin response and an increased acid secretion. Whisky and ethanol, in comparable concentrations, bring about this effect only to a lower extent or not at all (43, 68, 75).

Stimulation of gastric acid secretion by the ingestion of beer or wine is presumably caused by an enhanced release of gastrin. The mechanism through which alcohol solutions below 5% stimulate gastric acid production is not known yet.

11.1.5. Influence of alcohol on the intestines

The effects of alcohol on the intestines have been the subject of many studies and reviews (51, 56, 86). The structural and functional changes found in the intestines of alcoholics and surveyed in these reviews will be summarized below.
Morphology. Acute ingestion of alcohol causes haemorrhagic erosion of the jejunal villi. Among the effects of chronic consumption of alcohol are shortening of the villi and a decrease of the number of epithelial cells lining the villi (59). This may be due to mitotic inhibition (40). These morphologic changes can presumably be reversed by alcohol abstinence (74).

Permeability. An increased permeability to a number of substances as a consequence of chronic alcohol consumption has been demonstrated. Intact molecules with a molecular weight < 5000 can pass the intestinal mucosa of alcoholics (1). The consequences of this increased permeability may be that small toxic, ordinarily non-absorbable, compounds might be absorbed by alcoholics and accelerate extra-intestinal damage. Otherwise, increased losses of substances from the blood to the gut lumen may occur (86).

Absorption of nutrients by the small intestine. Malabsorption of nutrients in the small intestine is an important factor in malnutrition of alcoholics. Damage to the intestinal mucosa interferes with the physiological absorption of vitamins (especially the B vitamins and folate), minerals, proteins and amino acids (51, 56, 86). Fat malabsorption frequently occurs in alcoholics. This defect disappears when these alcoholics receive an adequate diet, irrespective of whether they continue to use alcohol. Alcohol reduces the absorption of glucose by inhibiting the glucose transport mechanisms (55).

Intestinal enzymes. Alcohol consumption results in a decreased activity of disaccharidases and ATPase activity, and in increased microsomal enzyme activities (51).

11.2. The pancreas

The pancreas is a gland with both external and internal secretion. The external secretion, containing digestive enzymes and bicarbonate, contributes to the digestion of food in the duodenum. The internal secretion, containing insulin and glucagon among other substances, regulates carbohydrate metabolism.

11.2.1. Harmful doses of alcohol in relation to pancreas injury

In several studies (66) it was found that most patients suffering from pancreatitis consumed over 150–175 g alcohol per day. It is, however, not possible to establish a lower 'safe' limit; there appears to be a large individual variability in susceptibility to the development of pancreatitis. The relative risk increases logarithmically with the quantity of alcohol ingested. The duration of the period of (excessive) alcohol consumption until clinical symptoms become apparent is estimated at 3 to 20 years (men 18 years, women 11 years on average). The mean total alcohol consumption before disease becomes manifest has been calculated to be 1166 ± 176 litres for men, and 605 ± 195 litres for women (79).

Although acute doses of alcohol have been found to induce changes in volume and composition of the pancreatic secretion in vitro (31, 76), in animal experiments (3, 83) and in human studies (12, 18, 44), there are no indications that moderate quantities of alcohol are harmful to the pancreas. This will still have to be confirmed by specific investigations, however.

11.2.2. Pancreatitis

Alcoholism is an important cause of pancreatitis, in which both a chronic and an acute inflammation may be found. More than 75% of patients with chronic pancreatitis have a history of heavy alcohol consumption (82).

Three major hypotheses have been posed to explain the pathogenesis of pancreatitis (72).

– *Flow/reflux hypothesis*

Duodenopancreatic reflux. According to this hypothesis, alcohol reduces the tone of the sphincter of Oddi, thereby facilitating reflux of duodenal contents into the pancreatic duct. Enterokinase, present in the duodenal contents, activates intraductal pancreatic enzymes that diffuse into the pancreatic interstitium, thereby causing acute pancreatitis.

Biliary pancreatic reflux. Proponents of this theory postulate that ethanol causes spasms of the sphincter of Oddi, thereby creating a common channel between bile and pancreatic duct. This permits reflux of bile into the pancreatic duct and causes pancreatitis.

– *Ductal-plug hypothesis*

This hypothesis proposes that secretion of pancreatic juice rich in proteins may 'plug' the small ductules, leading to obstruction, inflammation and, consequently, degeneration of the area drained by these ductules, thus inducing chronic pancreatitis.

– *Toxic metabolic hypothesis*

This hypothesis suggests that ethanol has a direct injurious effect, leading ultimately to pathophysiologic changes culminating in pancreatitis.

Because none of the concepts outlined above are mutually exclusive, it may be true that most or all of the changes proposed by these hypotheses occur at different stages during progression of ethanol injury to the pancreas (72).

The role of nutritional factors in chronic pancreatitis is uncertain. Sarles (66) has hypothesized that a diet rich in fat and proteins predisposes alcoholics to pancreatitis. However, Mezey et al. (52), who studied pancreatitis in subjects who consumed more than 50% of their energy as alcohol, concluded that a high dietary intake of protein and fat is not a factor in the development of chronic pancreatitis in alcohol-dependent individuals.

In Stockholm county the incidences of hospitalization for chronic and acute pancreatitis have declined during the past decades. Moreover, the apparent national and Stockholm prevalence of chronic pancreatitis declined in the same period. A concomitant decline in the Stockhol sales figures of spirits was also observed, whereas the sales for wine and beer increased. From these data Schmidt (67) concludes that the consumption of spirits, but not of beer and wine, appears to be a risk factor for pancreatitis. This interpretation indicates the population at high risk to be the chronic alcoholics, who are known to consume predominantly spirits.

11.3. The eye

11.3.1. Effect of acute alcohol consumption on the eye function

The inhibition of brain activity caused by alcohol also affects vision. In acute alcohol consumption motor disturbance will occur, including reduction of velocity of eye movements, impairment of convergence and induction of nystagmus (33, 53, 77). At a blood alcohol concentration (BAC) of 0.8 g/l accommodation and convergence is disturbed to such a degree that depth vision is impeded (26). Double vision, well known to be attendant on alcohol intoxication, may also occur. Gramsberg-Danielsen & Lindner (22) have investigated the effect of alcohol consumption on vision in relation to justifiable participation in traffic. Vision is affected most before the maximum BAC has been reached.

11.3.2. Effect of chronic alcohol consumption on the eye

Alcoholism may be accompanied with disturbances of the motor activity of the eye. Chronic consumption of alcohol may result in histological changes or in premature ageing of the brain including the oculomotor system (39). The 29 patients with a slowly progressing decrease of vision examined by Henkes (26) all reported to drink at least 150 g alcohol daily. Heavy smoking and poor dietary habits were also observed in most of them. It is not clear whether alcohol has a direct toxic effect on the optic nerve or whether vitamin deficiency (vitamin A (63), vitamin B-12 (4, 26)) is a causal factor. Cyanide, present in tobacco smoke and beer, may also cause this opticoneuropathy, formerly also referred to as alcohol- or tobacco-induced amblyopia (4, 26). Local retinol deficiency going with alcohol consumption plays a part in causing night blindness (see Chapter 6).

Heavy drinking (more than 4 glasses/day) may be a risk factor for cataract (24). There is a risk of confounding between heavy drinking, cigarette smoking, and work on a military base, but analyses have indicated that these three risk factors are independent (25).

11.4. The lung

Since alcohol is excreted in significant quantities by the lung and since alcohol is toxic to several organs is seems plausible that alcohol affects the lung as well. However, since most heavy drinkers are also smokers it is to be expected that their pulmonary function will be abnormal. Therefore it is difficult to assess the effects of alcohol per se on the lung.

No significant difference in pulmonary function between alcoholics and their matched controls was shown by Lyons et al. (45). The authors conclude that the high prevalence of respiratory disease in alcoholics is largely attributable to their smoking habits. No evidence of a specific pulmonary toxic effect of alcohol was identified. The relationship between alcohol consumption and respiratory symptoms has also been studied by Lebowitz (42). In this study heavy drinkers were stratified by smoking status in order to clarify the independent effect of alcohol consumption. The author observed that heavy drinkers (> 17 glasses/week) had more productive cough, wheeze, and reported more acute respiratory illness than other drinkers. In this study smoking was found to be a far more important contributor to respiratory symptoms, but alcohol is a significant risk factor as well. No relationship was observed between alcohol consumption and lung function. In a recent longitudinal study alcohol was found to be a significant independent predictor of decline in forced expiratory volume (FEV). The authors of this study also looked after the interaction between alcohol consumption and smoking. Alcohol consumption in smokers led to a slower decrement in FEV than expected as a result of smoking alone (41). These results have been confirmed in a case-control study (19).

11.5. The muscles

11.5.1. Myopathy

During the past century muscle dysfunction resulting from chronic alcohol abuse accompanied by weakness of the limbs was described for the first time. Later on, more cases of myopathy in alcoholics were reported (65). Chronic alcohol consumption causes morphological changes of the skeletal muscle. Atrophy of fibres (especially type IIb fibre), reduction of the protein/DNA ratio, and reduction of myofibrillary Ca^{2+}-ATPase activity have been observed in alcoholics. According to Martin et al. (47), these results suggest a selective loss of type II fibre myofibrillary protein. Athrophy of type II fibres after alcohol consumption in alcoholics has been observed in several other studies (54, 62)). Type II muscle fibre atrophy is often reversible and is directly related to alcohol consumption, not to malnutrition, vitamin deficiency or peripheral neuropathy (28, 48, 60). In animal studies it has been found that alcohol feeding significantly reduces the RNA content. This loss of protein synthetic apparatus may be an important event in the development of skeletal muscle myopathy (49).

Perkoff (58a) classified alcoholic myopathy into three forms: a subclinical, an acute and a chronic form. The first two forms are presumably entirely, and the last one is partly reversible upon abstinence (65).

11.5.2. Harmful amounts of alcohol in relation to myopathy

Research on alcoholic myopathy has almost exclusively been done with alcoholics. Roughly speaking, excessive alcohol consumption may be said to go with alcoholic myopathy almost without exception. In Rubin's experiment (64) 225–260 g/day was consumed, which corresponds with a daily consumption of 22–26 drinks. These are clearly no examples of 'moderate' alcohol consumption!

From literature reviews it becomes apparent that the development of reversible selective atrophy of type IIb fibres may result from the consumption of amounts exceeding 100 g alcohol/day for at least three years (30).

There are no indications of moderate alcohol consumption causing muscle damage. On the other hand, no studies reporting harmlessness of alcohol to muscle have come to our knowledge either. Also in this field studies on the effects of moderate doses of alcohol are needed, e.g. in sportsmen.

11.6. Blood and bone marrow

The effects of chronic alcoholism on bone marrow, erythrocytes and leucocytes have currently been summarized by Ballard (2) and by others (21, 56). It has been noted that the occurrence of vacuolated pronormoblasts (precursor cells of erythrocytes) is the most striking effect of alcohol on marrow cells. This vacuolization disappears during the first week of abstinence. The rapid reversibility of vacuolation is consistent with a direct toxic effect of alcohol. An elevated erythrocyte mean corpuscular volume (MCV) occurs frequently in heavy alcohol consumers. This may be the consequence of membrane damage to primitive bone marrow cells. Several forms of anaemia have been noted to be associated with alcoholism, including haemolytic anaemia (increased rate of red cell attrition or destruction). Thrombocytopenia (decreased platelet count) and impaired platelet function have also been associated with excessive alcohol consumption. Little is known about the effects of alcohol on white blood cells. It has, however, een noted that an ineffective immune system may be a consequence of alcoholism.

11.7. Other organs

In case of severe continuous alcohol abuse the total function of the body will be disturbed to such an extent that in addition to the dysfunctions mentioned in other chapters, damage may be caused to almost every other organ. This damage will not be a consequence of the direct toxicity of alcohol as a rule, but of secondary disturbances of the organism. These disturbances are beyond the cope of this literature review and will not be discussed here.

The quantity of alcohol causing this damage to occur depends on individual sensitivity, but generally exceeds or equals the quantity which may injure the liver, heart or brain.

Summary

Alcohol consumption affects the function and structure of the gastrointestinal tract in many direct and indirect ways. Alcohol decreases gastric motility, delays gastric emptying and possibly affects the secretion of gastrin and gastric juice (indirectly).

Acute alcohol consumption may cause damage to the oral, oesophageal, gastric and duodenal mucosa ('barrier breaker'). The degree of the damage caused depends on the alcohol concentration of the solution that comes into contact with the mucosa, and consequently on the concentration and quantity (with respect to dilution) of the beverage consumed.

Alcoholism is one of the main causes of pancreatitis. In many instances the injury manifests itself as acute pancreatitis which may be considered the first shift of chronic relapsing pancreatitis. Three hypotheses with regard to the pathogenesis of pancreatitis in alcoholics have been posed. An acute dose of alcohol alters the secretion of pancreatic juice. Evidence of the harm, if any, of the consumption of moderate quantities of alcohol caused to the pancreas is not available.

Alcohol causes a reversible disturbance of visual motor activity. Chronic excessive alcohol consumption affects the optic nerve and may cause night blindness. Cataract may be a consequence of heavy drinking.

A decline in forced expiratory volume and occurrence of respiratory symptoms may be related to excessive consumption of alcohol. However it is difficult to establish the independent role of alcohol consumption since most heavy drinkers are also smokers.

Prolonged excessive alcohol consumption may lead to morphological changes of the muscles. Atrophy of type IIb fibres has frequently been observed. This may cause subclinical, acute or chronic myopathy. There is no evidence that moderate alcohol consumption leads to muscle injury.

The effects of alcohol on blood are diverse. Abnormalities of erythrocytes, platelets and leucocytes may occur separately or in various combinations.

Functioning of the body in its entirety may be disturbed by chronic excessive alcohol consumption to such an extent that indirect damage may be caused to almost every organ. The quantity of alcohol which may cause this damage depends on the individual's susceptibility, but usually corresponds to those quantities which cause injuries of the liver, the heart or the brain.

References

1. Anonymous. The leaky gut of alcoholism. Nutr Rev 1985;43: 72-4.
2. Ballard HS. Hematological complications of alcoholism. Alcohol Clin Exp Res 1989;13: 706-20.
3. Bayer M, Rudick J, Lieber SC, Janowitz HD. Inhibitory effect of ethanol on canine exocrine pancreatic secretion. Gastroenterology 1972;63: 619-26.
4. Brontë-Stewart J, Pettigrew AR, Foulds WS. Toxic optic neuropathy and its experimental production. Trans Ophtalmol Soc UK 1976;96: 355-8.
5. Cohen MM, Young R, Kilam S, Wang HR. Aspirin-induced human antral injury is reduced by vodka pretreatment. Dig Dis Sci 1988;33: 513-7.
6. Cooke AR. The simultaneous emptying and absorption of ethanol from the human stomach. Am J Dig Dis 1970;15: 449-54.
7. Cooke AR. Ethanol and gastric function. Gastroenterology 1972;62: 501-2.
8. Davenport HW. Ethanol damage to canine oxyntic glandular mucosa. Proc Soc Exp Biol Med 1967;126: 657-62.

9. Davenport HW, Arbor A. Salicylate damage to the gastric mucosal barrier. N Eng J Med 1967;276: 1307-12.
10. Davenport HW. Gastric mucosal hemorrage in dogs: effects of acid, aspirin, and alcohol. Gastroenterology 1969;56: 439-49.
11. Demol P, Singer MV, Hotz J, Eysselein V, Goebell H. Different actions of intravenous ethanol on basal (= interdigestive) secretion of gastric acid, pancreatic enzymes and bile acids and gastrointestinal motility in man. Alcohol Alcoholism 1985;20: 19-26.
12. Demol P, Singer MV, Hotz J, Hoffman U, Hanssen LE, Eyssellein VE, Goebell H. Action of intragastric ethanol on pancreatic exocrine secretion in relation to the interdigestive gastrointestinal motility in humans. Arch Int Physiol Biochem 1986;94: 251-9.
13. Dinoso VP, Chey WY, Braverman SP, Rosen AP, Ottenberg D, Lorber SH. Gastric secretion and gastric mucosal morphology in chronic alcoholics. Arch Intern Med 1972;130: 715-9.
14. Domzalski CA, Wedge BM. Elevated serum amylase in alcoholics. Am J Clin Path 1948;18: 43-7.
15. Dutta SK, Douglass W, Smalls UA, Nipper HC, Levitt MD. Prevalence and nature of hyperamylasemia in acute alcoholism. Dig Dis Sci 1981;26: 136-41.
16. Dutta SK, Dukehart M, Narang A, Latham PS. Functional and structural changes in parotid glands of alcoholic cirrhotic patients. Gastroenterology 1989;96: 510-8.
17. Eastwood GL, Kirchner JP. Changes in the fine structure of mouse gastric epithelium produced by ethanol and urea. Gastroentrology 1974;67: 71-84.
18. Fried GM, Ogden WD, Zhu XG, Greeley GH, Thompson JC. Effect of alcohol on the release of cholecystokinin and pancreatic enzyme secretion. Am J Surg 1984;147: 53-7.
19. Garshick E, Segal MR, Worobec TG, Salekin CMS, Miller MJ. Alcohol consumption and chronic obstructive pulmonary disease. Am Rev Resp Dis 1989;140: 373-8.
20. Glass GBJ, Slomiany BL, Slomiany A. Biochemical and pathological derangements of the gastrointestinal tract following acute and chronic ingestion of ethanol. In: Majchrowicz E, Noble EP, eds. Biochemistry and pharmacology of ethanol. New York: Plenum Press, 1979;1: 551-86.
21. Goudsmit R. Haematologische afwijkingen ten gevolge van overmatig alcoholgebruik. Tijdschr Alc Drugs 1989;15: 210-2.
22. Gramsberg-Danielsen B, Lindner HJ. Zur Beziehung zwischen Blutalkoholkurve und Lichtunterschiedsempfindlichkeit, Blendempfindlichkeit und Readaptionszeit. Klin Monatsbl Augenheilk 1972;161: 710-5.
23. Green PHR. Alcohol, nutrition and malabsorption. Clin Gastroenterol 1983;12: 563-74.
24. Harding JJ, Van Heyningen R. Drugs, including alcohol, that act as risk factors for cataract, and possible protection against cataract by aspirin-like analgesics and cyclopenthiazide. Br J Ophtalmol 1988;72: 809-14.
25. Harding JJ, Van Heyningen R. Beer, cigarettes and military work as risk factors for cataract. In: Hockwin O, Sasaki K, Leske MC, eds. Risk for cataract development. Dev Ophtalmol 1989;17: 13-6.
26. Henkes HE. Stoornissen in het gezichtsvermogen als gevolg van ethylalcoholgebruik. Ned Tijdschr Geneesk 1979;123: 1260-64.
27. Hirschowitz BI, Pollard HM, Hartwell SW, London J. The action of ethyl alcohol on gastric acid secretion. Gastroenterology 1956;30: 244-53.
28. Hodges DL, Kumar VN, Redford JB. Effects of alcohol on bone, muscle and nerve. Am Family Physician 1986;34: 149-56.
29. Horowitz M, Maddox A, Bochner M, et al. Relationships between gastric emptying of solid and caloric liquid meals and alcohol absorption. Am J Physiol 1989;257: G291-G298.
30. Hudgson P. Alcoholic myopathy. Br Med J 1984;288: 584-5.
31. Iwatsuki K, Chiba S. Effects of ethanol on the isolated blood-perfused canine pancreas. J Stud Alcohol 1981;42: 811-3.
32. Jian R, Cortot A, Ducrot F, Jobin G, Chayvialle JA, Modigliani R. Effect of ethanol ingestion on postprandial gastric emptying and secretion, biliopancreatic secretions, and duodenal absorption in man. Dig Dis Sci 1986;31: 604-14.
33. Katoh Z. Slowing effects of alcohol on voluntary eye movements. Aviat Space Envir Med 1988;59: 606-10.

34. Kawashima K, Glass GBJ. Alcohol injury to gastric mucosa and its potentiation by stress. Gastroenterology 1973;64: 752A.
35. Kawashima K, Glass GBJ. Alcohol injury to gastric mucosa and its potentiation by stress. Am J Dig Dis 1975;20: 162-172.
36. Keshavarzian A, Iber FL, Greer P, Wobbleton J. Gastric emptying of solid meal in male chronic alcoholics. Alcohol Clin Exp Res 1986;10: 432-5.
37. Keshavarzian A, Iber FL, Ferguson Y. Esophageal manometry and radionuclide emptying in chronic alcoholics. Gastroenterology 1987;92: 651-7.
38. Kölbel CBM, Singer MV, Dorsch W, et al. Pancreatic and gastric responses to gastric versus jejunal beer in humans. Pancreas 1988;3: 89-94.
39. Kotabake K, Yoshii F, Shinohara Y, Nomura K, Takagi S. Impairment of smooth pursuit eye movement in chronic alcoholics. Eur Neurol 1983;22: 392-6.
40. Landsdown ABG, Dayan AD. Alterations in crypt cell populations in the small intestine as an early toxic response to sub-acute ethanol administration. Arch Toxicol 1987;59: 448-52.
41. Lange P, Groth S, Mortenson J, et al. Pulmonary function is influenced by heavy alcohol consumption. Am Rev Respir Dis 1988;137: 1119-23.
42. Lebowitz MD. Respiratory symptoms and disease related to alcohol consumption. Am Rev Respir Dis 1981;123: 16-9.
43. Lenz HJ, Ferrari-Taylor J, Isenburg JI. Wine and 5 percent ethanol are potent stimulants of gastric secretion in humans. Gastroenterology 1983;85: 1082-7.
44. Llanos OL. Swierczek JS, Teichman RK, Rayford PL, Thompson JC. Effect of alcohol on the release of secretin and pancreatic secretion. Surgery 1977;81: 661-7.
45. Lyons DJ, Howard SV, Milldge JS, Peters TJ. Contributions of ethanol and cigarette smoking to pulmonary dysfunction in chronic alcoholics. Thorax 1986;41: 197-202.
46. Maier H, Born IA, Mall G. Effect of chronic ethanol and nicotine consumption on the function and morphology of the salivary glands of alcoholic cirrhotic patients. Gastroenterology 1989;96: 510-8.
47. Martin FC, Slavin G, Levi AJ, Peters TJ. Investigation of the organelle pathology of skeletal muscle in chronic alcoholism. J Clin Pathol 1984;37: 448-54.
48. Martin F, Ward K, Slavin G, Levi J, Peters TJ. Alcoholic skeletal myopathy, a clinical and pathological study. Q J Med 1985;55: 233-51.
49. Marway JS, Preedy VR, Peters TJ. Experimental alcoholic skeletal muscle myopathy is characterized by a rapid and sustained decrease in muscle RNA content. Alcohol Alcoholism 1990;25: 401-6.
50. Mezey E, Halsted CH. Effects of alcohol on gastrointestinal and pancreatic function in alcoholics. In: Gastineau CG, Darby WJ, Turner TB, eds. Fermented food beverages in nutrition. New York: Academic Press, 1979: 277-392.
51. Mezey E. Effect of ethanol on intestinal morphology, metabolism and function. In: Seitz HK, Kommerell B, eds. Alcohol related diseases in gastroenterology. Berlin: Springer-Verlag, 1985: 342-60.
52. Mezey E, Kolman CJ, Diehl AM, Mitchell MC, Herlong HF. Alcohol and dietary intake in the development of chronic pancreatitis and liver disease in alcoholism. Am J Clin Nutr 1988;48: 148-51.
53. Miller RJ, Pigion RD, Takahama M. The effect of ingested alcohol on accommodative, fusional, and dark vergence. Perception Psychophys 1986;39: 25-31.
54. Mills KR, Ward K, Martin F, Peters TJ. Peripheral neuropathy and myopathy in chronic alcoholism. Alcohol Alcoholism 1986;21: 357-62.
55. Money SR, Petroianu A, Kimura K, Jaffe BM. The effects of short-term ethanol exposure on the canine jejunal handling of calcium and glucose. Surgery 1990;107: 167-71.
56. Moore DT. Reversal of alcohol effects, acute and chronic conditions. Alcohol Health Res World 1986;11: 52-9.
57. Nazer H, Wright RA. The effect of alcohol on the human alimentary tract: a review. J Clin Gastroenterol 1983;5: 361-5.
58. Palmer ED. Gastritis: a revaluation. Medicine 1954;33: 199-290.
58a. Perkoff GT. Alcoholic myopathy. Ann Rev Med 1971;22: 125-32.

59. Persson J, Berg NO, Sjölund K, Stenling R, Magnusson PH. Morphologic changes in the small intestine after chronic alcohol consumption. Scand J Gastroenterol 1990;25: 173-84.
60. Peters TJ, Martin F, Ward K. Chronic alcoholic skeletal myopathy, common and reversible. Alcohol 1985;2: 485-9.
61. Pihan G, Regillo C, Scabo S. Free radicals and lipid peroxidation in ethanol- or aspirin-induced gastric mucosal injury. Dig Dis Sci 1987;32: 1395-401.
62. Preedy VR, Bateman CJ, Salisbury JR, Price AB, Peters TJ. Ethanol-induced skeletal muscle myopathy: biochemical and histochemical measurement on type I and type II fibre rich muscles in the young rat. Alcohol Alcoholism 1989;24: 533-9.
63. Roe DA. Drug-induced nutritional deficiencies. Westport, CT: AVI, 1976.
64. Rubin M, Katz AM, Lieber CS, Stein EP, Puszkin S. Muscle damage produced by chronic alcohol consumption. Am J Path 1976;83: 499-516.
65. Rubin E. Metabolic and pathological changes in muscle during acute and chronic administration of ethanol. In: Majchrowicz E, Noble EP. Biochemistry and pharmacology of ethanol. New York: Plenum Press, 1979;1: 623-40.
66. Sarles H. An international survey on nutrition and pancreatitis. Digestion 1973;9: 389-403.
67. Schmidt DN. Apparent risk factors for chronic and acute pancreatitis in Stockholm county. Int J Pancreatol 1991;8: 45-50.
68. Singer MV, Eysselein V, Goebell H. Beer and wine but not whisky and pure ethanol do stimulate release of gastrin in humans. Digestion 1983;26: 73-9.
69. Singer MV, Calden H, Eysselein VE, Leffmann C, Goebell H. Low concentrations of ethanol stimulate gastric acid secretion independent of gastrin release in humans. Gastroenterology 1984;86: 1254.
70. Singer MV, Leffmann CJ, Schnober D, Goebell H. Beer and wine but not whisky and cognac stimulate gastric acid secretion in humans. Gastroenterology 1986;88: 1588.
71. Singer MV, Leffmann C, Eysselein VE, Calden H, Goebell H. Action of ethanol and some alcoholic beverages on gastric acid secretion and release of gastrin in humans. Gastroenterology 1987;93: 1247-54.
72. Singh M, Simsek H. Alcohol and the pancreas; current status. Gastroenterology 1990;98: 1051-62.
73. Sjödahl R, Andersson PÅ, Tagesson C. Alcohol and gastric mucosas permeability to different-sized molecules. Scand J Gastroenterol 1984;19: 75-7.
74. Sjölund K, Persson J, Bergman L. Can villous atrophy be induced by chronic alcohol consumption? J Intern Med 1989;226: 133-5.
75. Snelten H, te Wierik E, Veenstra J, Schaafsma G. De invloed van bier en wijn op de gastrinespiegel. Voeding 1991;52: 86-9.
76. Solomon N, Solomon TE, Jacobson ED, Shanbour LL. Direct effect of alcohol on in vivo and in vitro exocrine pancreatic secretion and metabolism. Am J Dig Dis 1974;19: 253-60.
77. Stapleton JM, Guthrie S, Linnoila M. Effects of alcohol and other psychotropic drugs on eye movements: relevance to traffic safety. J Stud Alcohol 1986;47: 426-32.
78. Stuyt PMJ, van Tongeren JHM. Alcohol en maagdarmziekten. Ned Tijdschr Geneesk 1979;123: 1255-9.
79. Teichmann W, Zastrow R. Alkohol und Pankreas. Z Ges Inn Med 1981;36: 567-71.
80. Terano A, Hiraishi H, Ota S, Shiga J, Sugimoto T. Role of superoxide and hydroxyl radicals in rat gastric mucosal injury induced by ethanol. Gastroenterol Jpn 1989;24: 488-93.
81. Thomson AD, Majumdar SK. The influence of ethanol on intestinal absorption and utilization of nutrients. Clin Gastroenterol 1981;10: 263-93.
82. Van Thiel DH, Lipsitz HD, Porter LE, Schade RR, Gottlieb GP, Graham TO. Gastrointestinal and hepatic manifestations of chronic alcoholism. Gastroenterology 1981;81: 594-615.
83. Walton B, Shapiro H, Woodward ER. The effect of alcohol on pancreatic secretion. Surg Forum 1960;11: 365-6.
84. Williams AW. Effects of alcohol on gastric mucosa. Br Med J 1956;1: 256-9.
85. Willson CA, Bushnell D, Keshavarzian A. The effect of acute and chronic ethanol administration on gastric emptying in cats. Dig Dis Sci 1990;35: 444-8.
86. World MJ, Ryle PR, Thomson AD. Alcoholic malnutrition and the small intestine. Alcohol Alcoholism 1985;20: 89-124.

CHAPTER 12

Alcohol and cancer

E. te Wierik and W.G. Vrij-Standhardt

The results of the various studies on the relation between alcohol and cancer are not unanimous. A large group of experts, invited by the International Agency of Research on Cancer, have reviewed the literature on this subject and concluded that the occurrence of malignant tumours of the oral cavity, pharynx, larynx, oesophagus and liver is causally related to the consumption of alcoholic beverages (16). However, this conclusion has met with much criticism because it does not account for a dose-effect relationship.

There is some evidence of a relationship between the consumption of alcohol and cancer of the lungs, the pancreas and the breast and between the consumption of beer and intestinal cancer and cancer of the rectum. However, these relationships need not all be causal.

12.1. Epidemiology

12.1.1. Cancer of the upper digestive tract

In reaction to the monograph issued by the International Agency for Research on Cancer (16) mentioned above, Turner et al. (53) reviewed studies into the relationship between moderate drinking and the risk of cancer of the oropharynx, larynx, oesophagus and liver. They concluded that there is no solid and consistent scientific evidence that moderate use (< 0.7 g ethanol per kg body weight per day) of alcoholic beverages is associated with an enhanced risk of these types of cancer. Moreover, they noted that most of the studies reveal cancer of the oropharynx and the larynx to be positively associated with very heavy alcohol use, but that similar associations were noted between these types of cancer and environmental factors like smoking and other tobacco use, and ingestion of very hot drinks. The separate effects of heavy alcohol use and smoking are hard to establish unambiguously.

A case control study of cancer of the oral cavity and cancer of the oropharynx was performed in Torino, Italy. Among men, an effect of alcohol was obvious only in those consuming 120 g alcohol or more per day. For women, a dose-effect relationship was suggested. In both sexes an elevated risk was found for subjects who drank beer whether or not in combination with other alcoholic beverages. However, there was no effect of alcohol in non-smokers. Among heavy users of both alcohol and tobacco the risk of both oral and oropharyngeal cancer was very high. This study corroborates the suggestion of a combined effect of alcohol and tobacco on the risk of oral cancer (31). A study of 178 cases of cancer of the oesophagus in New York confirmed these findings. Both alcohol and smoking independently increased the risk of oesophageal cancer. In a case-control study Tuyns et al. (56) have paid special attention to the risk of various sites of the larynx and hypopharynx. The consumption of more than 80 g alcohol increased the relative risk for cancer of the epilarynx or hypopharynx by 4.3, and the relative risk for cancer of the endolarynx by 2.1. These relative risks were adjusted for smoking and other factors. The combined relative risk among heavy drinkers and smokers was 135.5 for cancer of the epilarynx/hypopharynx and 43.2 for cancer of the endolarynx.

12.1.2. Liver cancer

About 60 to 90% of all cases of liver cancer are associated with cirrhosis (55). Epidemiological studies are equivocal with regard to the causal relation between alcohol consumption and liver cancer (16, 46).

12.1.3. Breast cancer

Epidemiological data regarding the relation between alcohol consumption and the risk of breast cancer are inconsistent. A meta-analysis (a quantitative review) of the available data has been performed by Longnecker et al. (26). The authors found strong evidence for a dose-response relationship in both the case-control and follow-up data. The risk of breast cancer at an daily alcohol intake of 24 g (about 2 drinks) relative to the risk for non-drinkers was 1.4 in the case-control and 1.7 in the follow-up data. Although these findings are not a proof of causality, they support a strong association between alcohol consumption and risk of breast cancer. Stampfer et al. (49) reviewed studies into the relationship between the consumption of alcoholic beverages and breast cancer. They concluded that epidemiological data demonstrate an association between alcohol intake and the risk of breast cancer. This association appeared to hold for all forms of alcoholic beverages. The authors also noted a dose-response effect, but were not able to delineate the shape of the curve. The evidence was insufficiently strong to conclude that the association is of a causal nature. Although some studies do not support the association between alcohol consumption and breast cancer (3, 43, 47) other studies do (14, 35, 42). It has been suggested that the risk of breast cancer may be higher in women who have begun drinking at a later age (35). Assessment of the influence of the age at which alcohol consumption began, the duration of exposure as well as the identification of plausible mechanisms of action would clarify the role of alcohol consumption in breast cancer pathogenesis (20).

12.1.4. Lung cancer

The observed relation between alcohol consumption and lung cancer (39, 40) may largely be attributed to the fact that heavy drinkers often also smoke heavily (25, 50).

Schmidt & Popham (44) found no difference in lung cancer mortality figures between alcoholics and non-alcoholic veterans with similar smoking habits. A direct influence of alcohol on the genesis of lung cancer is considered to be unlikely. In another study in which patients with lung cancer were compared to healthy controls the patients were found to consume significantly more alcohol (beer in particular) and to be more likely to smoke. When the analysis was restricted to cigarette smokers and adjusted for other variables, beer consumpion remained a significant factor. The authors conclude that the association of beer consumption with lung cancer does not appear to be explained solely by confounding by cigarette smoking (14).

12.1.5. Pancreas cancer

Pancreatitis has often been observed in conjunction with pancreatic cancer. Heavy alcohol consumption is strongly associated with pancreatitis. Therefore, it has been suggested that pancreatitis is causally related to pancreatic cancer and that alcohol leads to pancreatic cancer via pancreatitis. Meanwhile, it has been shown that alcohol may alter pancreatic function in various ways not mediated by alcohol-induced pancreatitis. A carcinogenic effect seems plausible. An association between alcohol consumption and pancreas cancer could not been demonstrated consistently and the available evidence is inconclusive (57). The IARC Working Group concluded that the consumption of alcoholic beverages is unlikely to be causally related to cancer of the pancreas (16).

12.1.6. Other types of cancer

Alcohol consumption has been found to be associated with stomach cancer (13), rectal cancer (6, 12, 18, 50) and cancer of the sigmoid colon (12). No relation between prostate cancer (50) and stomach cancer (12, 36, 50) on the one hand and alcohol use on the other was observed in other studies. The associations between alcohol consumption and rectal cancer observed may be attributable to an incomplete control for confounding variables such as socio-economic status and dietary practices (18). With regard to the risk of stomach cancer Boeing et al. (2) observed a positive association with beer consumption and a negative association with wine and liquor consumption. These associations of alcoholic beverages with stomach cancer may reflect a particular life-style rather than a causal relationship.

12.2. Differences between various types of beverages

There are several, mainly epidemiological, indications that the various types of beverages differ in their influence on carcinogenesis. Several epidemiological studies have revealed a relation between the consumption of beer and the incidence of several forms of cancer (2, 10, 39, 40), but this has not been confirmed by other investigations (3, 36, 43). In some studies cancer of the upper digestive tract has been found to be associated specifically with the consumption of either spirits or beer. The IARC Working Group concludes that drinking of alcoholic beverages is causally related to cancers of the oral cavity, pharynx, larynx, oesophagus and liver and that there is no evidence that the effect is dependent on type of beverage. The IARC Working Group notes that some epidemiological data are suggestive of a causal role of alcoholic beverages in rectal cancer. Moreover, this role of alcoholic beverages in rectal cancer could often be explained by beer consumption (16). It is by no means certain, however, whether the observed relationship is causal, i.e. whether beer does actually cause this type of cancer. If this proves to be the case, it should be investigated which substance present in beer is responsible for this effect. The difference between types of beer (e.g. dark/light) may be of importance.

The presence of nitrosamines, known to be carcinogenic, in alcoholic beverages will be discussed in this chapter. In maize beer and its distillates, as well as in calvados, whisky and some brands of beer these substances have been demonstrated, albeit in minor quantities. Their presence in wine and some distilled beverages such as gin and vodka seems unlikely.

12.3. Mechanisms

It is not clear how the influence of alcoholic beverages on the pathogenesis of cancer of the mouth, pharynx, larynx, oesophagus and liver must be explained; the beverages may be carcinogenic in themselves, or be active as a cocarcinogen; either ethanol or other additives may cause those effects.

Indirect effects of alcoholism may also play a part. Regarding carcinogenesis in sites that are directly exposed to alcoholic beverages, a local effect has been suggested. The following mechanisms have been mentioned in order to explain the relationship between alcohol consumption and carcinogenesis (8, 24, 46).

12.3.1. Direct carcinogenic effect of alcoholic beverages

One or more of the many congeners which may be present in alcoholic beverages in greater or smaller quantities, might have a carcinogenic effect. Some seem to do so, other ones do not. Some of these substances will be discussed below.
– *Ethanol.* Ethanol itself is unlikely to be carcinogenic. Experiments performed to induce cancer with ethanol in well-fed laboratory animals without liver injury have failed so far (19).
– *Fusel alcohols.* Fusel alcohols, i.e. the higher alcohols present in all alcoholic beverages in varying quantities, contribute to the specific taste and flavour of these beverages. Several animal experiments have shown that fusel alcohols (in far greater quantities than usual) are carcinogenic (9, 19, 37, 54).
– *Tannins.* On the basis of correlation studies Morton (33) has suggested that the absorption of certain types of tannin from sorghum (used in preparing kaffir beer) is associated with a high incidence of cancer of the oesophagus. The action and activity

of different types of tannin vary widely. Consequently, it would not be justified to consider tannins (found in wine and other beverages) as carcinogenic.

– *Nitrosamines.* Small quantities of nitrosamines, known to be carcinogenic, have been found in African maize beer and distilled maize beverages (29) while in the areas where these beverages are consumed, the incidence of cancer of the oesophagus is high (4, 37). In cider distillates, too, slight quantities of nitrosamines have been found. Some suggest this as a possible explanation of the fact that mortality from cancer of the oesophagus in France is highest in Normandy and Brittany (54, 55). Whisky and some types of beer (especially the dark types), contain (as does cigarette smoke (30)) traces of nitrosamines (19, 48, 58) which are formed when the malt is roasted. Furthermore, the presence of ethanol seems to catalyse the production of nitrosamines from their constituent compounds, namely nitrites and secondary amines, under the conditions encountered in the upper gastrointestinal tract (38).

– *Other congeners or contaminants.* Alcoholic beverages are extremely complex mixtures in which hundreds of substances may be present. It can by no means be excluded that one or more of these contribute to carcinogenesis. Mutagenetic investigations have revealed mutagenic activity in several commercially produced and home-made apple brandies (27) as well as in red wine and in red grape juice (52). A variety of carcinogens such as polycyclic hydrocarbons (28) and asbestos fibres (59) have been detected in alcoholic beverages.

12.3.2. Dietary deficiencies and alcohol abuse

Nutritional deficiencies are very common in alcoholics. A poor nutritional status seems to be a risk factor for oesophageal cancer (61).

Chronic alcohol consumption has striking effects on tissue vitamin A levels (see also Chapter 6). Considering the enhancing effects of vitamin A deficiency on carcinogenesis, it is conceivable that part of the interaction between alcohol and cancer may be caused at least indirectly by the alteration of the vitamin A status induced by chronic alcohol abuse (21).

Deficiencies of some other vitamins (riboflavin, vitamin B-6, vitamin E) have also been implicated in the pathogenesis of cancer (24).

12.2.3. Hepatitis B virus and hepatocellular carcinoma

Hepatitis B virus (HBV) is associated with an increased risk of hepatocellular carcinoma (cancer affecting the liver cells) (22, 41), and alcoholics have an increased risk of HBV infection (32). HBV-DNA is capable of integrating into host genomic DNA and thus may induce chromosomal alterations important in hepatic carcinogenesis (11). In addition to the association between HBV infection and hepatocellular carcinoma, an association between cirrhosis and hepatocellular carcinoma has been observed (17).

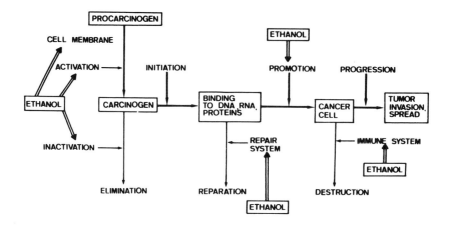

Fig. 12.1. Simplified scheme of two-step carcinogenesis and possible sites of action of ethanol (46).

12.2.4. Ethanol as a cocarcinogen

There is no evidence from animal studies that ethanol per se is carcinogenic (19). However, alcohol may increase the susceptibility of various tissues to chemical carcinogens by a variety of mechanisms (51). Among these are alteration of the metabolism and/or distribution of carcinogens, interfering with the repair of carcinogen-mediated DNA alkylation and the immune response, and stimulation of cellular regeneration. Fig. 12.1 shows the possible sites for application in carcinogenesis.
- Alcohol causes changes in the permeability of membranes ('barrier breaker') (30, 37) which may increase the diffusion of carcinogenic substances from tobacco smoke and other sources into the cells.
- By acting as a solvent for carcinogens alcohol may increase its activity. Polycyclic aromatic hydrocarbons, dissolved in alcohol, may initiate cancer in laboratory animals, whereas the same quantity dissolved in water does not (54).
- As a possible mechanism MEOS induction has been suggested. Regular alcohol consumption induces the microsomal ethanol-oxidizing system (MEOS) in the liver and in the intestine (see Chapter 3) which is associated with an increase in various constituents of proteins of the microsomal system, namely phospholipids, cytochrome P-450 reductase and cytochrome P-450. Other substances (such as drugs) are also metabolized via this system. The conversion of procarcinogenic substances into carcinogens in the intestine and/or liver may also be stimulated via induction of microsomal enzymes (23, 37, 45).
- There are two effects of alcohol on DNA metabolism that might be associated with cocarcinogenic activity, namely its effect on sister chromatid exchanges (SCEs) and on DNA repair. Obe & Ristow (37) have reported that acetaldehyde, the first metabolite of alcohol, induces an abnormal exchange of genetic information from

one sister chromatid to another in cells grown in culture. In addition, they found an elevation of chromosomal aberrations in the lymphocytes of alcoholics. A second mechanism by which alcohol abuse may increase the risk of developing cancer is by inhibiting the capacity of cells to repair carcinogen-induced DNA damage. Results of animal studies indicate that chronic alcohol consumption inhibits the activity of the DNA repair enzyme O^6-methylguanine transferase. However, it remains to be determined whether alcohol also inhibits human O^6-methylguanine transferase (7).

– DNA demethylation might be related to tumour development by enhancing gene expression or inhibiting chromatin condensation. Inoue et al. (15) noted that demythylation of exon 2 of the *c-myc* gene of the hepatocellular carcinoma is more (albeit not significantly more) frequently encountered in groups of heavy drinkers.

– Since several studies have shown an association between heavy drinking and decreased immune response (34), alcohol-associated immunosuppression has been considered a possible factor contributing to an increased risk of cancer (5). For the most part, however, studies in which decreased immune response has been associated with alcohol abuse involved patients who already had alcoholic liver disease. Therefore, it is difficult to assess whether the immunological defects observed were due directly to alcohol or reflected other aspects of the ongoing disease process. There is reason to question the significance of immunosuppression in general carcinogenesis. Although cancer incidence is higher among immunosuppressed patients or animals, the cancers observed are mostly cancers of the immune system itself (1). Nevertheless, the immune system may play a vital role in the defence against virally induced tumours, particularly in HBV-associated hepatocellular carcinoma (8).

12.3. Safety limits

Considering the fact that the mechanism through which alcoholic beverages promote or initiate carcinogenesis remains largely unknown as yet, as is the complexity of factors which appear to influence this (e.g. the composition of the beverage itself, interaction with tobacco consumption), it is not possible to determine a border-line between safe and harmful quantities of alcohol consumption in relation to cancer. Presumably no well defined limit can be established and a continuous increase of the risk with increasing alcohol consumption appears more likely, the rate of increase depending on modifying factors such as type of beverage and tobacco consumption. As to this, everyone should weigh enjoyment against risk, as in many other matters in life. In social alcohol consumption the risk hadrly increases, or does not increase at all, particularly in non-smokers. Heavy smokers who are also heavy drinkers, particularly drinkers of spirits, run a highly increased risk for cancer of the mouth, throat, larynx and oesophagus. This risk may be decreased by restricting either the use of alcohol, or smoking less, or (which is even more effective) both.

Liver cancer is associated with liver injury (cirrhosis). Consequently, the quantity of alcohol consumed which initiates cancer of the liver is equal to or larger than the quantity that causes liver injury.

Summary

There is a relationship between (excessive) consumption of alcoholic beverages and the incidence of cancer of the oral/pharyngeal cavity, the oesophagus, the pharynx and the liver. There is probably a relationship between alcohol consumption and breast cancer, lung cancer and rectal cancer. This compares with the notion that not only alcohol, but also other substances (such as those present in tobacco smoke) play a role in the pathogenesis of cancer of the mouth, throat, oesophagus and larynx. The increased mortality as a consequence of these types of cancer among alcoholics may be partly attributed to the fact that they often smoke heavily, and possibly also to the fact that they differ from moderate drinkers in other respects (life-style, (mal)nutrition). Since most cases of liver cancer occur among cirrhotics, it appears to be caused by liver injury. The increased mortality caused by lung cancer among alcoholics may be attributed almost entirely to tobacco.

Via what mechanism the cancer-promoting action of alcoholic beverages proceeds is unkown. Presumably there are distinctive mechanisms underlying the pathogenesis of cancer of the liver, and of cancer of the organs and tissues which are in direct contact with alcoholic beverages (local effect). Pure ethanol is probably not carcinogenic considering the results obtained in several animal experiments. However, fusel alcohols, nitrosamines and tannins (only certain types) have been shown to be carcinogenic. Possibly ethanol or another substance present in alcoholic beverages acts as a cocarcinogen. Dietary deficiencies, commonly observed in alcoholics, may also contribute to carcinogenesis. Infection with hepatitis B virus, impairment, induction of microsomal enzymes that activate procarcinogens, impairment of DNA metabolism, and immunosuppression may also contribute to carcinogenesis.

As yet it has not proved possible to determine limits below which alcoholic beverages have no effect on carcinogenesis. It is also difficult to differentiate between wine, beer and spirits, albeit beer consumption has been associated with the incidence of cancer of the colon and the rectum. The IARC Working Group concludes that drinking of alcoholic beverages is causally related to cancers of the oral cavity, pharynx, larynx, oesophagus and liver and that there is no evidence that the effect is dependent on type of beverage.

A great deal of research will still have to be done to arrive at more definite conclusions about the mechanism and to develop, on the basis of that knowledge, more effective preventive policies and treatment.

References

1. Baird SM, Beattie GM, Lennon RA, Lipsick JS, Jensen FC, Kaplan NO. Induction of lymphoma in antigenically stimulated athymic mice. Cancer Res 1982;42: 198-206.
2. Boeing H, Frenzel-Beyme R, Berger M, et al. Case-control study on stomach cancer in Germany. Int J Cancer 1919;47: 858-64.
3. Chu SY, Lee NC, Wingo PA, Webster LA. Alcohol consumption and the risk of breast cancer. Am J Epidemiol 1989;130: 867-77.
4. Cook P. Cancer of the oesophagus in Africa: a summary and evaluation of the evidence for the frequency of occurrence, and a preliminary indication of the possible association with the consumption of alcoholic drinks made from maize. Br J Cancer 1971;25: 853-80.
5. Daynes RA, Harris CC, Connor RJ, Eichwald EJ. Skin cancer development in mice exposed chronically to immunosuppressive agents. J Natl Cancer Inst 1979;62: 1075-81.
6. Freudenheim JL, Graham S, Marshall JR, Haugey BP, Wilkonson G. Lifetime intake and risk of rectal cancer in Western New York. Nutr Cancer 1990;13: 101-9.
7. Garro AJ, Espina N, Farinati F, Lieber CS. Ethanol and the repair of deoxyribonucleic acid (DNA). Alcohol Health Res World 1986;10(3): 26-7.
8. Garro AJ, Lieber CS. Alcohol and cancer. Ann Rev Pharmacol Tocicol 1990;30: 219-49.
9. Gibel W, Lohs K, Wildner GP. Experimentelle Untersuchungen zur karzinogenen Wirkung von Lösungsmitteln am Beispiel van Propanol-1,2-Methylpropanol und 3-Methylbutanol-1. Arch Geschwulstforsch 1975;45: 19-24.
10. Graham S, Marshall J, Haughey B, et al. Nutritional epidemiology of cancer of the esophagus. Am J Epidemiol 1990;131: 454-67.
11. Henderson AS, Ripley S, Hino O, Rogler CE. Identification of a chromosomal aberration associated with a hepatitis B DNA integration site in human cells. Cancer Genet Cytogenet 1988;30: 269-75.
12. Hirayama T. Association between alcohol consumption and cancer of the sigmoid colon, obervations from a Japanese cohort study. Lancet 1989;ii: 725-7.
13. Hoey J, Montvernay C, Lambert R. Wine and tobacco: risk factors for gastric cancer in France. Am J Epidemiol 1981;113: 668-74.
14. Howe G, Rohan T, Decarli A. The association between alcohol and breast cancer risk: evidence from the combined analysis of six dietary case-control studies. Int J Cancer 1991: 74: 707-10.
15. Inoue K, Aiba N, Nambu S, Sasaki H. Influence of ethanol on hepatocarcinogenesis, site specific demethylation of c-myc oncogen in hepatocellular carcinoma (HCC) of heavy drinkers. In: Kuriyama K, Takada A, Ishii H. Biomedical and social aspects of alcohol and alcoholism. Amsterdam: Excerpta Medica, 1988: 817-22.
16. International Agency for Research on Cancer. IARC Monographs on the evaluation of carcinogenic risk to humans: alcohol drinking. Lyon: International Agency for Research on Cancer, 1988;44.
17. Johnson PJ, Krasner N, Eddleston ALWF, Williams R. Hepatocellular carcinoma in Great Brittain: influence of age, sex, HBsAg status and aetiology of underlying cirrhosis. Gut 1978;19: 1022-6.
18. Kabat GC, Howson CP, Wynder EL. Beer consumption and rectal cancer. Int J Epidemiol 1986;15: 494-501.
19. Keller A. Alcoholic beverages and carcinogenesis. NY State J Med 1980: 1245-52.
20. Kelsey JL, Berkowitz GS. Breast cancer epidemiology. Cancer Res 1988;48: 5615-23.
21. Leo MA, Mak KM, Lieber CS. Interaction of ethanol with vitamin A in liver and in tumor development. Alcohol Health Res World 1986;10(3): 18-25.
22. Liaw YF, Tai DI, Chu M, et al. Early detection of hepatocellular carcinoma in patients with chronic type B hepatitis. Gastroenterology 1986;90: 263-7.
23. Lieber CS, Seitz HK, Garro AJ, Worner TM. Alcohol-related diseases and carcinogens. Cancer Res 1979;39: 2863-86.
24. Lieber CS, Garro AJ, Leo MA, Worner TM. Mechanisms for the interrelationship between alcohol and cancer. Alcohol Health Res World 1986;10(3): 10-7.

25. de Lint JEE. De invloed van het toenemende alkoholgebruik op het sterftepatroon. Tijdschr Soc Geneeskd 1980;58: 547-51.
26. Longnecker MP, Berlin JA, Orza MJ, Chalmers TC. A meta-analysis of alcohol consumption in relation to risk of breast cancer. JAMA 1988;260: 652-6.
27. Loquet C, Toussaint G, LeTalaer JY. Studies on mutagenic consituents of apple brandy and various alcoholic beverges collected in Western France, a high incidence area for oesophagal cancer. Mutation Res 1981;88: 155-64.
28. Matsuda Y, Mori K, Hirohata T, Kuratsune M. Carcinogenesis in the esophagus. III. Polycyclic aramatic hydrocarbons and phenols in whiskey. Gann 1966;75: 549-57.
29. McGlashan ND. Oesophagal cancer and alcoholic spirits in Central Africa. Gut 1969;10: 643-50.
30. McGoy GD, Wynder EL. Etiological and preventive implications in alcohol carcinogenesis. Cancer Res 1979;39: 2844-50.
31. Merletti F, Boffetta P, Cicconne G, Mashberg A, Terracini B. Role of tobacco and alcoholic beverages in the etiology of the oral cavity/oropharynx in Torino, Italy. Cancer Res 1989;49: 4919-24.
32. Mills PR, Pennington TH, Kay P, MacSween RNM, Watkinson G. Hepatitis B antibody in alcoholic cirrhosis. J Clin Pathol 1979;32: 778-82.
33. Morton JF. Tentative correlations of plant usage and esophagal cancer zones. Econ Bot 1970;24: 217-26.
34. Mufti SI, Darban HR, Watson RR. Alcohol, cancer and immunomodulation. CRC Crit Rev Oncol/Hematol 1989;9: 243-61.
35. Nasca PC, Baptiste MS, Field NA, et al. An epidemiological case-control study of breast cancer and alcohol consumption. Int J Epidemiol 1990;19: 532-8.
36. Nomura A, Grove JS, Stemmerman GN, Severson RK. A prospective study of stomach cancer and its relation to diet, cigarettes, and alcohol consumption. Cancer Res 1990;50: 627-31.
37. Obe G, Ristow H. Mutagenic, carcinogenic and teratogenic effects of alcohol. Mutation Res 1979;65: 229-59.
38. Pignatelli B, Castegnaro M, Walker EA. Effects of gallic acid and ethanol on formation of nitroso-diethylamine. Lyon: International Agency for Research on Cancer, 1976: 173-8. (IARC Science Publication 14.)
39. Pollack ES, Nomura AMY, Heilbrun LK, Stemmerman GN, Green SB. Prospective study of alcohol consumption and cancer. N Engl J Med 1984;310: 617-21.
40. Potter JD, Sellers TA, Folson A. Beer and lung cancer in older women, the IOWA Women's Health Study. Am J Epidemiol 1990;132: 784.
41. Qiao ZK, Halliday ML, Rankin JG, Goates RA. Relationship between hepatitis B surface antigen prevalence, per capita alcohol consumption and primary liver cancer death rate in 30 countries. J Clin Epidemiol 1988;41: 787-92.
42. Richardson S, de Vincenzia J, Pujol H, Gerber M. Alcohol consumption in a case-control study of breast cancer in Southern France. Int J Cancer 1989;44: 84-9.
43. Rosenberg L, Palmer JR, Miller DR, Clarke EA, Shapiro S. A case-control study of alcoholic beverage consumption and breast cancer. Am J Epidemiol 1990;131: 6-14.
44. Schmidt W, Popham RE. The role of drinking and smoking in mortality from cancer and other causes in male alcoholics. Cancer 1981;47: 1031-41.
45. Seitz HK, Garro AJ, Lieber CS. Effect of chronic ethanol ingestion on intestinal metabolism and mutagenicity of benzo(α)pyrene. Biochem Biophys Res Commun 1978;85: 1061-6.
46. Seitz HK, Simanowski UA. Alcohol and carcinogenesis. Ann Rev Nutr 1988;8: 99-119.
47. Simon M, Carman W, Wolfe R, Schottenfeld D. Alcohol consumption and the risk of breast cancer: a report from the Tecumseh Community Health Study. Am J Epidemiol 1990;132: 784.
48. Spiegelhalder B, Eisenbrand G, Preussmann R. Volatile nitrosamines in food. J Stud Alcohol 1981;42: abstr 2015.
49. Stampfer MJ, Colditz GA, Willett WC. Alcohol intake and risk of breast cancer. Comprehensive Therapy 1988;14: 8-15.
50. Stemmerman GN, Nomura AMY, Chyou PH, Yoshizawa C. Prospective study of alcohol intake and large bowel cancer. Dig Dis Sci 1990;35: 1414-20.

51. Swann PF. Effect of ethanol on nitrosamine metabolism and distribution: implications for the role of nitrosamine in human cancer and for the influence of alcohol consumption on cancer incidence. Lyon: International Agency for Research on Cancer, 1984: 501-512.(IARC Science Publications 57.)
52. Tamura G, et al. Fecalase: a model for activation of dietary glycosides to mutagens by intestinal flora. J Stud Alcohol 1982;42: abstr 307.
53. Turner TB, MacMohan B, Kimball AW, Chappel CI. Moderate drinking and the risk of cancer of the oropharynx, larynx, esophagus and liver. Maryland: Alcoholic Beverage Medical Research Foundation, 1989.
54. Tuyns AJ. Cancer and alcoholic beverages. In: Gastinau CF, Darby WJ, Turner TB, eds. Fermented food beverages in nutrition. New York: Academic Press 1979: 427-38.
55. Tuyns AJ. Epidemiology of alcohol and cancer. Cancer Res 1979;39: 2840-3.
56. Tuyns AJ, Estève J, Raymond L, et al. Cancer of the larynx/hypopharynx, tobacco and alcohol. IARC International case-control study in Turin and Varese (Italy), Zaragoza and Navarra (Spain), Geneva (Switzerland) and Calvadas (France). Int J Cancer 1988;41: 483-91.
57. Velema JP. Walker AM, Gold EB. Alcohol and pancreatic cancer, insufficient epidemiologic evidence for a causal relationship. Epidemiol Rev 1986;8: 28-41.
58. Webb KS, Gough TA. Human exposure to preformed environmental N-nitroso compounds in th U.K. J Stud Alcohol 1981;42: abstr 1669.
59. Wehman HJ, Plantholt BA. Astbestos fibrils in beverages I. Gin. Bull Envir Contam Toxicol 1974;11: 267.
60. Yu MC, Mack T, Hanisch R, Peters RL, Henderson BE, Pike MC. Hepatitis, alcohol consumption, cigarette smoking, and hepatocellular carcinoma in Los Angeles. Cancer Res 1983;43: 6077-8.
61. Ziegler RG, Morris LE, Blot WJ, Pottern LM, Hoover R, Fraumine F. Esophagal cancer among black men in Washington DC II. J Natl Cancer Inst 1981;67: 1199-206.

CHAPTER 13

Alcohol and pregnancy

E. te Wierik

Even in antiquity alcohol consumption during pregnancy has been associated with the incidence of malformations of the unborn child (45). In ancient Greece the marital couple was not permitted to drink alcohol during the wedding in order to prevent abnormal children (44).

When spirits were first introduced in England and alcohol consumption simultaneously increased dramatically (the gin epidemic, 1720–1750), several types of abnormalities were observed in the children of alcoholics as well as increased mortality and morbidity (44, 45).

An identifiable syndrome observed in babies of alcoholic mothers was first described 1968. Since 1973 this syndrome has been known as the foetal alcohol syndrome (FAS) (28).

13.1. Clinical symptoms

Maternal alcohol consumption during pregnancy may have a variety of effects on the unborn child. The most pronounced effect is the foetal alcohol syndrome (FAS). To standardize the criteria for FAS the Fetal Alcohol Study Group of the Research Society on Alcoholism formulated the criteria for diagnosis in 1980 as follows (5):
- pre- and/or postnatal growth retardation (weight, length, and/or head circumference < 10th percentile, when corrected for gestational or postnatal age);
- central nervous system involvement (signs of neurologic abnormality, developmental delay or intellectual impairment, e.g. mental retardation);
- characteristic facial dysmorphology (at least 2 of 3 symptoms):
  · microcephaly (head circumference < 3rd percentile)
  · microphthalmia and/or short palpebral fissures
  · poorly developed philtrum, thin upper lip, and flattening of the maxillary area.

To satisfy a diagnosis of FAS, an individual must exhibit an abnormality in each of the three categories. In addition to these, abormalities of other organs and systems (cardiovascular, skeletal, genital systems) may occur (56). If only one or two of these characteristics are evident and the mother is suspected of alcohol use during pregnancy, a diagnosis of 'possible FAS', 'partial FAS', 'foetal alcohol effects' (FAE) or 'alcohol-related birth defects' may be made (2).

13.2. Incidence

In a review of 20 relevant epidemiological papers in which a total of 88 236 births had been evaluated Abel & Sokol (3) observed 164 cases of FAS. They conclude that the world-wide incidence of the FAS is 1.9 per 1000 births. Incidence rates vary considerably depending on study site, ranging from 0.6 to 2.6 per 1000.

The incidence of FAS in children of mothers identified as 'problem drinkers' is considerably higher. Among alcoholic women in the USA the incidence ranges from 24 to 42 per 1000, and in Europe from 66 to 259 per 1000 live births (5).

13.3. Specific foetal alcohol effects

Alcohol use during pregnancy has been associated with a number of specific foetal effects. Since 1973, hundreds of studies into the relationship between alcohol and foetal effects have been published (1, 30, 47, 60).

13.3.1. Growth retardation

Birth weight is an important factor in future development. Reduced birth weight is one of the more reliably observed effects associated with in utero alcohol exposure in man and animals.

In a prospective study in California, involving 5093 mother-infant pairs, it has been observed that frequent consumption of beer (> 20 beers/month) is a significant determinant in that it decreases intra-uterine growth by 100 g. This effect on birth weight was independent of the effects of cigarette smoking and caffeine intake, both of which also appeared to reduce infant birth birth weight independently and interactively (31). Barr et al. (9) noticed that maternal alcohol use during early pregnancy was significantly related to infant weight and length both at birth and at the age of 8 months. Maternal alcohol consumption was significantly related to head circumference at birth only. A positive relationship between heavy drinking and growth retardation has been observed in several other studies (19, 23, 32, 50).

The effect of moderate alcohol consumption on foetal growth was studied by Little et al. (35). Average daily consumption of 10 g alcohol in the week prior to recognition of pregnancy was related to a decrease in infant birth weight of 225 g. This relationship between alcohol consumption and birth weight was stronger in male than in female infants. Consumption of more than 10 g per day, in early pregnancy, has been found to be associated with growth retardation in other studies (23, 62). Fried & O'Connell (19) reported both birth weight and length at birth to be negatively correlated with an average alcohol use exceeding 28 g per day. Mills et al. (40) have demonstrated that drinking an average of one to two drinks per day reduces infant birth weight by 83 g.

In the Netherlands the effect of moderate alcohol use on intra-uterine growth retardation has been studied. This study involved 3400 pregnant women (mean gestation ca. 18 weeks) who were questioned about their alcohol consumption. This consumption was categorized into 0, 1–50, 51–120 and >120 g alcohol per week. Intra-uterine growth retardation did not occur more frequently in drinking women than in non-drinking women (58). In a prospective study at Boston City Hospital (46) no differences in growth were found between offspring of women who were rare drinkers (abstaining or drinking less than once a month) and offspring of moderate drinkers (mean consumption 8.6 g/day). Walpole et al. (59) also failed to show a significant relationship between light or moderate (<2 drinks/day) maternal alcohol consumption and foetal growth.

The growth retardation in offspring of heavily drinking mothers may be reduced when drinking has been moderated during the first trimester (15, 50)

In conclusion, there appears to exist a negative relationship between heavy alcohol consumption during early pregnancy and foetal growth. There may be a dose-response relationship, but it is not clear whether there is such a thing as safe drinking during pregnancy.

13.3.2. Effects on the central nervous system

Infants exposed to alcohol prenatally, even when they do not suffer from FAS, may be at high risk for many of the negative outcomes typically found among children of alcoholics including hyperactivity and other behavioural and learning problems.

Smith et al. (50) observed that infants of women who used to take two or more drinks per day throughout pregnancy had significantly lower scores on orientation (i.e. they were less able to maintain a quiet alert state and attend to the visual and auditory stimuli in their environment) than infants of non-drinking women. Infants of women who continued to drink during pregnancy scored significantly lower on autonomic regulation than infants of mothers who stopped drinking or who abstained. Streissguth et al. (51) observed in a follow-up study that maternal alcohol use during pregnancy at a level of about four drinks per day or above has an adverse effect on mental and motor development in their offspring. In a series of studies Coles et al. (15) investigated the incidence and persistence of central nervous system (CNS)-related behavioural alterations in three groups of infants ($n = 103$) born to (i) those women who never drank in pregnancy, (ii) those women who had an average of 5 drinks per day, and (iii) those women who drank an equivalent amount initially but stopped doing so by the second trimester of pregnancy. The neonates were examined on three separate days. Those who had been exposed to alcohol were found to be less optimal in neurobehavioural responses. Infants whose mothers had continued to drink had significantly lower scores on their orientation towards auditory and visual stimuli, motor performance and autonomic regulation than the non-exposed infants. Although a second study found that some of these effects were related to the neonatal withdrawal syndrome, a follow-up to 30 days of age found that there were persistent behavioural alterations. Over the first month, infants in the 'stopped drinking' group showed more recovery than those in the 'continued drinking' group in reflexive behaviour and autonomic control. A reassessment at six months of age among 60 of the infants indicated that differences in orientation, motor performance, reflexive behaviour and autonomic control at three days of age were predictive of six-month mental and motor performance in the Baylay Scales of Infant Development. More recently, in a longitudinal prospective study Streissguth et al. (52) examined the long-term effects of moderate prenatal alcohol exposure. Consumption of two drinks per day or more on the average was related to a 7-point decrement in intelligence quotient (IQ) in 7-year-old children. Communication skills in children aged 4½ to 9½ years with diagnosed forms of FAS were assessed by Becker et al. (10). As compared to the matched non-FAS controls, FAS children demonstrated abnormalities of the speech mechanism articulation abilities inconsistent with mental age. Further, the FAS children demonstrated a reduced capacity to process and store critical elements as compared with non-FAS children.

Ioffe & Chernick (27) reported that the power of an EEG at birth both during REM (rapid eye movements) and quiet sleep is a sensitive index of alcohol effects on the foetal brain and may be used to predict future motor and mental development.

13.3.3. *Malformations*

Children of alcoholic mothers appeared to have an increased risk of congenital anomalies (23, 25, 42, 48, 61). These defects include craniofacial, cardiac, renal and skeletal abnormalities (38).

O'Connor et al. (42) noticed that an increased incidence of anomalies was noted in the infants of mothers who reported a higher alcohol consumption during pregnancy. The data suggest that mothers who were used to drink large quantities of alcohol, even af they did it infrequently, had infants with a greater proportion of anomalies. This suggestion is consistent with the finding that social drinkers who 'binged' had infants with increased EEG power compared to infants of chronic alcoholics who drank in a more continuous pattern (26).

In a study into the relationship between moderate drinking during pregnancy and the risk of malformations Mills et al. (41) collected information from 32 870 women about the first trimester of pregnancy. The authors observed total malformation rates not to be significantly higher among offspring of women who had an average of less than one drink per day (77.3 per 1000) or one or two drinks per day (83.2 per 1000) than among non-drinkers (78.1 per 1000). However, the prevalence of sex organ malformations and genito-urinary malformations taken together increased significantly with increasing alcohol consumption.

13.4. Mechanisms of action

The mechanisms of action are still unknown. There are, however, various suggestions of the way alcohol can induce foetal abnormalities. The hypotheses relating to the cause of the FAS and other alcohol-induced abnormalities may be classified into three categories (38, 39):

– direct and indirect mechanisms of action of alcohol
– nutrition-related effects of alcohol exposure
– cellular mechanisms of alcohol-induced growth retardation.

13.4.1. Direct and indirect mechanisms of action of alcohol

Both alcohol and its major metabolite, acetaldehyde, readily cross the placenta and have been shown to be teratogenic (29). Because the foetus lacks the enzymes required to degrade these substances, it is exposed long after these substances have been cleared from the maternal system (29). Both alcohol and acetaldehyde have shown to impair nucleic acid and protein sythesis directly in vitro (16). In particular, the pharmacological effects of ethanol are the result of cell membrane narcosis, leading to general depression of cell function (29).

13.4.2. Nutrition-related effects of alcohol exposure

A poor nutritional status of the mother in addition to the toxic effects of alcohol may be more detrimental to the developing foetus than either factor alone. In addition, impairment of placental transport of amino acids by alcohol may result in transient or chronic deprivation of essential amino acids, resulting in intra-uterine growth retardation, and amplify any toxic effects of ethanol on embryogenesis. Reduced placental transfer of zinc and folate (18, 20) and an altered vitamin A metabolism (22) has been demonstrated in animal models of alcohol-induced teratogenesis.

13.4.3. Cellular mechanisms of alcohol-induced growth retardation

According to Kennedy (29) all alcohol-related developmental abnormalities result from derangements in protein synthesis occurring during periods of rapid organ growth. This hypothesis is based on three basic observations.
– Alcohol-related abnormalities include a cluster of physical malformations involving many organ systems, which are highly variable in their frequency and severity of expression. This suggests that alcohol is a non-specific teratogen that affects some developmental process common to all cell types.
– The most common consistent manifestation of gestational alcohol abuse is intra-uterine growth impairment (reductions in head circumference, body length, body weight).
– Normal growth and development are dependent upon the accumulation and organization of protein. Alcohol directly and indirectly impairs protein synthesis in foetal and adult tissues through reductions in RNA and DNA, as well as in total and subcellular processes that are vital to foetal growth (29, 39).

The current state of our knowledge of the interactions of these events is schematized in Fig. 13.1.

Fig. 13.1. Relationship between maternal alcohol intoxication and placental nutrient transport deficiencies, protein synthesis inhibition and malnutrition (39).

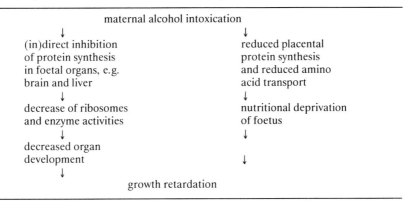

13.5. Breast-feeding

Several physicians prescribe one beer a day to increase lactation for new mothers who are to breast-feed their babies. Although some experts are in favour with this prescription (17, 21), others are not (7, 12).

After birth maternal consumption patterns may still have detrimental effects on the child via breast-feeding. The alcohol concentration of the milk equals the mother's the blood alcohol concentration (33). In case of high BACs (< 2 g/l) the milk yield will be reduced (14). The inhibitory effect of alcohol on milk yield may be attributable to disruption of transmission of the suckling stimulus (53).

Binckiewicz (11) has reported a case of the pseudo-Cushing syndrome (characterized by retarded linear growth and increased weight) affecting the baby of a mother who consumed some 2.5 litres of beer a day in addition to other alcoholic beverages. The mother's milk contained 1 g/l alcohol. The relation between the mother's alcohol use during breast-feeding and the infant's development at an age of one year has also been investigated by Little et al. (37) in a study covering 400 infants. After correction for alcohol exposure during gestation, motor development was still significantly lower in infants exposed regularly (1 drink or more per day) to alcohol in breast milk. No effects on mental development were observed. The authors postulated that other environmental factors could have a greater effect on motor development, and that any effect of alcohol on the infant would not be permanent.

However, Lawton (33) claimed that the strong dilution of the alcohol contained in the milk by the baby's body fluid renders the baby's resultant blood alcohol level very low. For example, if a mother's BAC is 1.19 g/l (after 7 drinks) her 6-month-old baby weighing 6.5 kg and drinking 180 ml of milk will have a BAC of approximately 0.06 g/l (33). It is uncertain that exposure to alcohol of that magnitude affects the baby.

13.6. Paternal alcoholism

Alcohol consumption impairs human sperm (57). Badr & Hussain found chromosome abnormalities among male alcoholics (8). There are only few scientific data available on the frequently assumed relation between paternal alcohol consumption and abnormalities of the child.

In animal experiments alcohol administration to the father before conception gave rise to a progeny lesser in number, size and brain weight (55). However, the findings are not consistent, which may be a consequence of differences in study design (4, 6, 13). Some preliminary reports suggest that prolonged paternal exposure to ethanol can result in behavioural effects in young rodent offspring (34, 54).

Little & Sing (36) reported a significant negative association between the father's alcohol use in the month before conception and the weight of their infants at birth. The authors studied 377 newborns and found that the birth weight of children of fathers who used to drink regularly (i.e. at least two drinks per day or 5 drinks on a single occasion at least once a month) was 137 g lower than that of children of fathers who drank less than that. This result was independent of the mother's drinking, smoking and marijuana use, and could not be ascribed to other factors such as the parents' height or weight. The infant's birth weight was less when the father drank regularly even when he did not smoke. This finding is hard to interpret because the biological mechanisms that might underlie it are obscure. Other authors do not support these results (49). To explore the role of parental alcohol consumption in miscarriage 80 women who had miscarried were interviewed about their own and their partner's drinking habits. Although this method employed for the assessment of alcohol use is open to criticism the authors concluded that light maternal (2 drinks or less per week) or paternal (4 drinks per week at most) alcohol consumption does not increase risk of miscarriage (24). Parazzini et al. (43) also conclude that light maternal and paternal alcohol use is unlikely to affect abortion rates.

Summary

Consumption of large quantities of alcohol during pregnancy may induce the foetal alcohol syndrome (FAS), characterized by growth disturbances, mental retardation and facial abnormalities. Moderate quantities of alcohol may also be injurious; the consumption of 1 drink per day appears to reduce slightly birth weight and mental and motor development. There may be a dose-response relationship between maternal alcohol use and foetal alcohol-related effects, but there is no such thing as a safe drinking level.

The incidence of FAS has been assessed at 1.9 per 1000 births for the total population, but at 66–259 per 1000 births for alcoholic mothers. The mechanism of its action is unknown although there are a few likely hypotheses. Alcohol my directly or indirectly inhibit protein synthesis in foetal organs, or reduce placental protein synthesis and amino acid transport both of which can lead to growth retardation.

After birth, maternal consumption may still have adverse effects on the child via breast-feeding.

The effect of paternal alcoholism is obscure.

References

1. Abel EL. Consumption of alcohol during pregnancy, a review of effects on growth and development of offspring. Human Biol 1982;54: 421-53.
2. Abel EL. Prenatal effects of alcohol. Drug Alcohol Depend 1984;14: 1-10.
3. Abel EL, Sokol RJ. Incidence of fetal alcohol syndrome and economic impact of FAS-related anomalies. Drug Alcohol Depend 1987;19: 51-79.
4. Abel EL. Duration of paternal alcohol consumption does not influence offspring growth and development. Growth Devel Aging 1989;53: 195-9.
5. Abel EL, Sokol RJ. Alcohol consumption during pregnancy, the dangers of moderate drinking. In: Goedde HW, Agarwal DP, eds. Alcoholism, biomedical and genetic aspects. New York: Pergamon Press, 1989: 228-37.
6. Abel EL, Bilitzke P. Paternal alcohol exposure, paradoxical effect in mice and rats. Psychopharmacology 1990;100: 159-164.
7. Auerbach K, Schreiber JR. Beer and the breast-feeding mom. JAMA 1987;258: 2126.
8. Badr FM, Hussain H. Chromosomal aberrations in chronic male alcoholics. Alcohol Clin Exp Res 1982;6: 122-9.
9. Barr HM, Streissguth AP, Martin DC, Herman CS. Infant size at 8 months of age, relationship to maternal use of alcohol, nicotine, and caffeine during pregnancy. Pediatrics 1984;74: 336-41.
10. Becker M, Warr-Leeper GA, Leeper HA. Fetal alcohol syndrome, a description of oral motor, articulatory, short-term memory, grammatical, and semantic abilities. J Commum Disord 1990;23: 97-124.
11. Binkiewicz A, Robinson MJ, Senior B. Pseudo-Cushing syndrome caused by alcohol in breast milk. J Pediatr 1978;93: 965-7.
12. Blume S. Beer and the breast-feeding mom. JAMA 1987;258: 2126.
13. Cake H, Lenzer I. On effects of paternal ethanol treatment on fetal outcome. Psychol Rep 1985;57: 51-7.
14. Cobo E, Quintero CA. Milk-ejecting and antidiuretic activities under neurohypophyseal inhibition with alcohol and water overload. Am J Obstet Gynecol 1969;105: 877-87.
15. Coles CD, Smith IE, Falek A. Prenatal exposure and infant behavior, immediate effects and implications for later development. In: Bean-Bayog M, Stimmel B, eds. Children of alcoholics. New York: The Haworth Press 1987: 87-104.
16. Dreosti E, Ballard FJ, Belling GB, et al. The effect of ethanol and acetaldehyde on DNA synthesis in growing cells and on fetal development in the rat. Alcohol Clin Exp Res 1981;5: 357-62.
17. Falkner F. Beer and the breast-feeding mom. JAMA 1987;258: 2126.
18. Fisher SE. Ethanol, effect on fetal brain growth and development. In: Tarter RE, van Thiel DH, eds. Alcohol and the brain, chronic effects. New York: Plenum Press, 1985: 265-81.
19. Fried PA, O'Connell CM. A comparison of the effects of prenatal exposure to tobacco, alcohol, cannabis and caffeine on birth size and subsequent growth. Neurotoxicol Teratol 1987;9: 79-85.
20. Ghishan FK, Patwardhan R, Greene HL. Fetal alcohol syndrome, failure of zinc supplementation to reverse the effect of ethanol on placental transport of zinc. Pediatr Res 1983;17: 529-31.
21. Grossman ER, Calif B. Beer, breast-feeding, and the wishdom of old wives. JAMA 1988;259: 1016.
22. Grummer MA, Zachman RD. The effect of maternal ethanol ingestion on fetal vitamin A in the rat. Pediatr Res 1990;28: 186-9.

23. Halmesmäki E. Alcohol counseling of 85 pregnant problem drinkers, effect on drinking and fetal outcome. Br J Obstet Gynaecol 1988;95: 243-7.
24. Halmesmäki E, Välimäki M, Roine R, Ylikahri R, Ylikorkala O. Maternal and paternal alcohol consumption and miscarriage. Br J Obstet Gynaecol 1989;96: 188-91.
25. Hollstedt C, Dahlgren L, Rydberg U. Outcome of pregnancy in women treated at an alcohol clinic. Acta Psychiatr Scand 1983;67: 236-48.
26. Ioffe S, Chernick V. Development of the EEG between 30 and 40 weeks of gestation in normal and alcohol-exposed infants. Dev Med Child Neurol 1988;30: 797-807.
27. Ioffe S, Chernick V. Prediction of subsequent motor and mental retardation in newborn infants exposed to alcohol in utero by computerized EEG analysis. Neuropediatrics 1990;21: 11-7.
28. Jones KL, Smith DW. Recognition of the fetal alcohol syndrome in early infancy. Lancet 1973;ii: 999-1001.
29. Kennedy LA. The pathogenesis of brain abnormalities in the fetal alcohol syndrome, an integrating hypothesis. Teratology 1984;29: 363-8.
30. Knipschild P. Is alcohol teratogeen? Epidemiologisch onderzoek naar het verband tussen alcoholgebruik in de zwangerschap en aangeboren afwijkingen bij het kind. Tijdschr Soc Gezondheidsz 1985;63: 471-6.
31. Kuzma JW, Sokol RJ. Maternal drinking behavior and decreased intrauterine growth. Alcohol Clin Exp Res 1982;6: 369-402.
32. Larsson G, Bohlin AB, Tunell R. Prospective study of children exposed to variable amounts of alcohol in utero. Arch Dis Child 1985;60: 316-21.
33. Lawton ME. Alcohol in breast milk. Aust N Z J Obstet Gynaecol 1985;25: 71-3.
34. Lee JA, Zajac C, Tracy NM et al. Paternal alcohol intake and offspring behavior in mice. Alcohol Clin Exp Res 1987;11: 204.
35. Little RE, Asker RL, Sampson PD, Renwick JH. Fetal growth and moderate drinking during pregnancy. Am J Epidemiol 1986;123: 270-8.
36. Little RE, Sing CF. Father's drinking and infant's birth weight, report of an association. Teratology 1987;36: 59-65.
37. Little RE, Anderson KW, Ervin CH, Worthington-Roberts B, Clarren SK. Maternal alcohol use durig breast-feeding and infant mental and motor development at one year. N Eng J Med 1989;321: 425-30.
38. Luke B. The metabolic basis of the fetal alcohol syndrome. Int J Fertil 1990;35: 333-7.
39. Michaelis EK, Fetal alcohol exposure, cellular toxicity and molecular events involved in toxicity. Alcohol Clin Exp Res 1990;14: 819-26.
40. Mills JL, Graubard BA, Harley EE, Rhoads GG, Berendes HW. Maternal alcohol consumption and birth weight, how much drinking in pregnancy is safe? JAMA 1984;252: 1875-9.
41. Mills JL, Graubard BI. Is moderate drinking during pregnancy associated with an increased risk for malformations? Pediatrics 1987;80: 309-14.
42. O'Connor MJ, Brill NJ, Sigman M. Alcohol use in primiparous women elder than 30 years of age, relation to infant development. Pediatrics 1986;78: 444-50.
43. Parazzini F, Bocciolone L, La Vecchia C, Negri E, Fedele L. Maternal and paternal moderate daily alcohol consumption and unexplained miscarriages. Br J Obstet Gynaecol 1990;97: 618-22.
44. Quelette EM. Alcohol in pregnancy; and its effect on offspring. In: Gastineau CF, Darby WJ, Turner TB, eds. Fermented food beverages in nutrition. New York: Academic Press, 1979: 439-456.
45. Rosett HL. Clinical pharmacology of the fetal alcohol syndrome. In: Majchrowicz E, Noble EP, eds. Biochemistry and pharmacology of ethanol. New York: Plenum Press, 1979;2: 485-505.
46. Rosett HL, Weiner L, Lee A, Zuckermann B, Dooling E, Oppenheimer E. Patterns of alcohol consumption and fetal development. Obstet Gynecol 1983;61: 539-46.
47. Rosett HL, Weiner L. Alcohol and pregnancy, a clinical perpective. Ann Rev Med 1985;36: 73-80.
48. Rostand A, Kaminski M, Lelong N et al. Alcohol use in pregnancy, craniofacial features, and fetal growth. J Epidemiol Commun Health 1990;44: 302-6.
49. Rubi DH, Leventhal JM, Krasilnikoff PA, Weile B, Berget A. Father's drinking (and smoking) and infant's birth weight. N Eng J Med 1986;315: 1551.

50. Smith IE, Coles CD, Lancaster J, Fernhoff PM, Falek A. The effect of volume and duration of prenatal ethanol exposure on neonatal physical and behavioral development. Neurobeh Toxicol Teratol 1986;8: 375-81.
51. Streissguth AP, Barr HM, Martin DC, Herman CS. Effects of maternal alcohol, nicotine, and caffeine use during pregnancy on infant mental and motor development at eight months. Alcohol Clin Exp Res 1980;4: 152-64.
52. Streissguth AP, Barr HM, Sampson PD. Moderate prenatal alcohol exposure, effects on child IQ and learning problems at age 7½ years. Alcohol Clin Exp Res 1990;14: 662-9.
53. Subramanian MG, Abel EL. Alcohol inhibits suckling-induced prolactin release and milk yield. Alcohol 1988;5: 95-8.
54. Tan S, Zajac CS, Moore C, Rudel D, Zajac C, Abel EL. Effects of paternal alcohol consumption on behavior of offspring in rats. Alcohol Clin Exp Res 1987;11: 193.
55. Tanaka H, Suzuki N, Arima M. Experimental studies on the influence of male alcoholism on fetal development. Brain Devel 1982;4: 1-6.
56. Tholen J, Siero S, Kok GJ. Gevolgen van alcoholgebruik tijdens zwangerschap. Tijdschr Alc Drugs 1988;14: 41-9.
57. Välimäki M, Ylikahri R. The effect of alcohol on male and female sexual function. Alcohol Alcoholism 1983;18: 313-20.
58. Verkerk PH, van Noord-Zaadstra BM, Verloove-Vanhorick SP. Alcohol-consumptie tijdens de zwangerschap en foetale groei. Tijdschr Soc Geneesk 1990;5: 26-7.
59. Walpole I, Zubrick S, Pontré J. Is there a fetal effect with low to moderate alcohol use before or during pregnancy? J Epidemiol Community Health 1990;44: 297-301.
60. Waterson EJ, Murray-Lyon IM. Preventing alcohol related birth damage, a review. Soc Sci Med 1990;30: 349-64.
61. Wright JT, Toplis PJ. Alcohol and the fetus in the west of Scotland. Br Med J 1983;287: 428.
62. Wright JT, Waterson EJ, Barrison IG, et al. Alcohol consumption, pregnancy and low birthweight. Lancet 1983;i: 663-5.

CHAPTER 14

Differences between alcoholic beverages

W.G. Vrij-Standhardt

14.1. Introduction

Alcoholic beverages may be classified into three groups: beer, wine and spirits.

Beer is brewed by fermenting malted barley and occasionally other cereals to which hops are often added in order to improve taste and keeping qualities. There are many types of beer, the most popular type in the Netherlands being the pilsener type.

Wine is made by fermenting grape juice (white wine) or crushed grapes (red wine and rosé wine). Sherry and port belong to the category of fortified wines, i.e. extra wine alcohol has been added.

For the preparation of distilled beverages (spirits) various sources of starch or sugar may be used: cereals and molasses (from sugar-beets) for the preparation of gin and Hollands (Hollands gin, geneva); grapes for cognac and brandy; barley, maize for bourbon; rye for whisky; potatoes (or sometimes cereals) for vodka and aquavit; cherries for kirsch; plums for slivovitz, and so on. When the sugar has fermented, the liquid is distilled – and sometimes dilluted again – usually until an ethanol[1] content of ca. 40% is achieved.

Some beverages are aromatized with herbs during or after distillation: gin and Hollands with juniper-berries, aquavit with caraway. Others get their flavour from the raw material (e.g. cognac and slivovitch).

A special group of distilled beverages are the liqueurs. These are a blend of neutral distilled ethanol, water, sugar and/or other sweeteners, natural (from herbs and spices) or artificial flavouring and colouring agents, and sometimes fruit juices, cream, eggs, and so on. The classical liqueurs have high ethanol contents (ca. 40%). Many modern, liqueur-like beverages, however, have an ethanol concentration below 15%. In the Netherlands a real liqueur legally must contain 20% ethanol at least.

In the past decade dealcoholized and low-alcohol beverages have increasingly gained popularity. Low-alcohol beers have been produced since many years, but a

---

[1] In contrast to other chapters, in which the terms 'alcohol' and 'ethanol' are used interchangeably, in this chapter the term 'ethanol' is used to discriminate between ethanol and other alcohols.

serious public demand for these beers is a recent development. Countries vary in definitions for low-alcohol and alcohol-free beers. In England alcohol-free beer must contain no more than 0.05% v/v ethanol, in Germany it must have less than 0.5% w/w (ca. 0.6 % v/v). In the USA alcohol-free beer does not exist: a beverage has to contain no alcohol to be called alcohol-free, and at least 0.5% ethanol to be called beer (55).

There are several methods currently used to produce beers with low alcohol contents, which can be classified into two procedures: limited production of alcohol (weak worts, poorly fermentable sugars, special yeasts, interruption of fermentation, conditions which do not allow fermentation to occur), and removal of alcohol (by steam distillation, distillation at reduced or normal pressure, dialysis, reverse osmosis and spray-drying) (55).

With regard to physiological effects it is not necessary, as a rule, to distinguish between different types of alcoholic beverages. Therefore, the statement 'alcohol is alcohol, irrespective of the beverage it is contained in' may safely be used as a starting point. For some aspects, on the other hand, it is not only the quantity of alcohol that matters, but also its concentration, and in these cases a distinction has to be made.

Moreover, alcoholic beverages are not just mixtures of ethanol and water, but may contain hundreds of substances which combine to constitute the nature of any particular beverage. These substances, present in small to minute quantities, may – individually or in combination – exert specific effects on the body or affect the action of ethanol.

Differences in type and quantity of these substances are not only found between the three classes of beverages, but also within these groups, which makes it possible, for instance, to distinguish between wines of different years or from different chateaus, and causes whisky to smell and taste differently from cognac.

In this chapter the following topics will be dealt with separately:
– ethanol
– macronutrients (carbohydrates, proteins and fats)
– vitamins
– minerals and trace elements
– congeners (higher alcohols, methanol, esters, etc.).

A number of references will be made to previous chapters of this book in which the differences between wine, beer and spirits concerning the effects of these substances have been touched on.

14.2. Ethanol

The quantity of alcohol contained in one drink does not only depend on the alcohol concentration of the beverage (6), but also on the capacity of the glass (see Table 14.1). If the capacity of glasses commonly used in the Netherlands is taken as a

Table 14.1. Average ethanol content of various alcoholic beverages in the Netherlands.

Beverage	Alcohol content		Capacity of a standard glass (ml)	Amount of ethanol per glass	
	% v/v	% w/w		ml	g
Beer (pilsener)	5	4	250	12	10
Wine	12	10	100	12	10
Distilled drinks					
(gin, cognac)	35	28	35	12	10
Sherry/port	18	14	50	9	7
Liqueurs	≥22	≥18	35	≥8	≥6
Whisky	44	35	35	15	12

standard (6), a glass of beer, wine or spirits will each contain about 10 g alcohol and a glass of sherry or port ca. 7.5 g. Several less current beverages such as blackcurrant gin (16% alcohol) contain less than 10 g alcohol per glass. It should be noted, however, that at home glasses may sometimes contain more than a standard glass as used in restaurants and the like (12). Roughly speaking, however, a glass may be considered to contain 10 g (ca. 12 ml) alcohol.

As regards the international literature, it should be taken into account that both the ethanol concentration of beverages and the actual capacity of a standard glass may differ from the values mentioned above. In Germany a standard glass of spirits is taken to contain 20 ml, and a standard glass of wine 200 ml, so a glass of 'Schnapps' will contain 6 g ethanol and a glass of wine circa 20 g. American measures also differ from Dutch ones. A typical American drink contains 12 g ethanol. The results of epidemiological studies concerning alcohol consumption are hard to compare because of these differences in ethanol content of standard glasses or units. It has been suggested to express alcohol consumption data in g/day, to increase compatibility of research results (53).

The ethanol concentration of a beverage only plays a role inasmuch it comes in direct contact with the body. This is the case in the upper digestive tract (mouth, oesophagus, stomach). Beverages with an ethanol concentration in excess of ca. 25% may cause reversible damage to the gastric mucosa depending on the quantity of alcohol and the presence of other substances in the stomach, whereas ethanol in beer, wine and long drinks is diluted to such a degree that these beverages can hardly cause mucosal damage (see Chapter 11). There is some controversy as to the question whether spirits are associated more strongly with carcinogenesis of the upper digestive tract than are other types of beverages.

Since it is not known whether moderate alcohol consumption also plays a part in carcinogenesis in the gastrointestinal tract, no distinction can be made between wine, beer and spirits, when used moderately, in relation to (co)carcinogenicity.

The concentration of ethanol is of importance to its absorption rate (see Chapter 2). The levels of other substances (particularly sugar and carbohydrates), however, also play a part since they slightly delay absorption. The blood alcohol curve consequently varies widely among the different beverages, particularly when ethanol is consumed on an empty stomach. When it is consumed in combination with food these differences largely disappear.

14.3. Macronutrients

Table 14.2 represents the macronutrient composition of a number of (groups of) alcoholic beverages derived from the NEVO table (Dutch nutrient database) (6). The data given in this table do not agree entirely with other sources. The famous handbook by Souci et al. (47), for instance, reports that beer does not contain carbohydrates whereas the NEVO table mentions the absence of protein. Obviously, the analytical data cannot be considered to be exact values but rather rough approximations. Values for the vitamin, mineral and congener contents should also be seen in this light. The most important cause of these divergences is the fact that there are large differences in composition between several batches of one particular type of beverage. *The* composition of whisky does not exist. Possible differences in methods of analysis may also partly account for the divergence in results obtained.

Wine and beer do not contain fat. Nor do spirits, except for eggnog and some cream liqueurs (Table 14.2). Beer and wine contain minute quantities of proteins

Table 14.2. Composition of alcoholic beverages (expressed per 100 g beverage) according to the Dutch NEVO table (6).

Beverage type	Energy	Protein	Fat	Mono- and disacch.	Poly- sacch.	Water	Alcohol
	kJ	g	g	g	g	g	g
Eggnog	1000	4	3	28	0	55	12
Blackcurrant gin	770	0	0	18	0	65	16
Beer (old brown)	126	0	0	2	2	94	2
Beer (pilsener)	184	0	0	2	1	91	4
Brandy	967	0	0	0	0	65	33
Campari	753	0	0	10	0	70	20
Lemon gin	753	0	0	3	0	73	24
Cognac	954	0	0	0	0	68	32
Hollands gin	820	0	0	0	0	70	28
Liqueur	1013	0	0	29	0	48	18
Port	628	0	0	13	0	71	14
Rum	967	0	0	0	0	66	33
Sherry	460	0	0	3	0	81	14
Vermouth (sweet)	619	0	0	16	0	71	12
Whisky	1025	0	0	0	0	65	35
Wine (red)	343	0	0	3	0	87	10
Wine (sweet)	690	0	0	22	0	66	11
Wine (white)	410	0	0	7	0	82	10

and amino acids (<0.5%); they are completely absent from spirits, again except for eggnog. Carbohydrates are present in some beverages as monosaccharides and disaccharides (sugars): in beer (2%), in wine (from practically nil in dry wine up to 22% in sweet Spanish wine), and in liqueurs and (spiced) bitters (10–30%). Besides, beer contains 1–2% polysaccharides.

The proteins and carbohydrates in alcoholic beverages delay the absorption of ethanol. The contribution of proteins and carbohydrates from alcoholic beverages to daily nutritional intake is negligible.

14.4. Vitamins

During their growth yeast cells produce B vitamins. Immediately after fermentation beer and wine consequently contain considerable quantities of B vitamins. Most of the vitamins are removed together with the yeast cells by filtration and clarification, so by the time these beverages are available for consumption they contain only minor amounts of vitamins (8). To cover the daily recommended intake of riboflavin, 5 litres beer are needed, and to satisfy one's vitamin B-6 requirement one should take at least 1 litre of wine a day (54). However, a positive correlation was found between alcohol consumption (90% of the alcohol was consumed as beer) and plasma folate and pyridoxin (B-6) levels in moderately drinking men (39).

The vitamin C present in grapes (10–180 mg/kg) is found in only negligible quantities in the final product wine (0–20 mg/kg) (8).

Distillation reduces the vitamin level even further. As a result, most distilled beverages contain hardly any vitamins, except for some liqueur-like beverages which may contain traces of some vitamins.

The contribution of alcoholic beverages to daily vitamin requirement may be neglected. That is why alcoholic beverages are sometimes said to supply 'empty' calories, i.e. merely energy (from ethanol and carbohydrates) and hardly any 'building' substances and 'protecting' substances such as proteins, vitamins and minerals. In case of moderate consumption this usually does not cause problems. Excessive consumption, however, may give rise to vitamin deficiency caused by substitution effects (see Chapter 6). Alcoholism is considered the main cause of malnutrition in Western societies (see Chapter 11 and Thomson & Makumdar (51)).

14.5. Minerals and trace elements

Beer and, in particular, wine contain a wide range of minerals and trace elements (see Table 14.3). The mineral content of wine may vary widely and is affected by many variables including soil type, grape variety, climatic conditions, harvesting practices and processing conditions (decrease due to clarification, increase through additives). The use of pesticides and fertilizers as well as accidental contamination during processing will also contribute to the content of metals.

Table 14.3. Average contents of some minerals and trace elements in wine and their contributions to dietary intake. After: Parr (36).[1]

Mineral or trace element	Concentration mg/l	Dietary intake mg/day	Proportion of dietary intake provided by 250 ml wine per day % w/w
Boron	5	1.3	96
Copper	0.6	3.5	4
Fluorine	0.3	1.8	4
Iodine	0.15	0.2	19
Iron	10	16	16
Lead	0.2	0.44	11
Potassium	700	3300	5
Magnesium	60	305	5
Manganese	1.5	3.7	10
Phosphorus	150	1400	3
Silicon	20	3.5	143
Tin	0.5	4	3
Zinc	2	13	4

[1] The contribution to dietary intake is < 1% for aluminium, arsenic, bromine, cadmium, chlorine, chromium, cobalt, lithium, molybdenum, nickel, sodium and vanadium.

Because the minerals precipitate in the residue during distillation the distilled drinks do not contain minerals unless these are added during processing.

The mineral most abundantly occurring in wine is potassium, followed by phosphorus, magnesium and calcium. The presence of some minerals may be regarded as favourable (potassium, calcium, magnesium, iron, chromium and silicium), whereas the presence of others should be considered as hazardous (cobalt, lead). The effects of the various minerals will not be discussed extensively here. Some minerals that are mentioned explicitly in the literature will be briefly dealt with.

Copper and zinc occur naturally in grapes, in certain enzyme complexes. Their presence in wine depends on the type and the pH of the soil, the ripeness of the grape and the general climatic conditions (11). However, the presence of these metals in wine can point at residues of insecticides and fungicides; other possible sources are industrial complexes nearby the vineyard and lead from exhaust gases of motor vehicles. High concentrations of lead (0.12 mg/l), copper (7.21 mg/l), and zinc (16.10 mg/l) have been found in Spanish 'Oloroso' wines from areas nearby highways and a city. In areas more remote from these sites concentrations of these metals were appreciably lower (0.06, 0.65 and 3.40 mg/l, respectively) (27).

Other sources of lead in wine and beer are lead-containing solder and bronze and brass fittings in processing equipment and draught beer installations, as well as the lead foil caps of wine bottles (29, 45, 46, 52). In a British investigation in 1982/83 into lead in wine and beer, 90% of canned and bottled beer contained 10 µg/l or less, whereas almost half the draught beers sampled contained more than 10 µg/l, and 4% even more than 100 µg/l (mean 20 µg/l) (45). As a reaction to these results, the lead sources for draught beer were identified and, where practicable, eliminated. A reinvestigation in 1985/86 showed a decrease in lead concentration in draught beer (mean 9.8 µg/l) (46). Not a single beer contained more than 100 µg/l.

An investigation of bottled wines showed a mean lead concentration of 88 µg/l (range 27–240 µg/l) (45). Pouring out lead-capped wine without cleaning the neck of the bottle resulted in a mean value of 565 µg/l (46). In 20% of the samples the lead concentration exceeded the (British) statutory limit for lead in wine (1 mg/kg). Thorough cleaning of the neck of lead-capped wine bottles, especially if corrosion is visible, decreases the lead content of poured wine (29). Smart et al. (46) estimated that consumption of 1 litre of wine containing 565 µg/l results in a 6 times greater uptake of lead than the weekly lead intake from the remainder of the diet (46). Wine drinkers have higher blood lead levels than drinkers of other types of beverages (10).

As a possible explanation of the consumption of port being associated with attacks of gout – as assumed in the 18th and 19th centuries – Gastineau et al. (12) suggest lead contamination of this beverage caused by the lead-containing solder of processing equipment. Contamination caused by the lead capsules sealing the bottles may also account for this. Daynes (9), on the other hand, supposed – based on his own experience – that a gouty attack afflicting individuals susceptible to it may be induced by oxidation products which are supposed to be generated in port exposed to the air for a long time.

The presence of silicium and/or chromium in wine is sometimes regarded as a possible explanation of the decreased risk of coronary heart disease in case of wine consumption, found in several studies (see Chapter 10). Deficiencies of these minerals are associated with the incidence of cardiovascular disease. According to Parr (36), 250 ml wine supplies ca. 1.5 times the daily silicium intake (see Table 14.3). Beer has practically the same silicium content (10–20 mg/l) (36).

Jennings & Howard (20) found an average chromium level of 450 µg/l in wine, 300 µ/l in beer, and 135 µg/l in spirits, while the daily requirement has been estimated at 20–500 µg. Parr (36) reports that 250 ml wine accounts for less than 1% of the mean daily intake of chromium.

The cardiomyopathy found among beer drinkers in Canada was caused by cobalt salts added to the beer as a head-retaining agent (2, 23). Arsenic poisoning, caused by contamination of raw materials, has been reported by Klatsky (23).

14.6. Congeners

The raw materials of alcoholic beverages (grapes, barley, etc.) are complex mixtures. During the fermentation process of alcoholic beverages small quantities of a number of compounds are generated in addition to ethanol. All these compounds – such as methanol, higher alcohols (alcohols with a longer hydrocarbon chain than ethanol which contains two C atoms), aldehydes, ketones, phenols, esters, hydrocarbons, lactones, and nitrogenous and sulphurous compounds – are designated as congeners in the English literature. The German literature uses the term 'Begleitstoffe''.

Congeners are also referred to as 'Fuselöl', which does not always include methanol because it is more volatile than ethanol. 'Fuselöl' is indicated as a mixture of substances produced during the distilling process, distinguishing itself from the ethanol/water system, which may include aliphatic alcohols, acids, esters, aldehydes and acetals of higher molecular weight which are formed during the metabolism of yeast as well as the volatile substances from the raw materials (24). In the latter category fall terpenes and furfural in the first place.

Generally, only the higher alcohols are regarded as fusel alcohols (German: 'Fuselalkohole'), although the term Fuselöl' is occasionally also used in this more restricted sense (24).

Since the analysis of minute quantities of substances became possible through the introduction of gas-liquid chromatography, a profusion of literature on the analysis of congeners in alcoholic beverages has appeared (1, 40, 42, 44). Several surveys have been devoted to this subject (21, 28, 38). The variation in levels of congeners reported in the literature is striking. Schreier et al. (44), for example, has found marked differences in levels of congeners in one particular kind of wine (Pinot Noir burgundy) over a number of years. Obviously, it does not make much sense to speak of *the* congener level, or of *the* methanol level, of wine. On the other hand, there are specific differences between types of beverages on the basis of which the genuineness and the purity of a beverage may be checked (40).

It is almost impossible to examine the effects of all of the approximately 400 congeners identified in alcoholic beverages (21). Of some substances, particularly ethanol and the fusel alcohols, some pharmacological effects are known (15, 19, 34, 35, 41). Congeners are said to play a role in hangover (41), carcinogenesis (33) and allergic reactions to alcoholic beverages (16).

14.6.1. Methanol

All alcoholic beverages contain methanol (48), varying in concentration from negligible quantities (4–50 mg/l in beer) to sometimes even 14 g/l in distilled beverages prepared from fruit such as 'Obstler', 'Williams', 'Kirsch' and 'Zwetschenwasser' (38, 40). Most methanol originates from pectins to which it is bound as an ester; it is released during maceration and fermentation of the grapes and fruits. Most other alcoholic beverages have ethanol levels not exceeding 400 mg/l (38, 48).

The lethal dose of methanol for someone weighing 70 kg has been estimated at 30–70 g (4, 17). The quantity of methanol consumed in case of moderate alcohol consumption is presumably harmless; when pectin-containing fruits and vegetables are digested in the gastrointestinal tract methanol bound to pectin as an ester is released, which does not cause any problems either (17).

Methanol poisoning is mainly characterized by – often fatal – metabolic acidosis and optic nerve damage ending in blindness. It is not the methanol itself, but its metabolites formaldehyde and formate, which cause the symptoms of poisoning (48). The formation of these metabolites is inhibited by the presence of ethanol (applied in the treatment of methanol poisoning), because alcohol dehydrogenase prefers ethanol to methanol. Alcoholics, however, are able to oxidize methanol despite high ethanol levels (7). The formaldehyde formed is highly reactive, forming condensation products with endogenous amines. These condensation products might partly be responsible for addiction and long-term toxicity (7, 48).

Methanol poisoning only occurs after the consumption of non-beverage methanol solutions or beverages which contain large quantities of methanol, either owing to mistakes or ignorance or if illicitly distilled ethanol has been mixed with methanol.

Until 1978, methanol poisoning was not uncommon among Swedish alcoholics, who spiked their regular drinks with cleansing solutions containing up to 80% methanol. Since 1978, only cleansing solutions with a maximum metanol concentration of 5% are allowed in Sweden. These solutions are quite popular among alcoholics. No acidosis or other signs of methanol poisoning have been found in alcoholics who had used these solutions excessively (30).

14.6.2. Fusel alcohols

Table 14.4 shows the average levels of higher alcohols and methanol reported in the literature, expressed as amount of congener (mg) per 80 g ethanol (about 8 glasses).

Not only the occurrence of fusel alcohols in alcoholic beverages has been investigated, but also the biological and toxic action of fusel alcohols. Since these two types of investigations have been carried out independently, a conclusion regarding the effect of fusel alcohols consumed during moderate alcohol consumption can only be drawn with great caution.

In common alcoholic beverages the fusel alcohol content is hardly one tenth of the toxic dose. There are only very few useful data available on reliable comparative studies relating to the toxicity and side-effects of beverages and the influence of fusel alcohols on human physiological functions.

There is no substantial difference in pharmacological action between ethanol and the higher alcohols. Differences between various types of fusel alcohols and differences between fusel alcohols and ethanol are mainly found in the degree and duration of their action (17, 24).

According to Prokop (41) fusel alcohols are known to be highly toxic substances with a specific effect on the nervous system, which affect a broad variety of biological functions. They have a distinct narcotic effect and cause many objectively and subjectively negative changes. Compared with ethanol, the narcotic effect of n-propanol is 3.9 times greater, n-butanol 14.3, isobutanol 11.7, secondary amyl alcohol 28, and isoamyl alcohol 52 times. Fusel alcohols, which are present in quantities varying between 0.1 and 2.0 g/l – depending on the base material of the beverage in question – may be associated with the well known fact that the effect of ethanol in specific beverages may occasionally increase or change, and unpleasant after-effects will be brought about, usually referred to as a hangover.

The higher alcohols are usually absorbed and metabolized more slowly by alcohol

Table 14.4. Some congeners in alcoholic beverages (in mg per 80 g ethanol). After: Piendl et al. (38).

Beverage type	Congener[1]								
	1	2	3	4	5	6	7	8	9
Pilsener beer	6	21	18	22	72	30	18	0?	37
White wine	54	30	72	34	122	30	52	122	65
Red wine	74	26	58	38	166	41	40	194	110
Obstler	798	204	59	61	141	3	10	40	107
Plum brandy	922	118	70	38	116	2	14	46	166
Gin	16	1	1	0	0	0	1	0	1
Hollands gin	12	6	19	13	25	1	3	0?	2
Brandy	54	38	82	48	237	2	18	12	60
Scotch whisky	9	38	93	29	129	3	7	1	42
Rum	8	53	22	18	135	1	11	0?	41
Vodka	38	0	0	0	0	0?	2	0?	1

[1] Congeners: 1, methanol; 2, n-propanol; 3, isobutanol; 4, 2-methylbutanol-1; 5, 3-methylbutanol-1; 6, 2-phenylethanol; 7, acetaldehyde; 8, ethyl lactate; 9, ethyl acetate.

dehydrogenase (ADH) than ethanol because they are less water-soluble and more fat-soluble. They are still detectable in the blood 14 hours after intake (41). Fusel alcohols are associated with hangovers, delayed reactions, nystagmus (jerky eye movement) and sleepiness (24, 41). It is as good as unknown what mechanism underlies these phenomena and after what quantities they occur.

14.6.3. Other congeners

In addition to the 37 alcohols identified in various alcoholic beverages, Kahn (21) reports the occurrence of 80 acids, 118 esters, 41 carbonyl compounds, 17 acetals, 41 phenols, 11 hydrocarbons, and 50 other compounds. Relatively little is known about the effects of these substances, particularly when they are consumed as a mixture of minute quantities, as is the case in moderate alcohol consumption.

A discussion of all the components of alcoholic beverages is beyond the scope of this literature review; only some of these will be discussed.

The histamines in some wines may account for some hangover symptoms (22). Grab (17) mentions that the narcotic effect of esters is highly dependent on their fat solubility. As regards the reported changes in behaviour and mood following the consumption of various alcoholic beverages (50) it should be investigated which substances bring them about.

Bourbon contains at least one biologically active phyto-oestrogen (plant-derived oestrogen-like substance) (13, 14, 43). It has been suggested that the effects of the consumption of alcoholic beverages, particularly as they relate to endocrine systems (e.g. feminization), are not resulting solely from ethanol.

Some people suffer from migraine attacks after consumption of some types of red wine. Although tyramine is known to cause migraine, the migraine-provoking wine in

Littlewood's experiment contained only 2 mg/l tyramine (26). He suggests phenolic flavanoids as responsible substances. Giesemann (16) suggests that migraine must be seen as an allergic reaction to fungi ('Schimmelpilzen', related to the ones used in the production of beer and wine.

Tannins (tannic acids), belonging to the group of phenols, are present in (mainly red) wine as well as in such products as tea and cocoa. Some tannins have exhibited measurable inhibitory activity towards various pathogenic organisms (3, 31). Some scientists (e.g. 18) regard the tannins from sorghum (used in the preparation of kaffir beer) as carcinogenic (see Chapter 12). However, it is not justified to regard the tannins in wine and tea as being carcinogenic too.

Tannin is a collective term for a group of substances each of which has an entirely different effect. Hydrolysable tannins are not only distinguished from condensed tannins nowadays, but the (most frequently occurring) condensed tannins differ from each other as regards activity. Some tannins prove to be harmful, other ones are almost or entirely harmless. Condensed tannins from grape pips proved not to be toxic when administered up to a dose of 2% in the food of laboratory animals (25). Further investigations into this matter are needed, also in view of the confusing nomenclature in older literature.

The Frenchman Masquelier (32) reports on his investigations into procyanidins in wine. Procyanidins are condensation products of 2 to 5 flavane-3-ols (catechins). With more than 5 flavane-3-ols the molecule is considered to be a tannin. Procyanidins in wine are said to have quite some positive physiological effects (e.g. antiviral effects, effects on collagens, cholesterol, free radicals, histamine), acting as an antidote of ethanol. To my knowledge, Masquelier's findings have not been confirmed by other studies into the role of procyanidins.

Congeners originating from the raw material or the fermentation process are not the only substances in alcoholic beverages. In the present-day production of most alcoholic beverages various additives are used. Many additives are of a cosmetic nature, used to improve the colour or body of the beverage, or are applied to retain the head in beer. There are at least 60 additives for beer and 20 for wines (37). Occasionally, additives give rise to accidents or affairs, like the case of ethylene glycol in Austrian wine in 1984, and the case of cobalt salts in Canadian beer in the 1960s.

Another group of congeners are the contaminants: substances which enter the beverages unintentionally during growth or harvest of the fruits and cereals or during processing. Traces of pesticides, asbestos, mycotoxins (such as aflatoxin), arsenic-containing insecticides and fungicides all have been demonstrated in alcoholic beverages, particularly in wine (5a).

Summary

Alcoholic beverages may be classified into three main groups: beer, wine (including fortified wine), and distilled drinks (including liqueur-like beverages). A recent development is the popularity of dealcoholized and low-alcohol wine and beer. In the

Netherlands standard glasses of wine, beer and spirits contain ca. 10 g ethanol each, except for glasses of sherry and port and some less current beverages which contain a little less alcohol. The ethanol concentration of a beverage is of importance in relation to damage to the gastrointestinal mucosa.

Besides ethanol and water (and sugar in sweet beverages), alcoholic beverages may contain hundreds of other substances which together characterize the beverage in question. These substances can be classified into macronutrients (carbohydrates and proteins), vitamins, minerals and trace elements, and congeners.

Small quantities of carbohydrates, proteins, vitamins and minerals may be present in alcoholic beverages, particularly in wine and beer. They do not contribute significantly to daily nutritional intake, possibly with the exception of some trace elements (e.g. silicium). However, even the significance of the contribution of solicium remains questinable. Consumption of wine and draught beer can result in a significant intake of lead (originating from lead capules on wine bottles and lead in processing and tap installations).

Congeners can originate from the raw material or be produced during fermentation (methanol, higher alcohols, phenols, acids). Other congeners are contaminants (pesticides, asbestos, mycotoxins) or additives (e.g. head-retaining agents, flavouring agents, preservatives).

The higher alcohols which may be present in alcoholic beverages, albeit in minute quantities, have an effect comparable to that of ethanol, but the intensiveness and the duration of the effects are more marked. Higher alcohols have also been associated with hangover symptoms. Methanol is present in small quantities in all alcoholic beverages, especially in distilled beverages made from fruit. In case of moderate consumption methanol intake does not cause intoxication. Methanol metabolites may play a role in the development of addiction.

Hardly anything is known about the physical effects of the other congeners, least of all of those caused by the mixture of minute quantities of these substances present in alcoholic beverages. An integrated approach by chemists, toxicologists and biomedical experts will be needed to be able to reach solid conclusions as to the biological activity of congeners in alcoholic beverages.

References

1. Adam L. Gaschromatographische Bestimmung flüchtiger Inhalsstoffe in alkoholischen Getränken. Dissertation, Technische Universität, München, 1973.
2. Alexander CS. Cobalt-beer cardiomyopathy. Circulation 1970;128(Suppl III): 470.
3. Anonymous. Wine and medical practice: a summary. San Francisco, CA: Wine Institute, 1979.
4. Anonymous. Methanol poisoning. Lancet 1983;i: 910-2.
5. Anonymous. Composition of foods: beverages (raw, processed, prepared). Washington DC: USDA, 1986. (Agriculture Handbook 8-14.)
5a. Anonymous. Chemical composition of alcoholic beverages, additives and contaminants. In: Berrino F, et al. Alcohol drinking. Lyon: WHO International Agency for Research on Cancer, 1988: 71-99. (IARC monographs on the evaluation of carcinogenic risks to humans 44.)

6. Anonymous. NEVO tabel: Nederlands voedingsstoffenbestand 1989/90. The Hague, Netherlands: Voorlichtingsbureau voor de Voeding/Zeist, Netherlands: NEVO Foundation, 1989.
7. Bonte W, Sprung R, Lesch OM. Ethanol-independent elimination of methanol in chronic alcoholics. In: Kuriyama K, Takada A, Ishii H, eds. Biomedical and social aspects of alcohol and alcoholism: proceedings of the Fourth Congress of the International Society for Biomedical Research in Alcoholism (ISBRA), Kyoto, Japan, 26 June–2 July 1988. Amsterdam: Elsevier Science Publishers, 1988: 637-40.
8. Darby WJ. The nutrient contributions of fermented beverages. In: Gastineau CF, Darby WJ, Turner TB, eds. Fermented food beverages in nutrition. New York: Academic Press, 1979: 61-82.
9. Daynes G. Gout and port. Br Med J 1979: 669.
10. Elinder CG, Lind B, Nilsson B, Oskarsson A. Wine: an important source of lead exposure. Food Addit Contam 1988;5: 641-4.
11. Fernandez C, Martin A. Contenido en elementos metalicos de vinos españoles. Alimentaria 1987;180: 75-83.
12. Gastineau CF, Darby WJ, Turner TB, eds. Fermented food beverages in nutrition. New York, Academic Press, 1979.
13. Gavaler JS, Rosenblum ER, Thiel DH, et al. Biologically active phytoestrogens are present in bourbon. Alcohol Clin Exp Res 1987;11: 399-406.
14. Gavaler JS, Imhoff AF, Pohl CR, Rosenblum ER, Van Thiel DH. Alcoholic beverages: a source of estrogenic substances? In: Lindros KO, Ylikahri R, Kiianmaa K, eds. Advances in biomedical alcohol research: third congress of the International Society for Biomedical Research on Alcoholism (ISBRA), Helsinki, Finland, 8–13 June 1986. Oxford: Pergamon Press, 1987: 545-9.
15. Gibel W, Lohs K, Wildner GP. Experimentelle Untersuchungen zur kanzerogenen Wirkung von Lösungsmitteln am Beispiel von Propanol-1,2-Methylpropanol-1 und 3-Methylbutanol-1. Arch Geschwulstforsch 1975;45: 19-24.
16. Giesemann G. Migräne nach Rotwein. Dt Med Wochenschr 1988;113: 1941.
17. Grab W. Pharmakologische Probleme bei Wein und Spirituosen. Arzneimittelforschung 1961;11: 73-79.
18. Heath A. Methanol poisoning. Lancet 1983;i: 1339-40.
19. Hillbom ME, Franssila K, Forsander OA. Effects of chronic ingestion of some lower aliphatic alcohols in rats. Res Commun Chem Path Pharmacol 1974;9: 177-80.
20. Jennings ME, Howard JMH. Chromium, wine, and ischaemic heart disease. Lancet 1080;ii: 90-1.
21. Kahn JH. Compounds identified in whisky, wine, and beer: a tabulation. In: Gastineau CF, Darby WJ, Turner TB, eds. Fermented food beverages in nutrition. New York: Academic Press, 1979: 505-17.
22. Kalish GH, Kaufman H. Headaches after red wine. Lancet 1981;i: 1263.
23. Klatsky AL. Effects of alcohol on the cardiovascular system. In: Gastineau CF, Darby WJ, Turner TB, eds. Fermented food beverages in nutrition. New York: Academic Press, 1979: 317-41.
24. Klausa A. Vorkommen und biologische Wirkung von Fuselöl in alkoholischen Getränken. Giessen, FRG: Institut für Ernährungswissenschaft, 1974.
25. Lang K. Biochemie der Ernährung. Darmstadt, FRG: Dietrich Steinkopf Verlag, 1974.
26. Littlewood JT, Glover V, Davies PTG, Gibb C, Sandler M, Rose FC. Red wine as a cause of migraine. Lancet 1988;i: 558-9.
27. López-Artíguez M, Grilo A, Soria L, Castro M, Repetto M. Levels of zinc, copper, and lead in wines from the area south of Seville. Bull Environm Contam Toxicol 1990;45: 711-7.
28. Maarse H, Visscher CA. Volatile compounds in food: Alcoholic beverages. Qualitative and quantitative data. Zeist, Netherlands: TNO Nutrition and Food Research, 1989.
29. Marks V, Taylor A. Lead foil on wine bottles. Lancet 1987;ii: 1473-4.
30. Martensson E, Olofsson U, Heath A. Clinical and metabolic features of ethanol-methanol poisoning in chronic alcoholics. Lancet 1988;i: 327-8.
31. Masquelier J. Vin et santé. Bull Office Int Vin 1979;52: 1025-35.

32. Masquelier J. Effets physiologiques du vin: sa part dans l'alcoolisme. Bull Office Int Vin 1988;61: 554-78.
33. Morton JF. Tentative correlations of plant usage and esophageal cancer zones. Econ Bot 1970;24: 217-26.
34. Murphree HB, Price LM. Cumpter time-series analysis of the EEG effects of alcoholic beverages. Fedn Proc 1966;25: 503.
35. Murphree HB, Greenberg LA, Carroll RB. Neuropharmacological effects of substances other than ethanol in alcoholic beverages. Fedn Proc 1967;26: 1468-73.
36. Parr RM. Silicon, wine and the heart. Lancet 1980;i: 1087.
37. Parratt T. Name your poison! A guide to additives in drinks. London: Robert Hale, 1990.
38. Piendl A, Geiger E, Hoffmann H. Über das Vorkommen von Äthanol und alkoholischen Begleitstoffen in Bier, Wein und Spirituosen. Brauwissenschaft 1977;30: 33-45.
39. van Poppel G, Schrijver J, Löwik MRH, Wedel M. Alcoholgebruik en vitaminestatus bij 35-jarige mannen. Voeding 1989;50: 118.
40. Postel W, Drawert F, Adam L. Aromastoffen in Branntweinen. In: Geruch- und Geschmackstoffe. Nürnberg, FRG: Verlag Hans Carl, 1975: 99-111.
41. Prokop L, Machata G. Höhere Alkohole und Äthanolwirkung beim Menschen. Blutalkohol 1974;11: 80-7.
42. Reinhard C. Über gaschromatographische Untersuchungen an Brennweinen. Branntweinwirtschaft 1970;110: 334-9.
43. Rosenblum ER, Van Thiel DH, Campbell IM, Eagon PK, Gavaler JS. Separation and identification of phytoestrogenic compounds isolated from bourbon. In: Lindros KO, Ylikahri R, Kiianmaa K, eds. Advances in biomedical alcohol research: third congress of the International Society for Biomedical Research on Alcoholism (ISBRA), Helsinki, Finland, 8–13 June 1986. Oxford: Pergamon Press, 1987: 551-5.
44. Schreier P, Drawert F, Abraham KO. Identification and determination of volatile constituents in Burgundy Pinot Noir wines. Lebensmittelwiss u. -technol 1980;13: 318-21.
45. Sherlock JC, Pickford CJ, White GF. Lead in alcoholic beverages. Food Addit Contam 1986;3: 347-54.
46. Smart GA, Pickford CJ, Sherlock JC. Lead in alcoholic beverages: a second survey. Food Addit Contam 1990;7: 93-9.
47. Souci SW, Fachmann W, Kraut H. Food composition and nutrition tables 1986/87, 3rd ed. Stuttgart, FRG: Wissenschaftliche Verlagsgesellschaft, 1986.
48. Sprung R, Bonte W, Lesch OM. Methanol: ein bisher verkannter Bestandteil aller alkoholischen Getränke: eine neue biochemische Annäherung an das Problem des chronischen Alkoholismus. Wien Klin Wochenschr 1988;100: 282-5.
49. Standhardt WG. Bier, wijn en gedestilleerd: verschillen in gebruik, bereiding, samenstelling en lichamelijke effecten. Tijdschr Alcohol Drugs 1983;9: 119-27.
50. Takala M, Pihkanen TA, Markkanen T. The effects of distilled and brewed beverages: A physiological, neurological and psychological study. Helsinki: Finnish Foundation for Alcohol Studies, 1957.
51. Thomson AD, Makumdar SK. The influence of methanol on intestinal absorption and utilization of nutrients. Clin Gastroenterol 1981;10: 263-93.
52. Todhunter EN. A historical perspective on fermentation biochemistry and nutrition. In: Gastineau CF, Darby WJ, Turner TB, eds. Fermented food beverages in nutrition. New York: Academic Press, 1979: 83-98.
53. Turner C. How much alcohol is in a 'standard drink'?: an analysis of 125 studies. Br J Addict 1990;85: 1171-75.
54. Vrij-Standhardt WG. Alcohol en nutriëntenvoorziening. In: Voeding en alcohol. Alphen aan den Rijn: Samsom Stafleu, 1986: 56-67. (Voeding en gezondheid 7.)
55. Wainwright T. The low alcohol revolution. Brew Guardian 1987;116: 16-8.

CHAPTER 15

Factors influencing behaviour in relation to alcohol

E. te Wierik

A great deal of attention has been paid to the aetiology of alcoholism. The real causes are still unknown. It must even be questioned whether a specific cause will ever be found. Various studies have tried to provide insight into the factors determining individual behaviour in relation to alcohol. The specific properties of alcohol as well as the individual's physical condition and personality play an important role. Factors influencing behaviour in relation to alcohol have been studied from the viewpoint of various disciplines. The following classification can be made (38):
– hereditary or constitutional studies,
– trait studies, also called 'psychological studies': investigations of the specific character features of the alcoholics,
– psychoanalytic studies, reconstructions of the alcoholic's development,
– learning theories,
– sociocultural studies,
– studies focusing on an interaction model,

15.1. Hereditary or constitutional studies

15.1.1. Family studies

The fact that alcoholism runs in families has been known for centuries. Aristotle, Plutarch ('Ebrii gignunt ebrios', Drunkards beget drunkards), old testament writers, doctors and preachers have commented on the familial nature of alcoholism. Alcoholism appears to occur more frequently among first-degree relatives of alcholics (161). Research on this subject has been carried out by many (2, 11, 114, 119). It has been estimated that about 40% of alcoholics have an alcoholic parent (70). Cotton (33) reviewed 39 family studies done over the preceding 40 years that involved a total of 6251 relatives of alcoholics and 4083 relatives of non-alcoholics. He concluded that an alcoholic is more likely than a non-alcoholic to have a mother, father or more distant relative who is an alcoholic. More recent studies continued to demonstrate familial aggregation in alcoholism. Guze et al. (59) observed that male relatives of alcoholics had higher rates of alcoholism than had female relatives. On the other hand, the sex of the proband did not influence the risk in his relatives. The observed transmission of liability to alcoholism does not fit the pattern one would expect when it was to be attributed to a single autosomal or sex-linked gene (4). Penick et al. (117) observed in a large multi-centre study comprising 568 male alcoholics that alcoholics with a positive family history (65%) had (1) an earlier onset of alcoholism, (2) greater alcoholic severity, (3) more medical and legal problems, (4) a broader range of treatments, (5) an increased life-time prevalence of additional psychiatric disorders and (6) a greater diversity of psychiatric disturbances among biological relatives. In their search for risk factors for alcoholism Nagoshi et al. (107) studied 35 subjects who reported to have an alcoholic parent or sibling (family history positive (FHP)) and who were matched with 35 controls (family history negative (FHN)). They found significant FHP decrements in cognitive performance prior to ingestion of alcohol. FHP subjects did not differ from FHN subjects in sensitivity and acute tolerance to alcohol or in perceived intoxication.

Despite the fact that family studies support the hypothesis that alcoholism is transmitted via kinship it cannot be established to what extent genetic or environmental factors contribute to this phenomenon. Studies among twins and adoptees have been carried out in attempts to distinguish between these aspects.

15.1.2. Twin studies

Monozygotic twins share all their genes whereas dizygotic twins share about 50% of their genes. Over 100 years ago, Francis Galton pointed out that twins provide natural experimental material enabling a separation to be made between the effects of nature and nurture.

The first study of alcoholism among twins was reported in 1960 by Kay (78). In that study alcoholism in monozygotic twins was found to occur twice as frequently as among dizygotic twins – a fact, according to Kay, clearly pointing to a genetic influence. Two other studies comparing the similarity in monozygotic twins with the similarity in dizygotic twins (58, 67) arrived at opposite conclusions as to the role of genetic factors in alcoholism.

The influence of heredity of normal drinking habits among non-alcoholic twins has been investigated by Partanen (115), Johnson & Nilson (74), Loehlin (88), Pederson (116), Kaprio (77) and others. The majority of studies agree that there is a substantial genetic effect on many aspects of normal drinking and that environmental influences increase with age. The similarity in drinking pattern of monozygotic twins may be related to the frequency of the mutual social contacts they have. To separate the effects of social interaction between monozygotic twins from their genetic similarity, Kaprio et al. (77) performed a stepwise multiple regression analysis, which showed that the greater social interaction between identical twins cannot account for their greater similarity in drinking pattern. The authors conclude that genetic variance significantly contributes to population variance in frequency, quantity and density of social drinking.

15.1.3. Adoption studies

One possibility of separating genetic and environmental factors is to study children who have been adopted by non-relatives immediately after birth.

Adoption studies began with Roe (134) who studied children who had been placed in institutions before the age of 10. No differences in drinking behaviour were found between children with and children without parental history of drinking problems. Roe concluded that alcoholism is presumably not hereditable. The incidence of alcoholism among children of alcoholics hence must result from environmental factors. The results of this study have been criticized because of the limited size and age range of the group of children studied.

The resurgence of the idea that inherited factors may be important determinants of alcohol abuse stems largely from studies performed by Goodwin and co-workers in the early 1970s in Denmark (52, 53, 54, 55). They investigated alcoholism and drinking patterns among adopted children with at least one parent who had been an alcoholic. The prevalence of alcoholism was found to be almost four times as high among adoptees with an alcoholic natural parent as in the control group of adoptees whose natural parents were non-alcoholics (52). Furthermore, the likelihood of the sons being alcoholic was related to the severity of parental alcoholism. No environmental factors predisposing to alcohol abuse were identified in this study. The increased susceptibility to alcoholism in sons raised by their alcoholic biological parents was about similar to that in their brothers who had been raised by non-alcoholic foster parents (53). For the daughters of alcoholic parents the findings were quite different (52, 54). There was no increase in prevalence of alcoholism compared with adopted-away daughters of non-alcoholic parents. The sisters of the adopted-away daughters of alcoholics who remained with their natural parents, however, had a higher incidence of depression, emphasizing the unfavourable effects of their remaining at home with an alcoholic parent (54, 55). These results seem, at first glance, to provide most convincing evidence that alcoholism has an important genetic basis in men, but not in women. The authors note that some evidence indicates that women develop alcoholism at a later age than do men, and possibly not all of these women in their thirties had entered the age of risk for alcoholism. Support for Goodwin's conclusions came from a Swedish study conducted by Bohman (14) who compared the alcoholism rate among 50 male adoptees who had an alcoholic father with that in a similar group of adoptees whose fathers had no history of alcoholism. Registrations for alcoholism were three-fold higher among the group with alcoholic fathers. When a similar analysis was made for female adoptees no difference was found between those who had an alcoholic biological parent and those who had not. Another adoption study was performed by Cadoret et al. (20) who studied a group of 127 male and 87 female adoptees. Antisocial personality and alcohol abuse were related to biological background and to environmental factors. In the men alcohol abuse increased by a background of problem drinking among first-degree biological relatives and by drinking problems in the adoptive home. An antisocial personality prevailed more frequently in men whose first-degree biological relatives had antisocial behaviour problems. Among the women the prevalence of alcohol abuse was elevated when first-degree biological relatives had a drinking problem.

15.1.4. The search for a biological marker

The evidence that alcoholism occurs more frequently among first-degree relatives of alcoholics, even when these relatives are adopted out of birth, and that the concordance rate is higher among monozygotic than among dizygotic twins, has prompted a search to identify possible genetically influenced factors that might contribute to the risk for alcoholism.

15.1.4.1. Electrophysiological markers of alcoholism
Ethanol ingestion has an effect upon the central nervous system (CNS). Two measures of CNS function are spontaneous brain activity measured by an electroencephalogram (EEG) and elicited activity via evoked potential (EP) or via event-related potential (ERP). The EEG patterns are known to be heritable and involve slow-wave (or alpha) activity (127). Male alcoholics and their sons have shown evidence of significantly decreased slow-wave activity on EEGs (48, 123, 126, 155). Begleiter and his co-workers have documented a phenomenon involving a positive polarity brain wave seen 300 milliseconds after an external stimulus has been applied. Among alcoholics and their sons this brain wave, termed P300, is significantly smaller in amplitude than among controls (7, 40, 108, 124). Begleiter notes that the possible predictive value of electrophysiological deficits in sons of alcoholics can only be assesed in longitudinal studies in which individuals at high risk and low risk for alcoholism are tested regularly over several years until they pass through the period of maximum risk for alcoholism (8). Other results suggest that the relationship between the P300 ERP and the inheritability of alcoholism is not very clear (121, 122).

15.1.4.2. Biochemical markers
Early investigations into the relationship between blood group and alcoholism revealed an association of blood group A with alcoholism (9, 109), but subsequent studies could not establish any relationship between ABO system and alcoholism (19, 23, 65, 131, 150).

Another approach to search for genetic factors in alcoholism is to study human lymphocyte antigens (HLA). The majority of studies, however, have not focused on alcoholism itself but on alcoholism-related diseases such as liver cirrhosis, steatosis and pancreatitis. Two studies defined alcoholism as the trait of interest. Robertson et al. (132) could not determine a difference in occurrence of HLA-A and HLA-B locus antigens between an alcoholic population and a control population. In another study HLA patterns of 30 alcoholics were compared with those of two control groups. HLA B40 and HLA DR4 occurred more frequently, and HLA DR3 less frequently among in alcoholics. The authors conclude that these findings are in agreement with the view that alcoholism has a genetic component (31).

One of the most cited markers in alcohol association studies is platelet monoamine oxidase (MAO). This enzyme is important in the degradation of brain neurotransmitters and hence may play a role in the regulation of mood and behaviour. The activity level of this enzyme is said to be low in alcoholics and in their sons (42, 91, 92, 93, 112). Knorring et al. (80, 81) applied the two types of alcoholism, Type I and Type II, classified by the Stockholm adoption study, to a clinical sample. Type I (milieu-limited) alcoholism was characterized by early onset and few social complications. Type II (male-limited) alcoholism was characterized by early onset and use and abuse not only of alcohol but also of other drugs. These two types of alcoholism could be distuinguished on the basis of platelet MAO activity. Platelet MAO activity was normal in Type I alcoholics as compared to healthy normals, but significantly lower in the Type II alcoholics. These results are consistent with other findings (149). In a recent study, however, significant differences in MAO activity between alcoholics and controls were observed, but the authors were unable to reproduce a platelet MAO activity among alcoholics specific for gender and subgroup (163). An increase in MAO activity has been noted upon alcohol withdrawal and abstinence (1, 104). In a recent study, the in vitro inhibition of monoamine oxidase by ethanol was found to be significantly higher in the platelets of alcoholics (151).

In conclusion, a link between MAO activity and genetic predisposition to alcoholism remains to be established unequivocally.

Another platelet enzyme investigated in relation to alcoholism is adenylate cyclase (AC). Tabakoff (151) and Watanabe (157) studied the activity of this enzyme in alcoholics and in non-alcoholic controls. No difference in basal activity was found, but platelet AC activity after stimulation was significantly lower in alcoholics and in alcoholics who had abstained for one to four years.

Schuckit and Gold (138) evaluated multiple aspects of the reaction to two doses of alcohol in 30 sons of alcoholics and 30 matched controls. A stepwise discriminant analysis showed that the hormones prolactin (after a low dose) and cortisol and the subjective feelings after a high dose of alcohol (discomfort, drug high, floating feelings, etc.) classified 83% of the controls and 70% of the sons of alcoholics correctly. These observations suggest that hormonal changes after a dose of alcohol may serve as a marker. In a study by Gianoulakis (49) differences in both the basal levels and the response of the cortisol and the β-endorphin system to an acute ethanol dose were noted between subjects at low and those at high risk for alcoholism.

The direct toxic effects of alcohol and alcohol-related physical alterations have been attributed to acetaldehyde rather than to alcohol itself. Acetaldehyde is the first catabolic product of alcohol oxidation and is much more toxic than alcohol itself. It has been suggested that the concentration of acetaldehyde, after drinking an acute dose of alcohol, in the blood of relatives of alcoholics rises to a significantly higher level than in controls, and this has been considered an indicator of a possible risk factor associated with alcoholism (137). But, as the authors themselves and others (41) emphasize, the method of blood acetaldehyde determination goes with many problems that can lead to erroneous values. Another method for the determination of blood acetaldehyde levels did not yield significant differences between alcoholics and their first-degree relatives (156).

Alcohol is broken down by alcohol dehydrogenase (ADH) to acetaldehyde which is metabolized by aldehyde dehydrogenase (ALDH). Multiple forms (isoenzymes) of ALDH have been found in the liver (61). A few years after the identification of these isoenzymes it was found that ALDH2 was missing in more than half of the Japanese people, whereas Germans appeared to have more isoenzymes (51). Orientals who are missing this isoenzyme are less likely than other orientals to drink heavily and appear to have a lower rate of alcoholism (62). In the absence of this isoenzyme ethanol oxidation results in significantly higher levels of blood acetaldehyde and consquently in a syndrome of facial flushing, tachycardia, sweating and nausea. Recently, the molecular and genetic basis of ALDH2 has been determined. The ALDH2 enzyme has two alleles (corresponding genes on the same locus of two homologous chromosomes) designated $ALDH2^1$ and $ALDH2^2$. The $ALDH2^1$ allele encodes information for the active form, whereas $ALDH2^2$ encodes information for the inactive form. The active and inactive forms of ALDH2 are different due to point mutation. In many cases, one allele encodes information that is dominant over the corresponding allele. In a study among Japanese subjects it was observed that all ALDH2-active livers had the $ALDH2^1/ALDH2^1$ genotype. The ALDH2-deficient livers had $ALDH2^1/ALDH2^2$ or $ALDH2^2/ALDH2^2$ genotypes. From these results it is concluded that the $ALDH2^2$ allele, the inactive allele, is dominant (34). Li et al. (87) investigated the ALDH2 genotypes of native Taiwanese and of Chinese living in Taiwan. The $ALDH2^2$ allele was found to occur in 50% of the Chinese and in only 15% of the native Taiwanese. The majority of Chinese alcoholics have the $ALDH2^1/ALDH2^1$ genotype for the active ALDH enzyme. Only a small number of the Chinese alcoholics possessed the other two genotypes for the inactive ALDH enzyme. These data are consistent with the results obtained in the Japanese study (140). So, one genetically determined enzyme seems to influence the chance of developing alcoholism among Orientals, thus providing a protective factor for heavy drinking.

15.1.4.3. DNA research
The aim of genetic analysis is to detect the location of the defective gene and to discover the nature of the mutation in the DNA molecule that has led to deviation in its function. Three strategies are commonly used to type the genetic markers (36). The first and most attractive is the candidate-gene approach. Numerous candidate genes have been mentioned some of which have been discussed as biochemical markers. A list of candidate genes in alcoholism is presented in Table 15.1. The second strategy is the so-called 'shot-gun' method (i.e. using large numbers of DNA probes to cover the entire genome). The third strategy is the use of hypervariable DNA probes to look for a close association with specified DNA sequences.

Table 15.1. Candidate genes in alcoholism (36).

Candidate gene	Chromosome location	References
Alcohol dehydrogenase 1	4q21–25	142
Alcohol dehydrogenase 2	4q21–25	142
Alcohol dehydrogenase 3	4q21–25	142, 113
Alcohol dehydrogenase 4	4q21–25	142
Alcohol dehydrogenase 5	4q21–25	142
Aldehyde dehydrogenase 1	9q21	130
Aldehyde dehydrogenase 2	12q24	142
Aldehyde dehydrogenase 3	17	136
Aldehyde dehydrogenase 4		
Formaldehyde dehydrogenase	4q21–25	142
Cytochrome P450IIE1	10	97
Fatty acid ethyl ester synthase		36
Esterase D	13q14	143
Transketolase		36
2,3-butanediol dehydrogenase		36
Monoamine oxidase A	Xp11–21	111
Monoamine oxidase B	Xp11–21	
Propiomelanocortin	2p23	43
Adenylate cyclase		
Phospholipase D		105
Arylsulphatase A	22q13–qter	69
MNS blood group	4q28–31	95
HLA-A,B,C,DR	6p21	118
Dopamine D2	11q22–23	13

However, the latter two strategies are not very useful.

Tanna et al. (152) performed a linkage study in which associations between alcoholism and 30 polymorphic marker loci were studied in 42 families. The authors suggest that a linkage may exist between a gene for alcoholism and the esterase-D locus on chromosome 13q.

15.2. Psychological studies

Alcoholics are supposed to have a type of personality predisposing them to turn into alcoholics. In other words, alcoholics are assumed to possess a common basis of personal characteristics. Many have studied the relationship between alcohol consumption and personality. Results of some of these studies are reviewed below. Most studies of personality factors bearing on the development of alcoholism are retrospective. Specific traits found may consequently both cause alcoholism and result from it (6). Heavy drinking has been related to:
– impulsiveness (64, 76, 39),
– tendency towards being easily upset (60),
– low self-esteem (18, 64, 71, 85, 102, 133, 162),

- poor hostility control and low tolerance towards frustration, anxiety and depressive feelings (64, 73, 102),
- low interpersonal orientation (25),
- higher extraversion scores (39, 85, 133)
- higher neurotism scores (85, 133)
- higher activity and sociability (39),
- bored temperament / longing for sensation (72, 100)
- low docility (100), and
- fear of failure (72).

It has been observed that Type A individuals (47), described as exhibiting impatience and a chronic sense of time pressure, enhanced competitiveness, aggressive drive and (often) some tendency towards hostility, drink more frequently than do Type B individuals who lack these traits (44). In another study (50) no relationship between Type A behaviour and alcohol consumption was found. Mayer (96) studied the relation between alcohol involvement and eighteen personality characteristics. The eight characteristics highly related to alcohol involvement as shown by regression analysis were, in descendant order: social presence, tolerance, achievement via conformance, socialization, flexibility, responsibility, self-control and sense of well-being. In a study of Forsyth and Hundleby the desire to drink among young adults was greater in (a) both stressful and convivial situations for those who scored higher on neurotism, (b) convivial situations for those who scored higher on depression, and (c) boring situations for those who scored higher on sensation seeking (45).

There are a few prospective studies which tried to solve the cause-effect relation. McCord & McCord (98) followed the possible development of alcoholism among a group of boys from deprived families ($n = 225$) and a control group ($n = 225$) over a period of 18 years. In these groups alcoholism manifested in 29 and 22 individuals, respectively. A number of shared personality traits such as aggressiveness, feelings of inferiority and hyperactivity were found. Jones (73) studied a group of individuals over a 30-year period, starting at school age. The results of this study indicate that alcohol-related behaviour, to some extent, reflects personality traits that were demonstrable even before the drinking pattern became manifest. These findings were confirmed in a study by Tarter et al. (153). Jones (73) found that eventual problem drinkers differed as children from moderate drinkers and non-drinkers by uncontrolled, impulsive and rebellious behaviour. Cloninger et al. (27) observed that novelty seeking, harm avoidance and reward dependence were predictive characteristics of later alcohol abuse. Schulsinger et al. (139) used for their study a Danish birth cohort. From the cohort 134 sons of alcoholic fathers (high-risk group) and 70 matched controls without parental alcoholism were selected. When the subjects were about 19 years old the high-risk group was characterized by poor verbal ability and impulsive behaviour. The authors indicate that these characteristics may result from paternal psychopathy rather than from alcoholism per se; a follow-up study has been planned. In a similar study, Sieber & Angst (141) questioned a birth cohort of 1577 men at the age of 19 and again at the age of 31. Personality, social background and substance use were connected with subsequent consumption of alcohol, tobacco and cannabis. Antisocial attitudes, depressiveness and psychosomatic complaints were found to be non-specific indicators. Specific predictors for the level of alcohol use were parental consumption and self-description as aggressive and extraverted. In another prospective study, in descendant order, antisocial personality disorder, gender and familial history of alcoholism were found to be predictors of alcoholism (144).

In conclusion, no single personality type is characteristic of all alcoholics although alcoholics share some common characteristics which may be already present before drinking starts.

15.3. Psychoanalytic studies

Most of these studies are founded on Freudian and neo-Freudian theories. According to Freud (46) the pregenital years are the most critical in the formation of personality structure. In the first five years of a child's life the pregenital (oral, anal and phallic) stages of development are believed to be experienced. In Freud's theories alcoholics have never passed beyond the oral phase of development in which the mouth maintains contact with the world. The oral character is taken to refer to a predilection to smoking, eating, drinking and munching sweets.

As a rule, psychoanalists define alcoholism in terms of dependence on a substance, an actitivy or a person believed to provide enjoyment on the one hand and relief of mental pain on the other (12, 24, 89). This dependence of the addict is the consequence of a failing development. In Dominicus' model (37) alcoholism is seen as a mixed bifactorial neurosis, composed of two personality characteristics: the neurotic primary disorder and the neurotic alcohol hunger component. These two neurotic characteristics develop according to a three-stage scheme (Fig. 15.1). In the first stage, the premorbidity stage, the primary disorder is manifest; the alcohol hunger component is still latent although a disposition to abuse is already present, which can be recognized retrospectivily. In this stage the premorbidic personality, i.e. neurotic symptoms, is expressed. During the second stage, the habituation stage, a raising alcohol abuse becomes evident. Normal social drinking is alternated with problem drinking. In the third stage, the depending stage, the hunger component dominates the primary disorder and alcohol abuse is there. These three stages can be preceded by a preliminary stage which reaches from birth until puberty. In this preliminary stage questions of guilt could arise.

Alcoholism is considered a substitute for an emotionally adult adaptation to living with conflicts. In this way addictions protect the individual against more serious consequences of his failures, such as suicide, psychosis, and antisocial and criminal behaviour (24). Other authors mention unconscious guilt as a cause of alcoholism and antisocial behaviour (99).

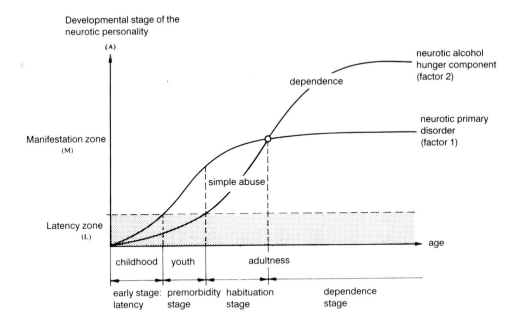

Fig. 15.1. Development of two personality characteristics in alcoholics, dependent on age (three-stage scheme) (37).

Some analysts relate behaviour with respect to alcohol to unsolved conflicts associated with the relationship with parents in early youth. The behaviour of the adult alcoholic is seen as a reflection of what is in fact a pathological remembrance of what/who parents were or is based on a distorted image, impressions from childhood and adolescence (12, 24). One problem inherent in this approach is the frequency at which similar personality syndromes may predispose towards many other types of deviant behaviour such as kleptomania, drug addiction and gambling (12). A critical treatise on this problem has been written by Blum (12).

15.4. Learning theories

In terms of learning or 'reinforcement' theories, behaviour is defined as a stimulus-response action. Learning is the interaction between the individual, the environment and specific stimuli in this environment (120). The way react to stimuli (e.g. alcohol) in various situations changes continuously. Personality traits are important factors. Part of the continuous learning process passes through its association with certain 'things'. An example regarding alcohol is that drinking behaviour may have different effects. If the response to a certain stimulus, such as alcohol, proves positive, this behaviour will be reinforced. This means that if drinking alcohol is experienced as enjoyable, it will be continued. If the response is negative, this inclination will be less strong. In other words, success leads to repetition and the individual is conditioned with respect to alcohol by a positive reward (relaxing or pleasant experiences), or by a negative effect (a hangover or a headache). Habituation resulting from a constantly repeated experience develops in many people. The effects become less strong and interest decreases. This may be the way excessive drinkers grow out of their drinking habits (120). In terms of learning theories alcoholism is regarded as a conditioned behavioural response which can be unlearned by appropriate modifications of environmental stimuli and reinforcement situations (148).

Bandura (5) developed a two-stage model to account for the causes of alcoholism. He finds the positive value of alcohol to derived first from central suppression and the stupefying properties of alcohol. Its pharmacological effect does in fact alleviate stress situations. Drinking behaviour will be stimulated because it is followed by a diminishment of the unpleasant experience. Repeated diminishment of fear, tension, or other disagreeable stimuli will result in a progressive reinforcement of drinking habits. Once the habit has become established, the excess of alcohol has adverse effects on the individual who in turn creates new stimulus conditions in order to continue drinking. Prolonged excessive drinking will eventually cause physiological changes in the body, resulting in physiological dependence.

The idea that alcohol can decrease anxiety dates back to antiquity. Kushner et al. (84) conclude that the relationship between alcohol problems and anxiety appears to be variable among the anxiety disorders. In agoraphobia and social phobia, alcohol problems appear more likely to follow from attempts at self-medication of anxiety symptoms, but panic disorder and generalized anxiety disorders may be more likely to follow from pathological consumption. The main starting point is that alcoholics start drinking and continue to drink while feelings of anxiety, fear, stress and tension exist (10, 75, 86, 103, 146, 147), or when events like 'loss' and 'danger' are present (56). Two possible explanations for drinking under these circumstances may be that intoxicated persons become less sensitive to negative social feedback by (1) isolating the emotional components aroused by such feedback from its cognitive content, or (2) isolating cognitive elements from other cognitive elements that in the sober state are connected to each other through association (57).

Besides these feelings, social variables play an important role in the onset of this behaviour (94, 164).

15.5. Socio-cultural studies

15.5.1 Cultural factors

An explanation of the development of differences in drinking behaviour is sought in the cultural sphere. In countries where proscription against alcohol use exists alcoholism is presumably more frequent than in countries where prescriptions for alcohol use are given. Cultures that teach children to drink responsibly, cultures that have ritualized when and where to drink tend to have lower rates of alcohol abuse than do cultures that forbid children to drink. Moreover, how a society socializes drunkenness is as important as how it socializes drinking. For example, both France and Italy learn their children responsible drinking practices, but public drunkenness is more socially acceptable in France than in Italy. This may be the reason why France experiences a higher rate of alcohol abuse. The Irish and American Indian forbid children and adolescents to drink, but they praise the capacity of men to drink large amounts of alcohol. In these cultures alcohol dependence occurs more frequently (154).

In a study among Australian, American and Papua New Guinean students, the majority of Australian and American students had drunk alcohol while the majority of Papua New Guinean young people had never drunk alcoholic beverages. The latter ones gave higher ratings of reasons for drinking and emphasized more than the other students that alcohol was related to feelings of being important and to friendship (158). It has been observed that West Germans consume more alcohol than Americans. This may be due to the cultural habit of Germans to support the consumption of alcohol (beer in particular) (30). Average consumption measured in Switzerland, Germany and the Netherlands was highest in Germany followed by Switzerland (79). Caetano (22) reported that U.S. Hispanics who are more acculturated embark more frequently on social events and drink more frequently in a number of social settings than less acculturated Hispanics.

15.5.2. Family relationships

Family studies show a higher prevalence of alcoholism among parents, brothers and sisters of alcoholics than in the entire population. An explanation may be the presence of a genetic factor. Another explanation is that the community transmits culture through a socialization process, a process in which parents and other educators play an important role, specifically through the transmission of drinking behaviour and the approval of drinking (63, 110, 145, 160, 165). Wilks & Callan (159) presented in their study 32 convivial situations and asked parents and their children in which situations they supported drinking. Daughters differed less in views from their parents than did sons. According to the authors differences between parents and their children are to be expected as older people reduce their number and variety of situations in which they drink, and older people may therefore make judgements according to their current drinking practices. On the other hand, similarities between children and their parents may reflect earlier socialization experiences and close contact.

15.5.3. Influence of the peer group

While an important part of what is learned is acquired at an early age, social learning is a continuous process. Initially learning is restricted to a relatively small (family) circle. Later on, an individual may participate simultaneously in several other groups, consisting of schoolmates or members of a sports club. New abilities acquired and new attitudes adopted as a participant in these groups may differ from those which the family teaches. The members of groups may influence each other in various spheres, such as education, clothes, drug use and also alcohol use. The influence of the peer group is held to be of particular importance during adolescence and continues the whole of one's life. From teen age onwards drinking will more and more take place away from home and the importance of the influence of friends parallels this changing situation. An explanation of social drinking behaviour might be social pressure exerted by peers. Rabow et al. (128) observed that the obligation to serve alcohol on social occasions influences both the frequency of drinking and the amount consumed. In other studies a relationship between drinking behaviour and perceived normative support for drinking has been found (63, 129, 160).

15.5.4. Personal variables

15.5.4.1. Sex and age
Results of a study in the USA (83) show that more women then men are lifelong abstainers, and that abstainers are older than alcohol consumers. This result shows a historical change since these young non-abstainers (18–34 years) cannot be lifelong abstainers when they are old. An increase in heavy drinking in the USA among people aged 21–34 has been observed before (66). More younger people than older people drink moderately or heavily and more men than women drink moderately or heavily (83). Aso a study in New York (3) indicated that more men than women drink moderately at least. In that study it was also observed that men reduce drinking when they are 50 years or over and that the younger one starts drinking, the more is consumed in later life. It has also been noted in Switzerland, Germany and the Netherlands (79) as well as in other studies (17, 32, 135) that alcohol consumption is higher among men than among women and higher among younger than among older people.

15.5.4.2. Socio-economic status (SES)
Education and/or profession and/or income of the breadwinner is generally taken as an index of social status. A profession is often associated with status or position of authority and may also be an indication of income. The income in turn indicates how large the financial means are which are disposable for the purchase of alcoholic beverages. That is to say, it is one of the factors playing a role in the availability of alcohol.

In the study of Knupfer (83) SES was based on income and education. He observed in both men and women that the lower the SES, the higher the proportion of abstainers. So, there are larger proportions of heavy drinkers in the higher SES groups. There were no differences in SES among the men who were frequently drunk. A positive association between drinking and SES has been observed in other studies (39, 50, 60, 68). Knibbe & Lemmens (79) found a difference between the sexes: in Germany, the Netherlands and Switzerland (albeit in the latter country the association was not significant) women consumed more alcohol when they had received a higher education. With the exception of the Netherlands, men with a higher education consumed less alcohol than those with a lower education. Higher alcohol consumption by educated women but not by their male counterparts was also observed by Braddon et al. (17). In a Swedish study (135) it was found that mean alcohol consumption was roughly the same among various socio-economic and educational categories for both sexes. However, among young people the proportion of heavy users was higher in both sexes among workers and those with a lower formal education. On the other hand, among older people the proportion of heavy users was generally higher among those with a higher formal education. In addition, Filipino males with a low income were most likely to be heavy drinkers (90).

15.5.4.3. Marital state
A relationship between single state and heavy drinking has been noticed in several studies (32, 73, 135). Knibbe & Lemmens observed that divorced German and Dutch men and women have a higher alcohol consumption than widowed and unmarried people. These findings are in agreement with other studies (35, 39, 50). Power & Estaugh (125) performed a study into the role of marriage and other partnerships and of the formation of families on the development of the drinking behaviour of young adults. Drinking was most strongly associated with the family formation variables: marital state, having children (moderate drinkers) and experiencing the breakup of a partnership (heavy drinkers). This finding was consistent for both sexes.

15.5.4.4. Religion
Studies relating to the differences in drinking behaviour between religious groups show that religion plays a part in the development of drinking behaviour. It appears that Protestants and Catholics consume more alcohol than Jews (3, 101). Protestants are more likely to abstain while Catholics are more permissive in their attitudes towards drinking (106). Atheists are found to have the highest alcohol consumption (35, 90). Cosper et al. (32) found that, among other variables, having a religion other than Protestant was a predictor of regularity in tavern visit. Irreligiosity was one of the other predictors. Cochrane & Bal (29) observed that Sikhs living in Britain were most likely to be regular drinkers, followed by white Britons and Hindus. The very few Muslim men who drank did not go to the mosque every week and consumed the highest amount of alcohol on average.

15.6. Interaction studies

The aetiology of alcoholism is best understood within the context of a framework that includes more than one factor: a multifactoral interaction model. Cloninger et al. (26) developed a model that captures the interactive relationship between the earlier

hereditary and environmental findings. Studying samples of adoptees they identified two distinct subtypes of alcoholism that are distinguished in terms of distinct alcohol-related symptoms, personality traits, ages of onset and patterns of inheritance:
– Type I ('milieu-limited') alcoholism is characterized by anxious (passive-dependent) personality traits and rapid development of tolerance and dependence on the anti-anxiety effects of alcohol. This leads to loss of control, difficulty in terminating binges, guilt feelings and liver complications following socially encouraged exposure to alcohol intake. This subtype demonstrates a significant gene-environment interaction.
– Type II ('male-limited') alcoholism is characterized by antisocial personality traits and a persistent longing for alcohol for its euphoriant effects. This leads to an early onset of inability to abstain entirely, as well as fighting and arrests when drinking. This subtype tends to be highly heritable from father to son across different social backgrounds and is associated with earlier onset of alcohol problems and criminality in the biological father.

In a companion study (15) these authors focused on a female sample and found, as predicted, that daughters with type I background were more likely to be alcohol abusers than were daughters of type II families. Furthermore, they found that type II daughters were more likely to have prominent somatic complaints as evidenced by their excessive physical complaints and a high level of absenteeism. In other words, given the same genetic predisposition, men usually express type II alcoholism and women express somatization. Somatization in women and type II alcoholism in men are associated with antisocial personality traits (16, 28). In another study these two types of alcoholism were compared to personality traits (82). Type II alcoholics had significantly higher scores than type I alcoholics on somatic anxiety and verbal aggression scales and significantly lower scores on socialization and inhibition of aggression scales. Furthermore, on the impulsive sensation-seeking psychopathy factor, type II alcoholics had significantly higher values. These results indicate that alcoholism accompanied with antisocial behaviour should be kept separate from alcoholism unrelated to antisocial behaviour. It is likely that type II alcoholics are similar to alcoholics with antisocial behaviour (82).

Another model is shown by Cadoret et al. (20) (Fig. 15.2). This model is based on the findings in the Lutheran Social Service data and has been verified by an adoption study done with Iowa Children's and Family Services (ICFS). This model shows that (1) biological family alcohol-related problems predict increased alcohol abuse in the adoptee, (2) biological family antisocial behaviour predicts increased antisocial personality diagnosis in the adoptee, and (3) environmental factors of alcohol-related problems in the adoptive family predict increased adoptee alcohol abuse.

Donovan (38) reviewed and critized studies into the aetiology of alcoholism and integrated hereditary, environmental and psychostructural risk factors into the design presented in Fig. 15.3. He observed four guideliness for future research:
– alcoholism is multifactorial in cause, and studies of alcoholic aetiology must be truly multifactoral;

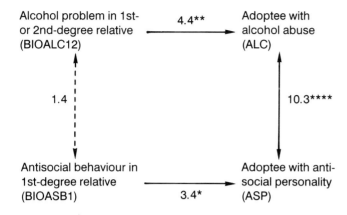

Fig. 15.2. Best-fitting four-factor model for ICFS data (solid lines show relationships found by analysis; broken lines show relationships forced into model for conditioning). P value of fit 0.42. *$P<0.05$; **$P<0.001$; ****$P<0.00001$. The model is based on a sample of 150 adult men. After: Cadoret et al. (20).

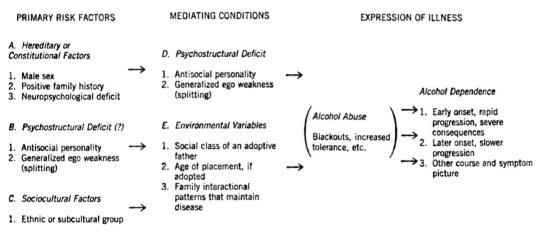

Fig. 15.3. An aetiological model of alcoholism (38).

– specific groups of patients should be studied separately because one may suspect that different risk factors lead to different forms of illness;
– personality variables must be presented in continuous versus categorial forms, irrespective of the design;
– because the morbidity of alcoholism can distort the perception of premorbid factors, research designs should be prospective whenever possible.

Summary

The study of factors influencing alcohol-related behaviour has been approached from various disciplines. A classification comprising hereditary, psychological, learning, socio-cultural and interactional studies has been made.

Heredity studies, mainly family, twin and adoption studies, emphasize the hypothesis that alcoholism is transmitted via kinship. Based on this evidence a search for a biological marker that might be a risk factor for developing alcoholism has arisen. Many potential markers have been suggested, but one single risk factor cannot yet be established.

Psychological studies used to be founded on the thesis that alcoholics have specific personal traits in common which were thought to be crucially important to the development of alcoholism. Formerly, attempts have been made to describe the so-called 'alcoholic personality', in other words, a constellation of characteristics and attitudes which together constitute a specific psychological susceptibility to develop alcoholism.

In psychoanalytical studies theories about development of alcoholics are presented.

Learning theories are based on the theory that learning is an interaction between the individual, the environment and specific stimuli in this environment (such as alcohol). According these theories alcoholics start drinking because alcohol can decrease negative emotions.

Socio-cultural investigations into drinking behaviour focus on various problems, such as drinking patterns and alcohol consumption, the development of problem drinking and alcoholism among various cultural, ethnic, social and regional groups. These theories refer to family, religion and demographic variables such as age, sex, social status, marital state and ethnicity.

Finally, interaction studies, which try to include more than one factor in one model – multifactoral interaction models – are described.

References

1. Agarwal DP, Philippen G, Milech U, Ziemsen B, Schrappe O, Goedde HW. Platelet monoamine oxidase and erythrocyte catechol-o-methyltransferase activity in alcoholism and controlled abstinence. Drug Alcohol Depend 1983;12: 85-91.
2. Amark C. A study in alcoholism: clinical, social-psychiatric and genetic investigations. Acta Psychiatr Neurol Scand 1951 (suppl).
3. Anonymous. Het alcoholgebruik in de staat New York. ANDO – Nuchter Bekeken 1989 (Sep).
4. Bandura A. Principles of behavior modification. New York: Holt, Rinehart and Winston, 1969: 528-54.
5. Barnes GE. The alcoholic personality: A review of the literature. J Stud Alcohol 1979;40: 571-634.
6. Begleiter H, Porjesz B, Bihari B, Kissin B. Related potentials in boys at high risk for alcoholism. Science 1984;225: 1493-6.
7. Begleiter H, Porjesz B. Neuroelectric processes in individuals at risk for alcoholism. Alcohol Alcoholism 1990;25: 251-6.

8. Bilal AM, Makhawi B, Al-Fayez G, Shaltout AF. Attitudes of a sector of the Arab-Muslim population in Kuwait towards alcohol and drug misuse: an objective appraisal. Drug Alcohol Depend 1990;26: 55-62.
9. Billington BF. Note on distribution of blood groups in bronchiectasis and portal cirrhosis. Australas Ann Med 1956;5: 20-2.
10. Birnbaum IM, Taylor TH, Parker S. Alcohol and sober mood state in female social drinkers. Alcoholism 1983;7: 362-8.
11. Bleuter M. Familial and personal background of chronic alcoholics. In: Diethelm O, Springfield, Charles C, eds. Etiology of chronic alcoholism. Thomas, 1955.
12. Blum EM. Psychoanalytic views on alcoholism. Q J Stud Alcohol 1966;27: 259-99.
13. Blum K, Noble EP, Sheridan PJ, Montgomery A, Ritchie T, Jagadeeswaran P, Nogami H, Briggs A, Cohn B. Allelic association of human dopamine D2 receptor gene in alcoholism. JAMA 1990;263: 2055-60.
14. Bohman M. Some genetic aspects of alcoholism and criminality: a population of adoptees. Arch Gen Psychiatry 1978;35: 269-76.
15. Bohman M, Cloninger CR, Sigvardsson S. Maternal inheritance of alcohol abuse: cross-fostering analysis of adopted women. Arch Gen Psychiatry 1981;38: 965-69.
16. Bohman M, Cloninger CR, Sigvardsson S, von Knorring AL. The genetics of alcoholics and related disorders. J Psychiat Res 1987;21: 447-52.
17. Braddon FEM, Wadworth MEJ, Davies MC, Cripps HA. Social and regional differences in food and alcohol consumption and their measurements in a national birth cohort. J Epidemiol Community Health 1988;42: 341-49.
18. Braucht GN, Brakarsh D, Follingstad D et al. Deviant drug use in adolescence: A review of psychosocial correlates. Psychosoc Bull 1973;79: 92-106.
19. Buckwalter JA, Pollock CB, Hasleton G, Krohn JA, Nance MJ, Ferguson JL, Bondi RL, Jaccobsen JJ, Lubin AH. The Iowa blood type disease research project II. J Iowa State Med Soc 1964;54: 58-66.
20. Cadoret RJ, O'Gorman TW, Troughton E, Heywood E. Alcoholism and antisocial personality. Arch Gen Psychiatry 1985;42: 161-7.
21. Cadoret RJ, Troughton E, O'Gorman W. Genetic and environmental factors in alcohol abuse and antisocial personality. J Stud Alcohol 1987;48: 1-8.
22. Caetano R. Acculturation, drinking and social settings among U.S. Hispanics. Drug Alcohol Depend 1987;19: 215-26.
23. Camps FE, Dodd BE, Lincoln PJ. frequency of secretors and nonsecretors of ABH groups substances among 1000 alcoholic patients. Br Med J 1969;4: 457-9.
24. Clare AW. The causes of alcoholism. In: Grant M, Gwinner M, eds. Alcoholism in perspective. Drug Abuse Alcohol Newsl 1975;4: 1-4.
25. Clark DC, Daughtery SR, Zeldow PB, Eckenfels EJ, Silverman CM. Alcohol use patterns of first-year medical students. II. Psychosocial characteristics associated with drinking level. Alcoholism Clin Exp Res 1986;10: 65-70.
26. Cloninger CR, Bohman M, Sigvardsson S. Inheritance of alcohol abuse. Arch Gen Psychiatry 1981;38: 861-8.
27. Cloninger CR, Sigvardsson S, Bohman M. Childhood personality predicts alcohol abuse in young adults. Alcoholism Clin Exp Res 1988;12: 494-505.
28. Cloninger CR, Sigvardsson S, Gilligan SB, von Knorring AL, Reich T, Bohman M. Genetic heterogeneity and the classification of alcoholism. In: Gordis E, Tabakoff B, Linnoila M, Stimmel B, eds. Alcohol research from bench to bedside. New York: Haworth Press, 1989: 3-16.
29. Cochrane R, Bal S. The drinking habits of Sikh, Hindu, Muslim and white men in the west Midlands: a community survey. Br J Addict 1990;85: 759-69.
30. Cockerham WC, Kunz G, Lueschen G. Alcohol use and psychological distress: a comparison of Americans and West Germans. Int J Addict 1989;24: 951-61.
31. Corsico R, Pessino OL, Morales V, Jmelninsky A. Association of HLA antigens with alcoholic disease. J Stud Alcohol 1988;49: 546-50.
32. Cosper RL, Okraku IO, Neuman B. Tavern going in Canada: a national survey of regulars at public drinking establishment. J Stud Alcohol 1987;48: 252-9.

33. Cotton NS. The familial incidence of alcoholism. J Stud Alcohol 1979;40: 89-116.
34. Crabb DW, Edenberg HJ, Bosron WF, Li TK. Genotypes for aldehyde dehydrogenase deficiency and alcohol sensitivity. The inactive ALDH^2 allele is dominant. J Clin Invest 1989;83: 314-6.
35. Crawford A. Self-reported alcohol consumption among population sub-groups in three areas of Britain. Drug Alcohol Depend 1988;21: 161-7.
36. Devor EJ, Reich T, Cloninger CR. Genetics of alcoholism and related end-organ damage. Semin Liver Dis 1988;8: 1-11.
37. Dominicus JE. Das psychobiologische Krankheitsmodell der bifaktoriellen Neurosenstrukturs des Alkoholismus. Drogalkohol 1990;14: 3-17.
38. Donovan JM. An etiologic model of alcoholism. Am J Psychatry 1986;143: 1-11.
39. Edward AM, Schork MA, Harburg E, Moll PP, Burns TL, Ozgoren F. Sources of variability in quantitive levels of alcohol use in a total community: sociodemographic and psychosocial correlates. Int J Epidemiol 1986;15: 82-90.
40. Elmasian R, Neville H, Woods D, Schuckit M, Bloom F. Event-related potentials are different in individuals at high risk for developing alcoholism. Proc Natl Acad Sci USA 1982;79: 7900.
41. Eriksson CJF. Elevated blood acetaldehyde levels in alcoholics and their relatives: a reevalution. Science 1980;207: 1383.
42. Faraj BA, Lenton JD, Kutner M, Camp VM, Stammers TW. Lee SR, Lolies PA, Chandora D. Prevalence of low monoamine oxidase function in alcoholism. Alcoholism Clin Exp Res 1987;11: 464-7.
43. Feder J, Gurling HMD, Darby J, Cavalli-Sforza LL. DNA restriction fragment analysis of the pro-opiomelanocortin gene in schizophrenia and bipolar disorders. Am J Hum Gen 1985;37: 186-94.
44. Folsom AR, Hughes JR, Buchler JF, Mittelmark MB, Jacobs DR, Grimm RH. Do Type A men drink more frequently than Type B men? Findings in the Multiple Risk Factor Intervention Trial (MRFIT). J Behav Med 1985;5: 27-235.
45. Forsyth G, Hundleby JD. Personality and situation as determinants of desire to drink in young adults. Int J Addict 1987;22: 653-69.
46. Freud S. The standard edition of the complete psychological works. Strachey J, ed. London: Hogarth 1953.
47. Friedman M, Rosenman RH. Association of specific overt behavior patter with blood and cardiovascular findings. JAMA 1959;169: 1286-95.
48. Gabrielli WF, Mednick SA, Volavka J, Pollock VE, Schulsinger F, Itil TM. Electroencephalograms in children of alcoholic fathers. Psychophysiology 1982;19: 404-7.
49. Gianoulakis C, Béliveau D, Angelogianni P et al. Different pituitary β-endorphin and adrenal cortisol response to ethanol in individuals with high and low risk for future development of alcoholism. Life Sci 1989;45: 1097-109.
50. Glynn RJ, de Labry LO, Hou DM. Alcohol consumption, Type A behavior and demographic variables. Am J Epidemiol 1988;127: 310-20.
51. Goedde HW, Harada S, Agarwahl DP. Racial differences in alcohol sensitivity: a new hypothesis. Hum Gen 1979;51: 331-4.
52. Goodwin DW, Schulsinger F, Hermansen L et al. Alcohol problems in adoptees raised apart from alcoholic biological parents. Arch Gen Psychiatry 1973;28: 238-43.
53. Goodwin DW, Schulsinger F, Moller N et al. Drinking problems in adopted and nonadopted sons of alcoholic. Arch Gen Psychiatry 1974;31: 164-9.
54. Goodwin DW, Schulsinger F, Knop J, Mednick S, Guze SB. Alcoholism and depression in adopted-out daughters of alcoholics. Arch Gen Psychiatry 1977;34: 751-5.
55. Goodwin DW, Schulsinger F, Knop J, Mednick S, Guze SB. Psychopathology in adopted-out daughters of alcoholics. Arch Gen Psychiatry 1977;34: 1005-9.
56. Gorman DM, Peters TJ. Types of life events and the onset of alcohol dependence. Br J Addict 1990;85: 71-9.
57. Gustafson R, Källmén H. Changes in the psychological defence system as a function of alcohol intoxication in men. Br J Addict 1989;84: 1515-21.

58. Gurling HMD, Murray RM. Genetic influence, brain morphology and cognitive deficits in alcoholic twins. In: Goedde HW, Agarwal DP, eds. Genetics and alcoholism. New York: Alan R. Liss, 1987: 71-82.
59. Guze S, Cloninger CS, Martin R, Clayton PJ. Alcoholism as a medical disorder. Compr Psychiatry 1986;27;501-10.
60. Hamlet K, Eaker ED, Stokes J. Psychosocial correlates of alcohol intake among women aged 45 to 64 years: The Framingham Study. J Behav Med 1989;12: 525-42.
61. Harada S, Agarwal P, Goedde HW. Isoenzyme variations in acetaldehyde dehydrogenase (E.C. 1.2.1.3) in human tissues. Hum Gen 1978;44: 181.
62. Harada S, Agarwal DP, Goedde HW et al. Aldehyde dehydrogenase isoenzyme variation and alcoholism in Japan. Pharmacol Biochem Behav 1983;18: 151-3.
63. Harford TC, Grant BF. Psychosocial factors in adolescent drinking contexts. J Stud Alcohol 1987;48: 551-7.
64. Hartocollis PC. Personality characteristics in adolescent problem drinkers. J Am Acad Child Psychiatry 1984;21: 348-53.
65. Hill SY, Goodwin DW, Cadoret R, Osterland K, Doner SM. Association and linkage between alcoholism and eleven serological markers. J Stud Alcohol 1975;36: 981-2.
66. Hilton ME. Trends in U.S. drinking patterns: further evidence from the past 20 years. Br J Addict 1988;83: 269-78.
67. Hrubec Z, Omenn GS. Evidence of genetic predisposition to alcoholic cirrhosis and psychosis. Alcohol Clin Exp Res 1981;5: 207-15.
68. Huffine CL, Folkman S, Lazarus RS. Psychoactive drugs, alcohol and stress and coping processes in older adults. Am J Drug Alcohol Abuse 1989;15: 101-13.
69. Hulgalkar AR, Nora R, Manowitz P. Arylsulfatase A variants in patients with alcoholism. Alcohol Clin Exp Res 1984;8: 337-341.
70. Institute of Medicine, Division of Health Science Policy. Heritable determinants of risk. In: Causes and consequences of alcohol problems. An agenda for Research. Washington DC: National Academy Press, 1987.
71. Jacobsen GR, Halikas JA, Morse C et al. Psychological characteristics and psychiatric diagnosis among juvenile offenders with substance use disorders. Abstract from Fourteenth International Institute on the Prevention and Treatment of Drug Dependence. Washington DC, 1984.
72. Johnson PB. Personality correlates of heavy and light drinking female college students. J Alcohol Drug Depend 1989;34: 33-7.
73. Jones MC. Personality correlates and antecedents of drinking patters in adult males. J Consult Clin Psychol 1968;32: 2-12.
74. Jonnson E, Nilsson T. Alkoholkonsumption hos monozygota och dizygota tvillingpar. Nord Hygienisk Tidsskr 1968;49: 21-5.
75. Kalodner CR, Delucia JL, Ursprung AW. An examination of the tension reduction hypothesis: the relationship between anxiety and alcohol in college students. Addict Behav 1989;14: 649-54.
76. Kandel DB. Drugs and drinking behavior among youth. Ann Rev Sociol 1980;6: 235-85.
77. Kaprio J, Koskenvuo M, Langinvainio H, Romanov R, Sarna S, Rose RJ. Social and genetic influences on drinking patterns of adult men: a study of 5638 Finnish twin brothers. Alcohol Alcoholism 1987(suppl 1)373-7.
78. Kay L. Alcoholism in twins. Stockholm: Almquist and Wiksell, 1960.
79. Knibbe RA, Lemmens PH. Korrelate des Alkoholkonsums in der Schweiz, Deutschland und den Niederlanden. Drogalkohol 1987;11: 27-41.
80. von Knorring AL, Bohman M, von Knorring L, Oreland L. Platelet MAO activity as a biological marker in subgroups of alcoholism. Acta Psychiatr Scand 1985;72: 51-8.
81. von Knorring L, Oreland L, von Knorring AL. Personality traits and platelet MAO activity in alcohol and drug abusing teenage boys. Acta Psychiatr Scand 1987;75: 304-14.
82. von Knorring L, von Knorring AL, Smigan L, Lindberg U, Edholm M. Personality traits in subtypes of alcoholism. J Stud Alcohol 1987;48: 523-7.
83. Knupfer G. The prevalence in various social groups of eight different drinking patterns, from abstaining to frequent drunkenness: analysis of 10 U.S. surveys combined. Br J Addict 1989;84: 1305-18.

84. Kushner MG, Sher KJ, Beitman BD. The relation between alcohol problems and the anxiety disorder. Am J Psychiatry 1990;147: 685-95.
85. Lester D, Rassas S. Personality and alcohol use: a preliminary study. Psychol Rep 1986;59: 694.
86. Lex BW, Mello NK, Mendelson JH, Babor TF. Reasons for use by female heavy, moderate and occasional social drinkers. Alcohol 1989;6: 281-7.
87. Li TK, Lockmuller JC. Why are some people more susceptible to alcoholism? Alcohol Health Res World 1989;6: 281-7.
88. Loehlin JC. An analysis of alcohol-related questionnaire items from the national merit twin study. Ann N Y Acad Sci 1972;197: 117-20.
89. Lolli G. Alcoholism as a disorder of the love disposition. Q J Stud Alcohol 1956;17: 96-107.
90. Lubben JE, Chi I, Kitano HHL. Exploring Filipino American drinking behavior. J Stud Alcohol 1988;49: 26-9.
91. Lykouras E, Moussas G, Markianos M. Platelet monoamine oxidase and plasma dopamine-β-hydroxylase activities in non-abstinent alcoholics. Relation to clinical parameters. Drug Alcohol Depend 1987;19: 363-8.
92. Lykouras E, Markianos M, Moussas G. Platelet monoamine oxydase, plasma β-hydroxylase activity, dementia and family history of alcoholism in chronic alcoholics. Acta Psychiatr Scand 1989;80: 487-91.
93. Major LF, Hawley RJ, Saini N, Garrick NA, Murphy DL. Brain and liver monoamine oxidase Type A and Type B activity in alcoholics and controls. Alcoholism Clin Exp Res 1985;9: 6-9.
94. Marlatt GA. Alcohol, stress and cognitive control. In: Sarason JG, Spielberger CD, eds. Stress and anxiety. New York: John Wiley, 1976;3: 271-96.
95. Mattei MG, London J, Rahuel C et al. Chromosome localization by in situ hybridisation of the gene for human erythrocyte glycophrin to region 4q28-31. Cytogenet Cell Genet 1987;46: 658.
96. Mayer JE. The personality characteristics of adolescents who use and misuse alcohol. Adolescence 1988;13: 382-404.
97. McBride OW, Umeno M, Gelboin HV, Gonzalez FJ. A Taq 1 polymorphism in the human P450IIE1 gene on chromosome 10 (CYP2E). Nucleic Acids Res 1987;15: 10071.
98. McCord W, McCord J. Origins of alcoholism. Stanford University Press, 1960.
99. McFadden J. Guilt is soluble in alcohol: an egoanalytic view. J Drug Issues 1987;2: 171-86.
100. Mehrabian A, O'Reilly E. Personality correlates of habitual alcohol use. Int J Addict 1988;23: 175-82.
101. Monteiro MG, Schuckit MA. Alcohol, drug and mental health problems among Jewish and Christian men at a university. Am J Drug Alcohol Abuse 1989;15: 403-12.
102. Mookherjee HN. Comparison of some personality characteristics of male problem drinkers in rural Tennessee. J Alcohol Drug Educ 1986;31: 23-8.
103. Mooney DK, Corcoran KJ. The relationship between assertiveness, alcohol related expectations for social assertion and drinking patterns among college students. Addict Behav 1989;14: 301-5.
104. Morinan A. Effects of chronic administration and withdrawal of ethanol and regional acetylcholinesterase and monoamine oxidase kinetics in brain. Neuropharmacology 1986;25: 661-2.
105. Mueller GC, Fleming MF, leMahieu MA, Lybrand GS, Barry KJ. Synthesis of phosphatidyl ethanol – a potential marker for adult males at risk for alcoholism. Proc Natl Acad Sci USA 1988;85: 9778-82.
106. Mullen K, Blaxter M, Dyer S. Religion and attitudes towards alcohol use in the western islets. Drug Alcohol Depend 1986;18: 51-72.
107. Nagoshi CT, Wilson JR. Influence of family alcoholism history on alcohol metabolism, sensitivity and tolerance. Alcoholism Clin Exp Res 1987;11: 392-8.
108. Neville HJ, Schmidt AL. Event-related brain potentials in subjects at risk for alcoholism. In: Chang NC, Chao HM, eds. Early identification of alcohol abuse. NIAAA Res Monogr 7, DHHS Publ (ADM) 85-1258, Government Printing Office, Washington DC, 1985: 228-39.
109. Nordmo SH. Blood groups in schizophrenia, alcoholism and mental deficiency. Am J Psychiatry 1959;1161: 460-1.

110. O'Connor I. The young drinkers. A cross-national study of social and cultural influence. London: Tarisstock, 1978.
111. Ozelius L, Hsu YP, Bruns G et al. Human monoamine oxidase gene (MAOA): chromosome position (Xp21-11) and DNA polymorphism. Genomics 1988;3: 53-8.
112. Pandey GN, Fawcett J, Gibbons R, Clark DC, Davis JM. Platelet monoamine oxidase in alcoholism. Biol Psychiatry 1988;24: 15 24.
113. Pandolfo M, Smith M. A Pvu II RFLP in the human ADH3 gene. Nucleic Acids Res 1988;16: 11857.
114. Parker F B. Sex-role adjustment in women alcoholics. Q J Stud Alcohol 1972;33: 587-666.
115. Partanen J, Bruun K, Markkanen T. Inheritance of drinking behaviour: A study of intelligence, personality and use of alcohol of adult twins. Helsinki: Finnish Foundation for Alcohol Studies, 1966.
116. Pederson N. Twin similarity for usage of common drugs. In: Gedda L, Parisi P, Nance W E, eds. Twin research 3 part c: Epidemiological and clinical studies. New York: Alan R. Liss, 1981: 53-9.
117. Penick EC. Powell BJ, Bingham SF, Liskow BI, Miller NS, Read MR. A comparative study of familial alcoholism. J Stud Alcoholism 1987;48: 136-46.
118. Pearson PL, Kidd KK, Willard MF et al. Report of the human gene mapping by recombinant DNA techniques. Cytogenet Cell Genet 1987;46: 390-566.
119. Pitts FN, Winokur G. Affective disorder. VII. Alcoholism and affective disorder. J Psychiatr Res 1966;4: 37-50.
120. Plant MA. Occupations, drinking patterns and alcohol-related problems: conclusions from a follow-up study. Br J Addict 1979;74: 267-73.
121. Polich J, Burns T, Bloom FE. P300 and the risk for alcoholism: Family history, task difficulty and gender. Alcoholism Clin Exp Res 1988;12: 248-54.
122. Polich J, Bloom FE. Event-related potentials in individuals at high and low risk of developing alcoholism: Failure to replicate. Alcoholism Clin Exp Res 1988;12: 368-73.
123. Pollock VE, Volovka J, Goodwin DW, Mednick SA, Gabrielli WF, Knop J, Schulsinger F. The EEG after alcohol administration in men at risk for alcoholism. Arch Gen Psychiatry 1983;40: 857-61.
124. Porjesz B, Begleiter H. Event-related potentials in individuals at risk for alcoholism. Alcohol (in press).
125. Power C, Estaugh V. The role of family formation and dissolution in shaping drinking in early adulthood. Br J Addict 1990;85: 521-40.
126. Propping P. Genetic control of ethanol action on the central nervous system. An EEG study in twins. Hum Genet 1977;35: 309-34.
127. Propping P, Kruger J, Mark N. Genetic predisposition in alcoholics and their relatives. Hum Genet 1981;59: 51-9.
128. Rabow J, Schwartz C, Stevens S, Watts RK. Social psychology dimensions of alcohol availability: the relationship of perceived social obligations, price considerations, and energy expended to the frequency, amount and type of alcoholic beverage consumed. Int J Addict 1982;17: 1259-71.
129. Rabow J, Neuman CA, Hernandez ACR. Contingent consistency in attitudes, social support and the consumption of alcohol, additive and interactive effects. Social Psychol Q 1987;50: 56-63.
130. Raghunatan L, Hsu LC, Klisak J, Sparkes RS, Yoshida A, Mohandas T. Regional localization of the human genes for aldehyde dehydrogenase-1 and aldehyde dehydrogenase-2. Genomics 1988;2: 267-9.
131. Reid NCRW, Brunt PW, Bias WB, Maddrey WC, Alonso BA, Iber FL. Genetic characteristics and cirrhosis: A control study of 200 patients. Br Med J 1968;2: 463-5.
132. Robertson DM, Morse RM, Moore SB, O'Fallon WM, Hurt RD. A study of HLA antigens in alcoholism.
133. Robson P. Psychological profile of social drinkers. Br J Addict 1988;23: 175-82.
134. Roe A. The adult adjustment of children of alcoholic parents raised in foster homes. Q J Stud Alcohol 1944/1945;5: 378-93.

135. Romelsjö A. The realtionship between alcohol consumption and social status in Stockholm. Has the social pattern of alcohol consumption changed? Int J Epidemiol 1989;18: 842-51.
136. Santisteban I, Povey S, West LF, Parrington JM, Hopkins DA. Chromosome assignment, biochemical and immunological studies on a human aldehyde dehydrogenase, ALDH3. Am Hum Genet 1988;49: 87-100.
137. Schuckit MA. Biological markers: metabolism and acute reactions to alcohol in sons of alcoholics. Pharmacol Biochem Behav 1980;13: 9-16.
138. Schuckit MA, Gold EO. A simultaneous evaluation of multiple markers of ethanol/placebo challenges in sons of alcoholics and controls. Arch Gen Psychiatry 1988;45: 211-6.
139. Schulsinger F, Knop J, Goodwin DW, Teasdale TW, Mikkelsen U. A prospective study of young men at high risk for alcoholism. Arch Gen Psychiatry 1986;43: 755-60.
140. Shibuya A, Yasunami M, Yoshida A. Genotypes of alcohol dehydrogenase and aldehyde dehydrogenase loci in Japanese alcohol flushers and nonflushers. Hum Genet 1989;82: 14-6.
141. Sieber MF, Angst J. Alcohol, tobacco and cannabis: 12-year longitudinal associations with antecedent social context and personality. Drug Alcohol Depend 1990;25: 281-92.
142. Smith M. Genetics of human alcohol and aldehyde dehydrogenase. Adv Hum Genet 1986;14: 249-90.
143. Squire J, Dryja TP, Dunn J et al. Cloning of the esterase D gene: A polymorphic probe closely linked in the retinoblastoma locus on chromosome 13. Proc Natl Acad Sci USA 1986;83: 6573-77.
144. Stabenau JR. Additive independent factors that predict risk for alcoholism. J Stud Alcohol 1990;51: 164-74.
145. Stacey B, Davies SD. Drinking bahaviour in childhood and adolescence: an evaluative review. Br J Addict 1979;65: 203-12.
146. Steele CM, Southwick L, Pagano R. Drinking your troubles away: the role of activity in mediating alcohol's reduction of psychological stress. J Abnormal Psychol 1986;95: 173-80.
147. Stockwell T, Smail P, Hodgson R, Canter S. Alcohol dependence and phobic anxiety stated. II. A retrospective study. Br J Psychiatry 1984;144: 58-63.
148. Stumphauzer JS. Learning to drink: adolescents and alcohol. Addict Behav 1983;5: 227-83.
149. Sullivan JL, Baenziger JC, Wagner DL, Rauscher FP, Nurnberger JI, Holmes JS. Platelet MAO in subtypes of alcoholism. Biol Psychiatry 1990;27: 911-22.
150. Swinson RP, Madden JS. ABO blood groups and ABH substance secretion in alcoholics. Q J Stud Alcohol 1973;34: 64-70.
151. Tabakoff B, Hoffman PL, Lee JM, Saito T, Willard B, De Leon-Jones F. Differences in platelet enzyme activity between alcoholics and nonalcoholics. N Eng J Med 1988;318: 134-9.
152. Tanna VL, Wilson AF, Winokur G, Elston RC. Possible linkage between alcoholism and Esterase-D. J Stud Alcohol 1988;49: 472-6.
153. Tarter RE, Hegedus A, Goldstein G, Shelly C, Alterman A. Adolescent sons of alcoholics: Neuropsychological and personality characteristics. Alcoholism 1984;8: 216-22.
154. Vailant G. Cultural factors in the etiology of alcoholism: a prospective study. In: Babor TF, ed. Alcohol and culture: comparative perspectives from Europe and America. Ann NY Acad Sci 1986;472: 1-239.
155. Vogel F, Propping P. The electroencephalogram (EEG) as a research tool in behavior genetics. In: Gerston ES, Matthyse S, Breakefield XO, Ciaranello RD, eds. Genetic research strategies for psychobiology and psychiatry. Pacific Groove: Boxwood Press, 1981: 269-80.
156. Ward K, Weir DG, McGrodden JM, Tipton KF. Blood acetaldehyde levels in relatives of alcoholics following ethanol ingestion. IRCS Med Sci 1983;11: 950.
157. Watanabe M, Tsuchiya F, Siraska T, Ideka H, Hatta Y, Katamura Y, Saito T. Platelet and lymphocyte adenylate cyclase activity in alcoholics. In: Kativama K, Tadaka A, Ishii H, eds. Biomedical and social aspects of alcohol and alcoholism. Amsterdam: Elsevier Science Publishers (Biomedical Division), 1988: 561-564.
158. Wilks J, Callan VJ. Alcohol related attitudes and stereotypes: teenagers in Australia, Papua New Guinea and the United States. J Drug Educ 1984;14: 119-32.
159. Wilks J, Callan VJ. Expectations about appropriate drinking contexts: comparisons of parents, adolescents and best friends. Br J Addict 1988;83: 1055-62.

160. Wilks J, Callan VJ, Austin D A. Parent, peer and personal determinants of adolescent drinking. Br J Addict 1989;84: 619-30.
161. Winokur G. et al. Alcoholism: diagnosis and familial psychiatric illness in 259 alcoholic probands. Arch Gen Psychiatry 1970;23: 104-7.
162. Yanish DL, Battle J. Relation between self-esteem, depression and alcohol consumption among adolescents. Psychol Rep 1985;57: 331-4
163. Yates WR, Wilcox J, Knudson R, Myers C, Kelly MW. The effect of gender and subtype on platelet MAO in alcoholism. J Stud Alcohol 1990;51: 463-7.
164. Young R, Oei TPS, Knight RG. The tension reduction hypothesis revisited: an alchol expectancy perspective. Br J Addict 1990;85: 31-40.
165. Zucker RA. Parental influence upon drinking patterns of their children. In: Greenblat M, Schuckit M, eds. Alcoholism problems in woman and children. New York: Seminars in Psychiatry, 1976: 211-39.

CHAPTER 16

Alcohol and traffic

E. te Wierik

Accidents are the third main cause of death in Europe 40% of which are road traffic accidents. One quarter of these deaths and 10% of the injuries are associated with alcohol (55). In 1989, 11.2% of fatal road traffic accidents and 6.5% of road accidents leading to injuries in the Netherlands were associated with the use of alcohol (9). Although alcohol plays a role in many traffic accidents it is not sure how many of these accidents would not have happened if the driver had not drunk.

Drivers can probably be roughly classified into two categories: a large group consisting of those who do not drink, or drink very little, before driving and a smaller one comprising those who regularly drink no matter whether they intend to drive. The former category can do without drinks, the latter one cannot. Various legislative measures with respect to the drinking-driving problem have been introduced. These measures can be divided into primary, secondary and tertiary prevention (12). Primary prevention strategies target young drivers with the goal of preventing the initiation of drinking-driving behaviour. Secondary prevention aims at the driving population as a whole and is founded on principles of general deterrence. Tertiary prevention focuses on convicted DWI (driving while intoxicated) offenders with the goal of reducing recidivism.

16.1. Motives for drinking before driving

Before starting measures to tackle the drinking-driving problem, more insight into the motives for drinking and driving after drinking is needed.

Among the reasons to drink drivers may have are: stress, worries, problems, need for comfort in social events and courage for tough situations, 'can't refuse' buddies, and boredom. These reasons for drinking form patterns related to psychological adjustment (46). Beck (5) noted that, as compared to non-driving drinking students, drunk driving students tended less to drink alcohol in their own home, but more to drink it in a dormitory. The most important drinking reasons for drunk driving students were: to get along better on dates, to get high, to get drunk, and to go to sleep. In a DWI offender sample, Snow & Wells-Parker (45) regressed four drinking reasons on alcohol consumption levels and frequency of drinking in seven types of locations. 'Escapism' reasons are related to quantity consumed per occasion, but are only weakly associated with specific locations. 'Sociability' reasons are associated with drinking in friends' home, but are not related to high consumption levels. Drinking 'for pleasure', 'to meet people of the opposite sex', and 'to get drunk' reasons are associated with both quantity consumed per occasion and away-from-home locations. This suggests a high traffic accident risk. The authors noted that opposite sex/drunkenness reasons are most important among young DWI offenders, and decline sharply in importance after the age of 25. Therefore, the authors recommend age-specific countermeasures rather than global strategies.

Rosenberg (38) interviewed 29 persons who were drinking regularly but who reportedly did not drive under the influence on 75% of drinking occasions and had never been convicted for DWI. The most frequently reported types of reasons for driving after drinking were:
– lack of perceived intoxication,
– perceived need or desire to go home or to some other location, often despite intoxication,
– geographical convenience ('I was in the middle of nowhere, if I ran off the road into a maize field, no big loss'),
– confidence in driving ability (e.g. 'I felt most fit to drive of the group').

The most frequently reported strategies to avoid detection/arrest when driving after drinking were:
– change of driving behaviour (e.g. driving slower, taking minor streets, being careful of speed limits, keeping an eye in the mirror),
– enhanced concentration and awareness (e.g. keeping alert, paying more attention, 'make myself conscious of what I'm doing'),
– self-talk (e.g. telling oneself 'you have to shake off the effect'),
– playing the radio, opening the car window, and talking/singing out loud (38).

16.2. Primary prevention

Young drivers represent a population at particular risk for accidents. Deaths and injuries due to road-crash involvement are a major health and safety problem, especially among the youth. Jonah (23), for example, noted that young people between the ages of 16 and 24 represented about 17% of the Canadian population and 21% of the licenced drivers. However, this age group accounted for 52% of all driver fatalities, for 31% of all traffic fatalities, and for 33% of all traffic injuries. Young people who drive after drinking have a greater risk of crash involvement than older drinking drivers at all blood alcohol concentrations (BACs). One explanation of this finding assumes that young people are inexperienced both with drinking and with driving. There is also evidence in favour of an alternative hypothesis, namely, that the higher crash risk among young drinking drivers is attributable to a subset of this group – those who engage in risky driving behaviour and who also happen to use alcohol (30). However, young people who do drive after drinking have a higher relative risk of total and fatal accident involvement than older drinking drivers (30). Young male drivers appear to have a higher risk than adolescent female drivers. Women seem to be more realistic and perhaps more responsible in the assessment of a dangerous drinking-driving situation (15).

Given the high relative risk of accidents among the young driving population, this group has become the focus of primary prevention strategies.

16.2.1. Education

Most educational programmes aimed at tackling the drinking-driving problem are school-based programmes. These programmes depart from the assumption that an increase in knowledge of the consequences of drinking and driving will result in a reduction of alcohol use, a change of drinking-driving behaviour, and a reduction of alcohol-related accidents (13)

The effectiveness of a Teams-Games-Tournament (TGT) alcohol education programme has been demonstrated by Wodarski (54). Two- year follow-up data indicated a permanent change of attitude towards drinking and driving after the TGT alcohol education programme. In his dissertation Kayser (24) evaluated an education programme offered by Dutch driving schools. The programme included a course book, slides, video tapes and a manual. The programme appeared to have a positive effect on knowledge, attitude and intended behaviour. With regard to self-reported driving after alcohol consumption no differences were observed between the two groups one year after completion of the programme. However, less car drivers in the experimental group drove with an assessed BAC higher than the legal limit of 0.5 g/l. More people who participated in the experimental group refused to be driven by a person who had consumed alcohol. In the Netherlands a publicity campaign focused on the combination of drinking and driving is conducted by 'Veilig Verkeer Nederland' (Dutch Road Safety Board). The evaluation of this campaign consists of the analysis of the results of a questionaire distributed among young male drivers aged 18–25. The results show no significant differences in number of non-drinking drivers, attitude, intention and normative beliefs towards drinking and driving. However, the number of glasses drunk before driving has decreased after the campaign. The decrease in BAC in drivers who had drunk was already observed in breath analyses (47). Mann et al. (29) have reviewed school-based programmes aimed at prevention of drinking in combination with driving. The most consistent finding is that that such programmes broaden the knowledge and understanding of alcohol and its effects as well as the risks of drinking combined with driving. An impact on attitudes concerning this behaviour has also been noted. Virtually no studies have evaluated the impact of education programmes on traffic safety measures. In the absence of an assessment of these variables it is difficult to determine the relative effectiveness of programmes with respect to their primary goal, namely, the improvement of traffic safety.

A public education campaign against drinking-driving directed towards 18–24-year-old Australian men has been evaluated by Reznik et al. (37). The authors arrived at the conclusion that the number of road traffic accidents in Wollongong fell significantly in the three-month period following the campaign as compared to other regions in New South Wales.

16.2.2. Increasing minimum purchase age

Another method to prevent drinking and driving among young people is to decrease the availability of alcohol by legislatively regulating the age at which it can be purchased.

As mentioned in Chapter 17, in recent times, many American states have changed the minimum legal purchase age (MLPA), with 29 states reducing the legal age between 1970 and 1975 (53), 24 of which raised the MPLA again within the next decade (7). In 1988, Wyoming became the last state to repeal a MLPA of 18, thus making the MLPA of 21 uniform throughout the country. The legal rise of the MLPA to the age of 21 in the USA was primarily based on studies relative to car accidents and fatalities in which youthful drivers were involved. Trends indicate that lowering the MLPA increases alcohol-related accidents and fatalities, while raising the legal age accomplishes the opposite in the affected age groups (4, 27). Reductions in fatal crash involvement as a consequence of raising the MLPA has been reported in other studies (2, 33, 51).

An alternative view suggests that the dangers of drinking and driving are increased not by a driver's youth per se, but by the yound driver's lack of experience both with drinking and with driving (3). In an examination of the long-term effects of raising the MLPA on fatal crashes it has been concluded that:
– a MLPA is associated with an increase of 5% on average in fatal crashes involving drivers in their first year after they reach the purchase age;
– when a state raises its MLPA, there is a corresponding, and unequal, shift in increased fatal crashes from a lower to a higher age group;
– states that did not raise their MLPA expierenced a slightly larger decrease in fatal crashes among drivers under 21 as compared to those states that did raise their MLPA (28).

Du Mouchel et al. (33) also determined whether the first year of legal purchase age was especially hazardous. The authors observed a 'negligible' effect, viz. a 2% increase of fatal crashes among beginning drinkers.

If reductions in alcohol consumption and motor vehicle accidents among young people are desired, both a MLPA of 21 and an increase in price of alcoholic beverages are suggested (11).

16.2.3. Lowering the maximum allowable BAC for traffic participants

A close relationship appears to exist between the height of the BAC on one hand and driving performance (16, 32) and the prevalence of traffic accidents (42) on the other.

Moskowitz & Robinson (32) have reviewed 41 findings of alcohol effects on driving performance, including 27 findings resulting from simulator studies. Results suggest that a BAC as low as 0.3 g/l can significantly impair the driver's performance; BACs of 0.8 g/l or less have shown to lower accuracy of steering, braking, speed control, lane tracking and gear changing, and also judgements of speed and distance in the driving situation. Smith (42) reported that lowering the legal BAC from 0.8 to 0.5 g/l in New South Wales had a beneficial effect on the casualty accident involvement of 17–20-year-old male drivers. A few years later he reported that a reduction of the legal BAC in several states of Australia had significantly reduced casualty accidents for drivers and motorcyclists (43, 44).

Although the relationship between alcohol and road safety has been the subject of numerous studies, much remains unknown about mechanisms by which alcohol contributes to traffic accidents. A number of other factors that can contribute to car accidents may interact with alcohol, possibly in a complex manner. Tolerance, age, driving experience, medicine, fatigue, stress and weather conditions may be among the factors that play a role in addition to alcohol (10). Since even very low BAC levels may impair the driver's capabilities, some authors recommend that licence holders should not be permitted to drive a motor vehicle after the ingestion of alcohol (1, 31).

16.3. Secondary prevention: general deterrence

Methods for intervening with individuals at risk for drinking in combination with driving, once the habit has taken root, are based on the model of deterrence. This model predicts that the threat of sanctions will prevent individuals from engaging in a specified behaviour (40).

16.3.1. Increasing risk of punishment and arrest

A significant inverse relationship between the perceived chance of apprehension and self-reported frequency of drinking when driving has been observed in males aged 17–30 (6). Increasing the probability of detecting incompetent drivers can be mediated by alcohol breath tests. The introduction of random breath testing in New South Wales in 1982 was immediately followed by a drop in road traffic casualties. The obviousness of police presence was considered as the cornerstone (19). Support for the hypothesis that increases in law enforcement leading to a greater probability of sanctions result in a lower incidence of fatal and serious injury accidents has been demonstrated in Scandinavia (50). Some studies have reported declines in night-time fatal crashes or other measures of alcohol-involved driving that preceded implementation of laws (14, 18, 22, 40). Studies on the mere effects of legislation have not uniformly demonstrated sustained declines in night-time fatal crashes. A typical pattern shows an immediate post-law decline in night-time fatalities followed by a return to the pre-law level. This trend has been observed in Great Britain, France and the Netherlands (40), as wel as in Canada (25), in the state of Maine (22), and in Stockton, a city in California (49). When the public becomes aware that police inforcement is not as rigid as initially perceived, the incidence of crashes returns to pre-law levels (40).

16.3.2. Increasing severity of punishment

It would be easier and less costly to punish DWI arrestees more severely than to tighten up the enforcement measures. Anincreased severity of punishment usually translates to more time in jail. Evaluations of such measures have not produced much support for their effectiveness (36, 39, 48). The failure of countermeasures based on increased severity of punishment lies in the possibility that the public disregards threats when it percieves a negligible likelihood of these threats being enforced. Another explanation is that members of the target population may not have perceived a jail as a severe punishment (41). It has been concluded that countermeasures based on severity of punishment may be more costly than initially anticipated and less effective than desired (36, 39).

16.4. Tertiary prevention

Since primary prevention and general deterrence strategies are not fully effective, individuals will continue to drive after drinking. Tertiary prevention focuses on those individuals who have been arrested for DWI. Some of the tertiary interventions represent deterrence stategies (fines, jail, licence revocation), other interventions involve educational or therapeutic approaches.

16.4.1. Licence revocation

It is assumed that the loss of driving privileges represents a punishment appropriate to the offence and that over the course of the revocation period the driver will have learned his lesson and will not repeat DWI offence (35). The loss of one's licence is viewed by the general public as a relatively severe sanction (26). Licence revocation does not lead to a complete elimination of driving by the convicted DWI offender. Between one-third and two-thirds of those people with licence suspensions or revocations have been found to drive during the sanction period (52). The results of studies that have evaluated the effectiveness of licence revocation have shown that it is the most cost-effective countermeasure identified as yet for reducing driving by drunken offenders (20, 21). For example, Hagen (20) noted that multiple DWI offenders who had their licence suspended had better subsequent records than comparable convictees who had not been punished with suspension. The benefits of licence suspension continued after the suspension period had expired.

16.4.2. Educational and therapeutic approaches

The rationale underlying a therapeutic approach is that a DWI offender signals a potential drinking problem that likely extends into many areas of the individual's life (35). The expectation is that treatment, by attenuating or eliminating drinking habits, will reduce the frequency of driving after drinking and thus reduce the probability of

DWI recidivism. On the other hand, educational approaches provide individuals with information about alcohol and its effect on behaviour without directly attempting to change drinking patterns.

Recently, the Dutch government has evaluated the 'Alcohol Verkeer Project Drenthe' (Drenthe Alcohol in Traffic Project), an education programme in which DWI offenders could volunteer. These volunteers received a milder punishment than DWI offenders who chose not to participate in this project. The project appeared to be effective; after the project, data showed that participants had a higher knowledge of alcohol and its effects on behaviour and driving performance. Moreover, the project seemed to reduce recidivism by 30% as compared to the matched control group (8).

Foon (17) reviewed drinking-driving education and treatment programmes reported in the literature. In his review he notes that the usefulness of educational programmes for drinking-driving offenders is questionable. While some of these programmes appear to increase knowledge about socially appropriate attitudes towards drinking and driving, others do not. Likewise, some programmes appear to reduce the number of drinking-driving and other alcohol-related traffic offenders, while other ones have a negative or indiscernable effect. It has also been noted that, as with education programmes, therapy programmes have not provided definitive evidence for their effectiveness. In a review Nichols (34) concludes that DWI offenders who have less severe drinking-related problems are more likely to benefit from an education programme than offenders with more severe drinking problems. According to Nichols (34), education and treatment programmes should be viewed as adjuncts, not as alternatives, to deterrent efforts.

Summary

Drinking impairs driving capabilities. Alcohol consumption is involved in many traffic accidents. Some reasons for drinking and driving after drinking are mentioned. Various legislative measures with respect to the drinking-driving problem are introduced. These measures can be classified into primary, secondary and tertiary prevention.

Primary prevention strategies target young drivers with the goal of preventing the initiation of drinking-driving behaviour. Educational programmes have shown to have a weak positive influence on traffic safety problems. Increasing the minimum purchase age or lowering the maximum allowable blood alcohol concentration for traffic participants have demonstrated to be more effective.

Secondary prevention strategies target the general population and are based on general deterrence. Increasing the perceived risk of punishment and arrest appears to have a short-lived effect. Increasing severity of punishment seems to be more expensive and less effective than desired.

Tertiary prevention focus on those individuals who have been arrested for DWI (driving while intoxicated) in order to prevent recidivism. Licence revocation appears to be the most cost-effective countermeasure identified for reducing driving by drunken offenders. Results of educational and treatment programmes are contradictary. These programmes may reduce DWI in less severe problem drinkers.

Overall, it can be concluded that none of these approaches is fully effective in reducing the drinking-driving problem. Integration of various of these approaches is probably the most effective countermeasure.

References

1. Anonymous. Alcohol and the driver: council on scientific affairs. JAMA 1986;255: 522-7.
2. Arnold R. Effect of raising the legal drinking age on driver involvement in fatal crashes, the experience of thirteen states. Washington, DC: National Center for Statistics and Analysis, 1985.
3. Asch P, Levy DT. Does the minimum drinking age affect traffic fatalities? J Policy Analysis & Management 1987;6: 180-92.
4. Ashley MJ, Ranking JG. A public health approach to the prevention of alcohol-related health problems. Ann Rev Public Health 1988;9: 233-71.
5. Beck KH, Summons TG. The social context of drinkign among high-school drinking drivers. Am J Drug Abuse 1987;13: 181-98.
6. Binns CW, Knowles SS, Blaze-Temple D. Is education enough? The drinking and driving practices of 17-30 year old males. Austr Drug Alcohol Rev 1987;6: 253-64.
7. Bonnie RJ. Regulating conditions of alcohol availability, possible effects on highway safety, J Stud Alcohol 1985(suppl 10): 129-43.
8. Bovens R. Evaluatie van het alcohol verkeer project Drenthe. The Hague: Staatsuitgeverij, 1987.
9. Centraal Bureau voor de Statistiek (CBS). Statistisch Zakboek. The Hague: Staatsuitgeverij, 1989.
10. Chan AWK. Factors affecting the drinking driver. Drug Alcohol Depend 1987;19: 99-119.
11. Coate d, Grossman M. Change in alcoholic beverage prices and legal drinking ages, effects on youth alcohol use and motor vehicle mortality. Alcohol Health Res World 1987;12: 22-5.
12. Donavan DM, Driving while intoxicated, different roads to and from the problem. Criminal Justice Behav 1989;16: 270-298.
13. Douglass RL. Youth, alcohol, and traffic accidents. In: Special population issues, alcohol and health, monograph No. 4. Rockville, MD: National Institute of Alcohol Abuse and Alcoholism, 1982. (Department of Health and Human Services publication [ADM] 82-1193.)
14. Epperlein T. Initial deterrent effects of the crackdown on drinking drivers in the State of Arizona. Accident Anal Prev 1987;19: 285-303.
15. Farrow JA, Brissing P. Risk for DWI, a new look for gender differences in drinking and driving influences, experiences, and attitudes among new adolescent drivers. Health Educ Quarterly 1990;17: 213-21.
16. Flanagan NG, Strike PW, Rigby CJ, Lochride GK. The effects of low doses of alohol on driving performance. Med Sci Law 1983;23: 203-8.
17. Foon AE. The effectiveness of drinking-driving treatment programs, a critical review. Int J Addict 1988;23: 151-70.
18. Forcier MW, Kurtz NR, Parent DG, Corrigan MD. deterrence of drunk driving in Massachusetts, criminal justice system impacts. Int J Addict 1986;21: 1197-220.
19. Gloag D. Drinking and driving, discretionary or random testing? BMJ 1988;297: 440-41.
20. Hagen RE. The efficacy of licensing controls as a countermeasure for multiple DUI offenders. J Safety Res 1978;10: 115-22.
21. Hagen RE, Rickey LW, McConnel EJ. The traffic safety impact of alcohol abuse treatment as an alternative to mandatory licensing controls. Accident Anal Prev 1979;11: 275-91.
22. Hingson RW, Heeren T, Kovenock D, et al. Effects of Maine's 1981 and Massachusetts's 1982 driving-under-influence legislation. Am J Publ Health 1987;77: 593-7.

23. Jonah BA. Accident risk and risk-taking behaviour among young drivers. Accident Anal Prev 1986;18: 255-71.
24. Kayser RE. Rijden onder invloed, effectiviteit van een voorlichtingsprogramma, een wetenschapppelijke proeve op het gebied van de sociale wetenschappen. Kampen: Mondiss, 1990.
25. Liben CB, Vingilis ER, Blefgen H. The Canadian drinking-driving countermeasure experience. Accident Anal Prev 1987;19: 159-81.
26. Little JW. A theory and emperical study of what eters drinking drivers, if, when, and why. Admin Law Rev 1971;23: 169-93.
27. Lotterhos JF, Glover ED, Holbert D, Barnes RC. Intentionality of college students regarding North Carolina's 21-year drinking age law. Int J Addict 1988;23: 629-647.
28. Males MA. The minimum purchase age for alcohol and young-driver fatal crashes: a long-term review. J Legal Stud 1986;15: 181-209.
29. Mann RE, Vingilis ER, Leigh G, Anglin L, Blefgen G. School based programmes for the prevention of drinking and driving, issues and results. Accident Anal Prev 1986;18: 325-37.
30. Mayhew DR, Donelson AC, Beirness DJ, Simpson HM. Youth, alcohol and relative risk of crash involvement. Accident Anal Prev 1986;18: 273-87.
31. McDermott FT, Hughes ESR. Driver casualties in Victoria. Med J Aust 1983;1: 609-11.
32. Moskowitz H. Robinson CD. Effects of low doses of alcohol driving-related skills: a review of the evidence. Alexandria: SRA Technologies, 1988.
33. duMouchel W, Williams AF, Zador P. Raising the alcohol purchase age, its effects on fatal motor vehicle crashes in twenty-six states. J Legal Stud 1987;16: 249-66.
34. Nichols JL. Treatment versus deterrence. Alcohol Health Res World 1990;14: 44-51.
35. Peck RC, Sadler DD, Perrine MW. The comparative effectiveness of alcohol rehabilitation and licensing control actions for drunk driving offenders, a review of the literature. Alcohol Drugs Driving 1985;1: 15-39.
36. Reed DS. Reducing the costs of drinking and driving. In: Moore MH, Gerstein DR, eds. Alcohol and public policy, beyond the shadow of prohibition. Washington, DC: National Academy of Science, 1981.
37. Reznik R, Morey S, Best JB. Evaluation of the Australian Medical Association drink-driving campaign in Wollongong. Med J Aust 1984;8: 818-21.
38. Rosenberg H. Coping strategies, reasons for driving, and the effect of self-monitoring in drinking-driving situations. Addict Behav 1988;13: 97-100.
39. Ross HL. Deterring drunken driving, an analysis of current efforts. J Stud Alcohol 1985(suppl 10): 122-8.
40. Ross HL. Deterring the drinking driver, legal policy and social control. Lexington, MA: Lexington Books, 1982.
41. Ross HL, McCleary R, LaFree G. Can mandatory jail laws deter drunk driving? The Arizona case. J Criminal Law & Criminology 1990;81: 156-70.
42. Smith DI. 0.05% or 0.08%? Med J Aust 1984;4: 176-7.
43. Smith DI. Effect of low proscribed blood alcohol levels (BALs) on traffic accidents among newly-licensed drivers. Med Sci Law 1986;26: 144-8.
44. Smith DI. Effect on traffic safety of introducing a 0.05% blood alcohol level in Queensland, Australia. Med Sci Law 1988;28: 165-70.
45. Snow RW, Wells-Parker E. Drinking reasons, alcohol consumption levels, and drinking locations among drunken drivers. Int J Addict 1986;21: 671-89.
46. Snowden RL, Campbell DR. Reasons for drinking among problem drinker-drivers, client and counselor reports during treatment. Addict Behav 1984;9: 391-4.
47. Söder JCM, de Bruin RA. 'Alcohol en verkeer, dat kun je niet maken', evaluatie onderzoek V.V.N.-campagne Alcohol in het verkeer, 1988-1989. Haren: Verkeerskundig Studiecentrum, Rijksuniversiteit Groningen, 1989.
48. Voas RB. Evaluation of jail as a penalty for drunk driving. Alcohol Drugs Driving 1986;2: 47-70.
49. Voas RB, Hause JM. Deterring the drinking driver, the Stockton experience. Accident Anal Prev 1987;19: 81-90.

50. Voley HL. Recent evidence from Scandinavia on deterring alcohol impaired driving. Accident Anal Prev 1984;16: 123-38.
51. Wagenaar AC. Preventing highway crashes by raising the legal minimum age for drinking: the Michigan experience six years later. J Safety Res 1986;17: 101-9.
52. Waller PF. Licensing and other controls of the drinking driver. J Stud Alcohol 1985(suppl 10): 150-60.
53. Wechsler H, Sands ES. Minimum age laws and youthful drinking, an introduction. In: Wechslet H, ed. Minimum-drinking-age laws, an evaluation. Lexington, MA: Lexington Books, 1980: 1-10.
54. Wodarski JS. Teaching adolescents about alcohol and driving, a two year follow-up. J Drug Educ 1987;17: 327-44.
55. World Health Organization, Regional office for Europe, Copenhagen: Working group on alcohol and accidents, ICP/APR 117/2, 1987.

CHAPTER 17

Prevention of alcohol abuse

E. te Wierik

Some authors assume a close relationship between average alcohol consumption and the prevalence of excessive drinking in a population to exist (9, 34, 59). Excessive drinking plays a distinct role in the development of all kinds of deviant behaviour and the pathogenesis of several syndromes. Alcohol-related problems are expected to increase with an increase in average alcohol consumption.

One way of tackling this problem may be the development of preventive policies. Dekker (12) considers prevention to encompass all measures and activities directed at preventing the occurrence of mental and physical disease and at promoting public health in general. We may distinguish primary, secondary and tertiary prevention. Primary prevention stands for the prevention of new cases. As regards alcohol-related problems this implies an effective governmental policy with regard to alcohol consumption, including promotion of activities like health education, and changing habits, norms and values so that another drinking pattern will be encouraged. Secondary prevention includes the early identification of prodromal or developing cases, their treatment and rehabilitation, in order to prevent the development of chronic cases – in other words, an intervention in the 'pathological process' at the earliest possible moment. Tertiary prevention is the treatment of problem drinkers by means of various methods in drying-out clinics or centres (72). This chapter will discuss primary prevention strategies. Secondary and tertiary prevention will be discussed in Chapter 18.

17.1. Models of control

The aims of tertiary prevention, and the measures to be taken, are relatively the least troublesome. The question is, however, what measures should be considered in primary and secondary prevention. The essential difference between these is that primary prevention concentrates on alcohol-related problems even before they have arisen. A number of prevention models have been described (31). The most popular model is the 'single distribution' model. Three models based on different starting points will be elucidated below.

17.1.1. The Ledermann or 'single distribution' model

The single distribution model is based on research done by the Frenchman Ledermann (34). Several aspects of Ledermann's theory have evolved and been modified in the course of time (59). The current model has been described by several authors (9, 31, 41, 54, 56, 62, 64). Epidemiological research has shown that the per capita consumption of alcohol and the prevalence of excessive alcohol consumption are associated: the larger the average quantity of alcohol consumed by the population, the higher the number of excessive drinkers (9). Theoretical speculations and empirical evidence indicate that a substantial increase in average consumption is almost always accompanied with an increase in number of excessive drinkers. When the per capita consumption increases, there will be concomitant increase of alcohol-related problems, such as mortality due to excessive drinking, cirrhosis of the liver, drunken driving, and hospitalization of alcoholics.

Ledermann's model (34) indicates the mathematical relation between mean alcohol consumption and the proportion of excessive drinkers in a homogeneous population. Ledermann classified all alcohol consumers in a given population according to the quantities of alcohol consumed during a particular period. Subsequently, he was able to study the distribution pattern in detail. He found that the frequency distribution of alcohol consumers as regards their individual

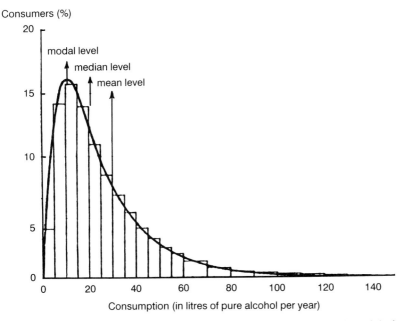

Fig. 17.1. Log-normal distribution of alcohol consumption among 30 million adults in France according to Ledermann (34).

consumption shows a highly positive skewed pattern. The distribution given in Fig. 17.1 shows a log-normal pattern. This is usually described by two parameters, namely the mean and the standard deviation (measure of dispersion). The mean in this case corresponds to the annual per capita consumption while the standard deviation was established by research. The latter may be a fairly constant value in various populations, so that the distribution may be described by applying only one parameter, viz. the mean alcohol consumption of a population.

Ledermann's model implies that, when the per capita consumption increases, the frequency of drinkers will always be distributed in accordance with a specific distributional pattern causing predictable proportions of drinkers to be classified in a higher consumption category and vice versa.

It has been questioned, however, whether the skewed distribution of alcohol consumption does really correspond with the fixed distribution Ledermann suggests and whether the mean consumption as a sole parameter indeed reflects the prevalence of excessive drinkers.

The interpretation of Ledermann's distribution curve has been discussed by Duffy (13, 14, 16), Duffy & Cohen (15) and Skog (64, 65, 66). Duffy & Cohen as well as Skog conclude that populations may deviate from the log-normal model. The hypothesis concerning the fixed relationship between mean alcohol consumption and the prevalence of excessive alcohol use is not tenable. Empirical investigations into the regularities in the distribution of alcohol consumption have experienced some disabilities that complicate attempts to draw precise, unequivocal conclusions. The most important disabilities met are: the underestimation of alcohol consumption in surveys, the small number of longitudinal studies (especially studies describing a decrease in mean alcohol consumption) and statistical problems related to sample size. Moreover, the theoretical foundations of Ledermann's model developed by Ledermann and by others are scanty, brief and incomplete (35). There is only one investigator, Skog (67), who has developed a theory about the distribution of alcohol consumption and who has tested the hypotheses drawn. His hypothesis, 'The use of alcohol by individuals is to such a degree a collective phenomenon that an increase in mean consumption is indicative of an increase in level of consumption for all drinkers (light and heavy)' is endorsed by Lemmens et al. (35). Fluctuations in mean alcohol consumption levels can thus be interpreted as being indicative of changes in consumption levels of all alcohol consumers. The description 'indicative of changes' means that it is not possible to give an exact, numerical prediction concerning the distribution of alcohol consumption.

17.1.2. The socio-cultural model

The socio-cultural model is characterized by the emphasis on social standards and on appreciation of the use of alcoholic beverages. Cultures may be classified according to their drinking patterns and attitudes. Pittman (53) made the following classification.

– *Abstinent cultures.* In an abstinent culture alcohol consumption is prohibited and drinking alcohol provokes strong negative feelings. Abstinence from alcohol is inherent in the life-style in this cultures. Members of these abstinent groups who start to drink have a greater risk of turning into an alcoholic because they have never learnt drinking. Examples of abstinent cultures are Mormones, Islamics and some Protestant groups.
– *Ambivalent cultures.* In an ambivalent culture the attitudes towards alcohol are contradictory in that there are two directly opposed value systems with regard to alcohol use. When there are no integrated control systems in a culture, the individual is in an ambivalent position which may lead to alcoholism. Examples of such ambivalent cultures are Ireland and Scotland.
– *Permissive cultures.* A permissive culture is defined as a culture in which attitudes towards alcohol consumption are favourable, but in which strong social sanctions exist regarding excessive drinking and intoxication or other types of deviant drinking behaviour. The number of alcoholics will be small when in a group or a community mutual consent exists on established habits, values, sanctions and attitudes towards alcohol use, and if these are compatible with the other norms in that culture. The Jewish and Italian cultures are often mentioned as being permissive.
– *Overpermissive cultures.* In an overpermissive culture attitudes towards alcohol use are favourable and types of deviant behaviour during drinking or resulting from drinking is accepted. This attitude is found, for instance, in France.

Some measures to prevent alcohol abuse in this model are:
– stimulation of the development and maintenance of prohibitive norms,
– encouragement of the use of alcohol in permissive situations and discouragement of situations in which alcohol drinking is the main activity,
– abolishment of the legal drinking age so that premises are open to families, and encouragement of consumption of alcohol in domestic situations.

The socio-cultural model may be open to criticism. The most pronounced criticism is expressed by Mäkelä (41) and Frankel & Whitehead (19). Mäkelä (41) has shown that information on differences in alcohol abuse among different cultures had not been interpreted correctly. The prevalence of alcohol abuse in the abstinent culture of Mormones, for example, is lower than in any of the other cultures defined above. Frankel & Whitehead (19) noted that factors such as permission of alcohol consumption and a prescriptive norm increase alcohol consumption and thus induce a higher prevalence of alcohol-related problems. They concluded that mean alcohol consumption most strongly affects the rate of negative consequences of alcohol consumption.

17.1.3. The mental-health model

Alcoholic beverages reduce stress. People who often have to face stressful situations are more inclined to drink to be relieved from these anxious feelings at least temporarily (55). Psychological studies support this hypothesis (see Chapter 15). However, cause and effect are hard to unravel: leads unemployment to alcohol use, or is it just the other way round? The 'mental-health model' assumes that social or psychological deprivation leads to excessive use of alcohol. The aim of prevention of alcohol abuse is to strengthen mental health. Plaut (55) mentioned, among other things, improvement of the quality of domestic life, advancement of the understanding of emotions and relations, reduction of poverty, discrimination and alienation, and the care of better housing. One can note that these measures are characterized by humanity and are not directed at alcohol consumption per se. Garretsen (22) noted that light and moderate drinkers also use alcohol to reduce stress. Moreover, not all problem drinkers consume alcohol to reduce anxious feelings. Therefore, prevention in line with the 'mental-health model' could only prevent a certain category from drinking excessively.

17.2. Measures to prevent alcohol abuse

Measures to prevent alcohol abuse which may be taken by the government include measures aimed at restricting the availability of or influencing the demand for alcohol. Among these measures are price control and a ban on or restrictions as to advertising.

17.2.1. Price control

One of the most fundamental economic laws states that increase of price tends to lead to a decrease in quantities bought (9, 33, 76).

The elasticy values of price and income are numerical expressions of the way consumers react to changing prices and incomes. All other factors remaining equal, a price increase will lead to a decrease in consumption while an increase in income will result in an increased consumption. The elasticy values will also be influenced by such variables as other goods, previous political controlling measures and marketing strategies which condition consumer preference. Moreover, in time elasticity values may change considerably in reaction to social, cultural and economic developments (68). Concerning price control the following observations have been made:
– alcohol consumption is more influenced by income than by price (17, 33),
– the calculated elasticities vary among types of alcoholic beverages: beer and wine appear to have a lower elasticity than spirits (17, 58).

Price regulation may be an effective alcohol policy element capable of forcing alcohol consumption down. However a number of objections have been raised. First, if the price increase merely leads to a switch to alcoholic beverages which do not increase in price, at least not to the same extent, then it is by no means certain that the price increase leads to a decline in overall alcohol consumption. Second, the econometric calculations in question are based on registered alcohol sales. A decline in registered consumption following upon a price increase can be counteracted by an increase in unregistered consumption. Third, it is conceivable that the great majority of moderate drinkers respond to price increases by reducing alcohol consumption, while the heavy drinkers sustain their old drinking pattern (55).

17.2.2. Restrictions bearing on the availability of alcohol

Some measures of control which are assumed to influence the prevalence of alcohol-related problems are:
– regulating the way alcohol is sold, i.e. whether there is a state monopoly, as in Scandinavian countries and in some states of the USA, or sales are in the hands of independent enterprises,
– limiting the number of outlets, i.e. the number of shops and public places where alcohol may be sold,
– fixing the times at which alcohol may be sold, i.e. the opening and closing times of bars and shops selling alcohol.

Popham et al. (56) were unable to show a significant difference in consumption level or liver cirrhosis mortality between US states where a licencing system was in force and the monopoly states. Similar findings are reported elsewere (29, 63).

According to Loeb (37) states in which licences are given out should have a higher consumption than states with a state monopoly. His conclusion is based solely on statistics relating to the consumption of spirits. The consumption of wine, beer and mixed beverages is not taken into account, thus resulting in a distorted picture. Pittman (53) quotes a survey in which seven US States in which the alcohol trade is government-controlled are compared with seven states in which alcohol is sold by private licencees. Consumption, as far as could be determined, was even higher in the states with a state monopoly. From a study on state monopoly systems and alcohol prevention in developing countries a relationship appear to emerge between the levels of alcohol consumption and State control: in countries with a high level of state intervention with regard to the availability of alcohol (Norway, Sweden, Finland, Poland) per capita consumption was lower than in countries with a low level of State control (USA, Ontario, FRG)(32). It should be noted that these data refer only to registered alcohol consumption, which may differ from real consumption.

A second possibility of limiting the availability of alcohol is restriction of the number of outlets. It is generally assumed that the more outlets of alcoholic beverages there are, the more will be drunk and the more often drunkenness will occur (56). Parker et al. (51) found that states with a high density of liquor outlets have higher levels of per capita consumption. Harford et al. (27) examined the effects of the number of on-premise outlets (i.e. where alcoholic beverages can be purchased for consumption on the premises) per 100,000 persons on alcoholism rates in the USA. They found that outlet density and urbanism, when controlled for income and consumption, were significantly related to rates of alcoholism in 38 states and the District of Columbia. In another study, the number of on-premise outlets per million population was found to be correlated significantly with and beer consumption (11). A study into the effects of allowing liquor-by-the-drink (LBD) (i.e. providing alcohol outlets the option to sell individual drinks for on-premise consumption), suggests a 6–7% increase in liquor sales attributable to LBD implementation. More recently, an association between alcohol availability (density of liquor outlets) and mortality has been demonstrated. The association was evident for heart disease, cancer, alcohol-related deaths and overall mortality. On the other hand, Smart (70) was not able to detect a significant correlation between availability and average consumption in a study covering 50 states. It was concluded that personal income may be a better predictor of variations in consumption than is availability. MacDonald & Whitehead (39) reviewed research into the relationship between per capita consumption of alcohol and density of alcohol retail outlets within a community. They conclude that a substantial increase in off-premise consumption is accompanied with an increase in consumption. Most of the studies reviewed can be interpreted alternatively. For example, the number of outlets within a community is affected by many factors including socio-demographic composition, disposable income and tourism as well as regulations controlling the number of outlets. A change of any one of these factors may lead to an increase in both the number of alcohol outlets and alcohol consumption. Although there seems to be an association between number of outlets and levels of alcohol consumption and alcohol-related diseases, the decision to restrict availability will not necessarily lead to lower alcohol consumption or less alcohol-related problems. This can be exemplified by the Polish experience. In October 1980, the Polish government decided to reduce the number of outlets purveying spirits and wine by one third. However, the Polish spirits sales in the last three months of that year were the highest ever recorded. This development was influenced, amongst other things, by a collapse in the markets for other products, a growth in national income and rumours of a planned increase of the price of alcohol. The closure of many outlets stimulated customers to buy 'just in case' whenever they passed a shop selling spirits. This example demonstrates that stable economic and political conditions are a prerequisite if control measures are to reduce consumption (32).

Limiting the hours during which alcohol is available would be more effective than restricting the number of outlets (9).

An attempt to reduce motor vehicle accidents associated with the early closing time for hotels (18.00) was made in Australia. Extension of the opening time to 22.00 on the premise was believed to reduce the urge to drink. The number of accidents was not reduced, but the peak accident time was shifted from 18.00–19.00 to 22.00–23.00 (57). The effects of closure of wine and spirits shops on Saturday have been measured in Norway. Both the number of reports of, and the number of incarcerations for, drunkenness, and the number of cases of domestic disturbance showed a reduction in the towns covered by the study relative to the control towns, particularly on Saturdays and Saturday nights, as might be expected. However, the number of reports of violence increased in the experimental towns relative to the control towns, especially on Friday nights (46). Insofar as they allow comparison, studies in Finland (30, 61) and Sweden (49) yield similar results. Frankel & Whitehead suggested that the hours during which alcohol could be sold determined where and when alcohol was consumed, but not the consumption rate or the total amount of alcohol consumed (19).

As regards the measures of control mentioned it may be pointed out that:
- seen in the wider context of a preventive policy, little is to be expected from changes in the legalization of alcohol sales,
- it has not been demonstrated convincingly that restriction of the number of outlets will result in lower sales,
- restrictions posed on opening and closing times may bring about side-effects, i.e. more may be drunk in the periods the shop is open.

17.2.3. Changing the minimum drinking age

In most countries sale of alcoholic beverages to persons under a specified minimum age is prohibited. In the Netherlands, for instance, the ages at which the purchase of weak and of strong alcoholic beverages is permitted differ, namely 16 and 18 years respectively.

In recent times, many states have changed the minimum legal purchase age (MLPA) for alcohol, with 29 states reducing that age between 1970 and 1975 (76) and 24 of these same states raising it again within the next decade (8). The decision in the USA to set the MLPA on 21 was primarily founded on studies into car accidents and fatalities among youthful drivers. A few authors have summarized previous studies on the impact of raising and lowering the MLPA on consumption, accidents and fatalities. Although results are somewhat conflicting, trends indicate that lowering the MLPA tends to increase the incidence of alcohol-related accidents and fatalities, whereas raising that age has the opposite effect in the affected age groups (5, 38).

Some studies have examined the effects of reductions in MLPA on consumption. Wagenaar (74) observed a short-term increase in draft beer sales after Michigan had reduced the MLPA in 1972, whereas package and total beer and wine sales remained unchanged. Barsby & Marshall (7) investigated liquor sales in 25 states one year before and one year after these states lowered their MLPA. Their analysis did not yield reliable evidence that liquor sales increased. However, the authors did not examine liquor consumption in the age group affected by the new law, but in the total market. Smart (70) examined per capita consumption for each beverage type separately in 25 states that had lowered their MLPA. Over a 3-year period, each state was compared with a neighbouring state that had not changed that age. Overall, beer and liquor sales increased in the states that had lowered the MLPA; wine sales remained stable.

Other studies have examined the effects of an increase of MLPA on consumption. Wagenaar (74) reported a decrease in package and total beer sales after the MLPA had been raised in Michigan, but draft beer sales increased as well. This increase was attributed to the simultaneous raise in the price of package beer. A short-term as well as a long-term reduction in frequency and quantity of drinking among age groups affected by a change of law (increase of MLPA) have been mentioned (78, 79).

Several negative effects of raising the MLPA for alcohol consumption have been mentioned. Students who are under age when the new drinking law is operative tend to change their location of alcohol use, tend to use other drugs more, try to obtain alcohol from their friends or tend to use a false or borrowed identification, but do not incline to teetotalism (24, 28, 38). Several studies have shown that most young people drink before reaching the legal minimum age. Usually this concerns drinking at home (18), sometimes permitted by their parents. Drinking without legal sanctions being applicable would promote the development of social ambivalence.

Suppression of juvenile alcohol use and abuse is accomplished best with both an increase in legal drinking age and an increase in prices of alcoholic beverages (10).

17.2.4. Mass media

Prevention of the abuse of alcoholic beverages can be mediated by mass media (i) by restricting or even prohibiting advertising of alcohol, and (ii) by media campaigns against alcohol abuse.

Restricting or prohibiting advertising of alcohol, as has happened in the Scandinavian countries, is regarded as a measure which might cause a decrease in alcohol-related problems. Smart & Cutler (69) studied the effect of a ban on advertising of alcohol over 14 months in British Columbia (Canada) in the early 1970s. This ban had no significant effect on the sales of beer, wine and spirits. Ogborne & Smart (48) investigated the effects of prohibition of advertising of beer through the press and the broadcasting media in Manitoba (Canada). This survey demonstrated, on the basis of a steady increase in the monthly per capita beer consumption statistics over the period 1974–1978, that the prohibition order had no effect whatsoever. A similar increase in alcohol consumption was observed in the control province Alberta. Another part of Ogborne & Smart's survey concerned the various restrictions applying to the advertising of alcoholic beverages in the press in several states of the USA. Here, too, no relation was found between the extent of official restrictions on the one hand and the per capita consumption of beer, wine and spirits on the other.

Most research assesses the effects of marginal changes in advertising expenditure on total consumption. Reviewers of this literature (20, 73, 77) conclude that the inability to find an effect on consumption suggests that the effect is unlikely to be substantial (45).

The Dutch liquor trade applies a fairly strict self-censorship regarding advertising of alcohol. It has, on its own initiative, adopted the German beer advertising code which has banned advertisements showing illustrations associating alcohol use with sportsmen, adolescents, cars, a profession, and so on. The complete code is essentially based on the fact that advertising should not stimulate alcohol consumption per se and should certainly not promote excessive consumption (4). A restriction of alcohol advertising is being considered: advertising may be prohibited on television in the future. This measure has been critized because some studies suggest that advertising has hardly any effect on total alcohol consumption.

Moreover, the measure would impede the introduction of low-alcohol beverages (2, 3).

Mass media can be used to advertise alcoholic beverages, but also to affect the public's attitude, and hence behaviour, towards the consumption of alcoholic beverages. The mass media are used because (23):
– many people can be reached that way
– mass media operate quickly
– the use of mass media is relatively simple
– the use of mass media is relatively cheap.

The Dutch government uses such slogans as 'Be honest ... how much do you really drink?' and 'Do you know, do you care?' to make the public aware of alcohol-related problems and to prevent moderate drinkers from becoming an alcoholic. Preliminary investigations suggest that the results of these campaigns are positive (41). Others dispute the success of media campaigns against drug and alcohol abuse (23). In the USA a similar campaign based on the slogan 'Be smart, don't start! – Just say no!' has started (21). It has been argued that personal contact is more credible than media contact. Future campaign efforts should explore ways of making people reflect on and, preferably, discuss their attitude and bahaviour prior to or perhaps in conjunction with specific drug/alcohol messages (6). Health education may be a good manner.

17.2.5. Alcohol education

Alcohol education programmes are often carried out in classrooms because a manageable number of young people are already together there.

The alcohol and drug education evaluation literature has been reviewed extensively (25, 43, 45, 60). These reviews are unanimous in their opinion that alcohol education significantly improves knowledge, but is hardly effective in preventing substance use or abuse. Therefore it is generally felt that school-based prevention programmes need to reconsider their conceptual bases, their logic and their strategies (26). An explanation for the lack of effect of these programmes is that schools do not know how to develop such programmes. Moreover, presenting programmes of this nature requires a kind of training that classroom teachers do not usually receive. Especially junior and senior highschool teachers who focus on presenting a content-orientated curriculum are less focused than elementary school teachers on juvenile learning processes (1, 50).

School-based programmes must be directed at all the major social influences en institutions that shape the youth. One possibility to change the prevention programmes is to concentrate more on factors such as parents and peers (42). Perry & Grant (52) noted that peer leadership is an important component of primary prevention programmes. That is why they performed a pilot study that was designed to elicit comparisons between peer-led and teacher-led instructions. Peer-leaders volunteered or were selected by their teachers. Schools in four countries were assigned to peer-led, teacher-led or control (no intervention) conditions. After intervention, peer-led conditions gave rise to significantly lower alcohol use scores than teacher-led and control conditions. There were no significant differences between peer-led and teacher-led programmes in knowledge scores. Students in the peer-led programme had significantly higher knowledge scores after the test than students in the control group. These outcomes provide some optimism for the potential of alcohol education.

Summary

Three models upon which prevention can be based are described. None of these models satisfies completely as a basis for alcohol prevention. According to Ledermann, the average alcohol consumption is regarded as an indication of the prevalence of excessive drinking and alcohol-related problems. Prevention of alcohol abuse in this model can be mediated by decreasing per capita consumption. The Ledermann model has been critized since the fixed relationship between mean alcohol consumption and the prevalence of excessive alcohol use is not tenable.

In the 'sociocultural model' cultures are classified according their drinking patterns and attitudes. In cultures in which alcohol abuse occurs frequently measures can be taken to influence behaviour. However, information on differences in alcohol abuse in different cultures seems to be interpreted incorrectly.

The 'mental-health model' assumes social or psychological deprivation to lead to excessive use of alcohol. The aim of prevention of abuse of alcohol is to improve mental health. Nevertheless, it is difficult to detect causes and effects: leads unemployment to alcohol use, or the other way round?

Several measures that can be taken by the government are described. Effects of the main policies can be summarized as follows:
– price control may decrease alcohol consumption under certain conditions;
– trends indicate that lowering the minimum legal purchase age results in an increase of alcohol consumption and alcohol-related accidents and fatalities, while raising the MLPA has the opposite effect in the affected age groups;
– alcohol advertisement on consumption has a small effect, if any, on total alcohol consumption;
– media used for campaigns against alcohol abuse may change the public's behaviour with regard to alcohol consumption, but firm conclusions cannot be drawn yet;
– alcohol education programmes may be a good policy to change the behaviour of the youth, but strategies have to be changed to be effective.

References

1. Amatetti SL. A prevention primer: putting theory to work in the classroom. Alcohol Health Res World 1987;11: 38-43.
2. Anonymous. Verbod tv-reklame voor alcohol zal wel doorgaan, al kan ieder weten dat het zinloos is. Adformatie 1987;6: 68-9.

3. Anonymous. Symposium reklamecode voor alcohol: overheid wil in feite alcoholreklame verbieden. Adformatie 1989;21: 63-4.
4. Anonymous. Code voor alcoholhoudende dranken. Stiva, etc., 1990.
5. Ashley MJ, Ranking JG. A public health approach to the prevention of alcohol-related health problems. Ann Rev Public Health 1988;9: 233-71.
6. Barber JJ, Grichting WL. Australia's media campaign against drug abuse. Int J Addict 1990;25: 693-708.
7. Barsbey SL, Marshall GL. Short-term consumption effects of a lower minimum alcohol purchasing age. J Stud Alcohol 1977;38: 1665-79.
8. Bonnie RJ. Regulating conditions of alcohol availability: possible effects en highway safety. J Stud Alcohol 1985(suppl 10): 129-43.
9. Bruun K, Edwards G, Lumio M et al. Alcohol control policies. In: Public Health Perspective. Helsinki: Finnish Foundation for Alcohol Studies, 1975;vol 25.
10. Coate D, Grossman M. Effects of alcoholic beverage prices and legal drinking ages on youth alcohol use. J Law Econ 1988;31: 145-71.
11. Colon I, Cutter HSG. The relationship of beer consumption and state alcohol and motor vehicle policies to fatal accidents. J Safety Res 1983;14: 83-9.
12. Dekker E. Elementen voor een beleid met betrekking tot de preventie van alkohol misbruik. Tijdschr Soc Geneesk 1978;56: 226-34.
13. Duffy JC. Estimating the proportion of heavy drinkers. In: Ledermann curve: report of a symposium. London: Alcohol Education Centre, 1977.
14. Duffy JC. Comment on the single distribution theory of alcohol consumption. J Stud Alcohol 1978;39: 1648-50.
15. Duffy JC, Cohen GR. Total consumption and excessive drinking. Br J Addict 1978;73: 259-64.
16. Duffy JC. The association between per capita consumption of alcohol and the proportion fo excessive consumers: a reply to Skog. Br J Addict 1980;75: 147-51.
17. Duffy M. The influence of prices, consumer incomes and advertising upon the demand for alcoholic drink in the United Kindom: an econometric study. Br J Alcohol Alcoholism 1984;16:200-8.
18. Farrel M. Young people and alcohol. Br J Addict 1988;83: 1137-8.
19. Frankel BG. Whitehead PC. Drinking and damage: theoretical advances and implications for prevention. New Brunswick, NY: Rutgers Centre of Alcohol Studies, 1984. (Rutgers Centre of Alcohol Studies Monograph 14.)
20. Frankena M, Cohen M, Daniel T, Ehrlich L, Greenspun N, Kelman D. Alcohol advertising, consumption and abuse. In: Recommendation of the staff of the Federal Trade Commission: omnibus petition for regulation of unfair and deceptive alcohol beverage marketing prices, Domacket no 209-46. Washington, DC: Federal Trade Commission, 1985.
21. Funkhouser JE. Before the cameras turn: the research base of the youth alcohol prevention campaign. Alcohol Health Res World 1987;11: 44-7.
22. Garretsen H. Probleemdrinken, preventiebepaling, beïnvloedende factoren en preventiemogelijkheden. Lisse: Swets & Zeitlinger, 1983.
23. Geinaert M. De effectiviteit van mediacampagnes voor alcohol en drug preventie. VAD Berichten 1989;2: 1-4.
24. George WH, Crowe LC, Abwender D, Skinner JB. Effects of raising the drinking age to 21 years in New York on self-reported consumption by college students. J Appl Soc Psychol 1989;19: 623-35.
25. Goodstadt MS, Caleekal-John A. Alcohol education programs for university students: a review of their effectiveness. Int J Addict 1984;19: 721-41.
26. Goodstadt MS. Alcohol education research and practice: a logical analysis of the two realities. J Drug Educ 1986;16: 349-64.
27. Harford TC, Parker DA, Paulter C, Wolz MW. Relationship between the number of on-premise outlets and alcoholism. J Stud Alcohol 1979;40: 1053-7.
28. Hughes SP, Dodder RA. Raising the legal minimum age: short-term effects with college student samples. J Drug Issues 1986;4: 609-20.
29. Jellinek EM. Recent trends in alcoholism and in alcohol consumption. Q J Stud Alcohol 1949;8: 1-42.

30. Kaski I. The experiment with Saturday closing of the ALKO shops. Alkoholpolitik 1978;41: 57-60.
31. Knibbe RA, van de Goor LAM. Modellen voor alcohol preventie. In: van der Stel JC, Buisman WR, eds. Alcohol preventie: achtergronden, praktijk en beleid. Alphen aan de Rijn/Brussel: Samson, 1988: 85-108.
32. Kortteinen T. State monopoly systems and alcohol prevention in developing countries: report on a collaborative international study. Br J Addict 1989;84: 413-25.
33. Lau HH. Cost of alcoholic beverages as a determinant of alcohol consumption. In: Gibbin J et al. eds. Research advances in alcohol and drug problems. New York: Wiley, 1975;2: 211-45.
34. Ledermann SC. Alcool, alcoolsime, alcoolisation. Paris: Institute National d'Etudes Démographiques, Traveaux et Documents, 1956. (Cahier No 29.)
35. Lemmens PHHM, Knibbe RA, Drop MJ. De verdeling van alcoholconsumptie in de bevolking: zin en onzin van het Ledermann-model. Gezondheid & Samenleving 1987;8:23-36.
36. de Lint J. Alcohol control policies as a strategy of prevention: a critical examination of the evidence. In: Madden JS, Walker R, Kenyon WH, eds. Alcohol and drug dependence: a multidisciplinary approach. New York: Plenum Press, 1977: 425-49.
37. Loeb B. Relationship of state law to per capita drinking. In: Ewing JA, Rouse BA, eds. Drinking: alcohol in American Society – issues in current research. Chigaco: Nelson-Hall, 1978; 219-38.
38. Lotterhos JF, Glover ED, Holbert D, Barnes RC. Intentionality of college students regarding North Carolina's 21-year drinking age law. Int J Addict 1988;23: 629-47.
39. MacDonald S, Whitehead P. Availability of outlets and consumption of alcoholic beverages. J Drug Issues 1983;13: 477-86.
40. Mäkelä K. Consumption level and cultural drinking patterns as determinants of alcohol problems. J Drug Issues 1975: 344-57.
41. Mäkelä K. Levels of consumption and social consequence of drinking. In: Israel Y et al. eds. Research advances in alcohol and drug problems. New York: Plenum Press, 1978;vol 4.
42. Mauss Al, Hopkins RH, Weisheit RA, Kearney KA. The problematic prospects for prevention in the classroom: should alcohol education programs be expected to reduce drinking by youth? J Stud Alcohol 1988;40: 51-61.
43. Meacci WC. An evaluation of the effects of college alcohol education on the prevention of negative consequences. J Alcohol Drug Educ 1990;35: 66-72.
44. Ministerie van Welzijn, Volksgezondheid en Cultuur, Bureau Alcohol Voorlichtings Plan. Alcohol opnieuw bekeken: een herhalingsonderzoek naar de rol van alcohol in de samenleving. Rijswijk: Ministerie van WVC, 1989.
45. Moskowitz JM. The primary prevention of alcohol problems: a critical review of the research literature. J Stud Alcohol 1989;50: 54-88.
46. Nordlund S. Effects of Saturday closing of wine and spirits shops in Norway. Oslo: National Institute for Alcohol Research, 1985. (SIFA-mineo 5/85.)
47. Nordlund S. Alcohol control policies in Norway and their effects. In: Skog OJ, Waahlberg R, eds. Alcohol and drugs: the Norway experience. Otta: Engers Boktrykkeri AS, 1988; 141-54.
48. Ogborne AC, Smart RG. Will restrictions in alcohol advertising reduce alcohol cosumption? Br J Addict 1980;75: 293-6.
49. Olsson O, Wikström PO. Effects of Saturday closing of wine and liquor stores in Sweden. Alkoholpolitik 1984;1: 84-95.
50. Palmer JH, Byrd V. North Carolina's implementation of the drug-free schools and community act of 1986. J Alcohol Drug Educ 1989;35: 16-22.
51. Parker DA, Wolz MW, Harford TC. The prevention of alcoholism: an experimental report on the effects of outlet availability. Alcoholism Clin Exp Res 1978;2: 339-43.
52. Perry CL, Grant M. Comparing peer-led to teacher-led youth alcohol education in four countries. Alcohol Health Res World 1988;12: 322-3.
53. Pittman DJ. Alcoholism. New York: Harper and Row, 1967: 5-20.
54. Pittman DJ. Primary prevention of alcohol abuse and alcoholism: an evaluation of the control of consumption policy. St. Louis, MO: Social Science Institute, Washington University, 1980.
55. Plaut IFA. Prevention of alcoholism. In: Golau et al. eds. Handbook of community health. New York: Appleton Century Crafts, 1972.

56. Popham RE, Schmidt W, de Lint J. The prevention of alcoholism: epidemiological studies on the effects of government control measures. Br J Addict 1975;70: 125-44.
57. Raymond A. Ten o'clock closing: the effect of the change in hotel bar closing time on road accidents in the metropolitan area of Victoria. Aust Road Res 1969;3: 3-17.
58. Ravn I. The control-of-consumption approach to alcohol abuse prevention. II: a review of emperical studies. Int J Addict 1987;22: 957-79.
59. Room R. Social science research and alcohol policy making. Paper given at the Conference on the Utilization of Research in Drug Policy Making, Washington, 1978.
60. Rundall TG, Bruvold WH. A meta-analysis of school-based smoking and alcohol use prevention programs. Health Educ Q 1988;15: 317-34.
61. Säilä SL. The experiment of Saturday closing and drunken misbehavior. Alkoholpolitik 1978;41: 61-9.
62. Schmidt W, Popham RE. The single distribution of alcohol consumption. J Stud Alcohol 1987;39: 400-19.
63. Simon JL. The economic effects of state monopoly of packed-liquor retailing. J Pol Econ 1966;74: 188-94.
64. Skog OJ. On the distribution of alcohol consumption. Oslo: National Institute for Alcohol Research, 1977.
65. Skog OJ. Is alcohol consumption lognormally distributed? Oslo: National Institute for Alcohol Research, 1979.
66. Skog OJ. Total alcohol consumption and rates of excessive use: a rejoinder to Duffy and Cohen. Br J Addict 1980:75:133-45.
67. Skog OJ. The collectivity of drinking cultures: a theory of the distribution of alcohol consumption. Br J Addict 1985:80: 83-99.
68. Skog OJ. An analysis of divergent trends in alcohol consumption and economic development. J Stud Alcohol 1986;47:19-25.
69. Smart RG, Cutler R. The alcohol advertising ban in British Columbia: problems and effects on beverage consumption. Br J Addict 1976;71: 13-21.
70. Smart RG. The relationship of availability of alcoholic beverages to per capita consumption and alcoholism rates. J Stud Alcohol 1977;38: 891-96.
71. Smart RG. Changes in alcoholic beverage sale after reductions in the legal drinking age. Am J Drug Alcohol Abuse 1977;4: 101-8.
72. Smart RG. Priorities minimizing alcohol problems among youth people. In: Blake HT, Chafetz ME, eds. Youth, alcohol and social policy. New York: Plenum Press, 1979: 229-62.
73. Smart RG. The impact of prvention measures: an examination of research findings. In: Institute of Medicine: an Inter-American workshop. Washington DC: National Academy Press, 1982: 224-46. (Publication IOM-82-003.)
74. Wagenaar AC. Aggregate beer and wine consumption: effects of changes in the minimum legal drinking age and a mandatory beverage container deposit in Michigan. J Stud Alcohol 1982;43: 469-88.
75. Walsh BM. Alcoholic beverages in Ireland: market forces and government policy. Br J Addict 1989;84:1163-71.
76. Wechsler H, Sands ES. Minimum age laws and youthfull drinking: an introduction. In: Wechsel H, ed. Minimum-drinking-age laws: an evaluation. Lexington, MA: Lexington Books, 1980: 1-10.
77. Whitehead P. Is advertising effective? Implication for public health policy. In: Rush B, Ogborne A, eds. Evaluation research in the Canadian addictions field. Ottawa, Canada: Health and Welfare, 1982.
78. Williams TP, Lillis RP. Changes in alcohol consumption by 18-year-olds following an increase in New York State's purchase age to 19. J Stud Alcohol 1986;47: 290-6.
79. Williams TP, Lillis RP. Long-term changes in reported alcohol purchasing and consumption following an increase in New York State's purchasse age to 19. Br J Addict 1988;83: 209-17.

CHAPTER 18

Measures to control alcohol abuse

E. te Wierik

As mentioned in Chapter 17, prevention of alcohol abuse can be classified into primary, secondary and tertiary prevention. Primary prevention has been discussed in Chapter 17. The aim of secondary prevention, i.e. early identification of developing cases, is to give an alcoholic access to help now, an alternative to the old strategy of allowing the alcoholic to 'hit bottom' before initiating efforts to change drinking behaviour. Tertiary prevention is the treatment of problem drinkers by means of various methods. Secondary and tertiary prevention will be elaborated below.

18.1. Early identification and treatment

The process of secondary prevention can be divided into screening, recruitment and implementation (31).

18.1.1. Screening

Screening is designed to differentiate among apparently healthy people, separating those who probably have the condition of interest from those who have not.

One of the first identified idicators of alcohol abuse is the level of serum gamma-glutamyltransferase (GGT). GGT is a liver enzyme whose level is elevated in case of excessive alcohol use (34). Another objective blood indicator used for screening and diagnosis is the mean corpuscular volume (MCV). However, the values of these two parameters are also affected by substances other than alcohol, as well as by physical conditions unrelated to drinking. Furthermore, they are not invariably elevated in heavy drinkers. Therefore, both GGT and MCV tests are considered as imperfect screening tests (7, 56), but for the time being they are the best available.

The level of serum ferritin, the major storage form of iron in the body, has been found to be significantly higher in heavy drinkers, but as a screening test it did not give additional information over a combination of GGT and MCV (12). Other new laboratory diagnostic markers have been developed, such as serum mitochondrial aspartate aminotransferase (mAST) activity (46, 47), carbohydrate-deficient transferrin level (58, 67), aldehyde dehydrogenase activity in red blood cells (70), platelet affinity to serotonin (36) and urinary dolichols (60). Further research is needed to confirm their usefulness in routine screening.

Besides these laboratorium tests self-report instruments like the Michigan Alcoholism Screening Test (55), the CAGE questions (40) and other diagnostic tests mentioned in Chapter 1 are being used. These interview tests are rapid and inexpensive. Additionally, they are generally more sensitive than routine biochemical tests (8, 33).

Until the new markers of alcohol use mentioned have proved their usefulness, GGT (perhaps in combination with MCV and AST) is still the best, cost-effective biochemical indicator of recent alcohol abuse. Self-administered or interview tests are known to have a high sensitivity and a reasonable specificity, but are prone to deliberate falsification and subconscious denial. Therefore, GGT combined with one or more self-administered or interview tests and with the full clinical picture is the method of choice for the assessment of alcoholism (26, 38, 48). Two screening procedures have combined the biological, clinical and verbal report methods into one screening procedure. These two screening approaches are the World Health Organization's Alcohol Use Disorders Identification Test (AUDIT) and the Alcohol Clinical Index (61, 64). Although these new screening tests have not been studied yet, the use of combined procedures presently holds great promise for early identification.

18.1.2. Recruitment

Once individuals at risk have been identified, the question is how they can be engaged in intervention intended to reduce risk? The global intensity of this intervention need not be determined by the measures required for the more damaged patient, which would imply that scarce clinical resources would be directed at a unnecessarily large number of people. Given the distinctive character – in the eyes of the potential client – of specialist units dealing primarily with alcohol-related problems, namely units dealing with 'alcoholics', it is unlikely that they will attract these potential clients. The solution to this dilemma, therefore, is to conduct the intervention in general hospitals, general practice, industry and the like (59).

Clinical studies have shown that problem drinkers consult their doctors more frequently than do others (10). So, health professionals in primary care and community health settings (e.g. family doctors, public health nurses) are often in a good position to identify individuals who drink excessively but do not consider themselves as 'alcoholics'. The results of a study in Sweden are encouraging (34): in a screening programme among middle-aged men, 585 problem drinkers were identified on the basis of elevated liver enzyme (GGT) level. These cases were randomly assigned to either brief advice and continuing follow-up consultations with a physician, or to a control group advised by letter to restrict their alcohol consumption. During a 4–6-year follow-up, the intervention group, as compared to the control group, gave a 80% reduction of absenteeism, a 60% reduction of hospital days and a 59% reduction of mortality. This study clearly demonstrates the potential impact of physician's advice to problem drinkers. The authors conclude that screening for early-stage problem drinkers linked with brief advice on drinking habits is an effective strategy that should be routinely used in general hospitals.

Recent reports suggest that the information obtained during screening can be used as feedback to motivate an individual's engagement in programmes aimed at a change of habits. Miller et al. (45) report a modest decrease of alcohol use and increased help-seeking in a population of problem drinkers for a 'drinker's check-up' that involved feedback regarding personal impairment related to alcohol use. The drinker's check-up is offered to individual drinkers as a means to discover what negative effects alcohol may have in their lives and consists of a battery of measures sensitive to alcohol's early effects on health and behaviour. Ojective feedback from these measures is given to the drinker with the intention to increase awareness of risk.

18.1.3. Implementation

Even when effective screening, recruitment and intervention strategies have been defined, there remain a number of logistical, technical and professional issues that must be addressed before promising findings are likely to be applied in clinical practice and public health settings. Evidence is mounting for the success of early intervention strategies for alcohol problems. Despite this promising prospect, early intervention programmes are not being implemented on any systematic basis. The following reasons are offered for this apparent lack of action (65):
– wide-spread pessimism among the physicians about being able to intervene effectively,
– the multidisciplinary nature of alcohol problems (social, medical, psychological problems, traffic accidents, crime) giving rise to confusion as to the question who is responsible for tackling alcohol problems,
– uncertainty with regard to the appropriate target population,
– inappropriate treatment programmes and philosophies,
– practical skills and techniques not being taught to health professionals.

Given the considerable promise that early intervention strategies hold for reducing alcohol and drug abuse, one must address the issue of how early intervention programmes can be implemented in a systematic way by health professionals. The following steps are proposed (65):
– adopt a basic strategy for early intervention,
– make practical techniques and training available,
– convince key people to implement the programme.

Secondary prevention is a promissory path, but there is still a long way to go.

18.2. Treatment of alcoholism

Recently a real explosion of studies have reported results from the multitude of disciplines involved in the treatment of alcohol abuse and dependence. Treatment seeks to bring about a change in the individual's use of alcohol and other drugs and to reduce problems accociated with that use. Typically, the goal of treatment programmes has been total abstention from alcohol and other drugs. Among the methods of treatment various approaches can be distinguished, such as detoxification, pharmacotherapies, mutual help groups, psychotherapy, hospitalization and relapse prevention.

18.2.1. Detoxification

A necessary first step in the treatment of alcoholism is detoxification. This procedure is designed to carry the individual safely through the process of alcohol withdrawal, minimizing the risks associated with the abstinence syndrome for those who are severely dependent on alcohol. Detoxification poses special problems for persons who experience an alcohol withdrawal syndrome. The cluster of symptoms ranging from tremulation and seizures to delirium tremens may manifest when alcohol consumption is discontinued after prolonged heavy consumption. According to Arnold & Feuerlein (6) a drug must ensure the following effects in treatment of delirium: sedation, elevation of the limit at which the patient decides to stop treatment, suppression of autonomic hyperexcitability, and antipsychotic effects.

Busch & Frings (11) have reviewed various frequently used pharmacological concepts. Chlomethiazole and benzodiazepines are established monotherapeutics against delirium. However, their somatic side-effects and, above all, the problems of dependence on this medication make it necessary to look for a lower-risk medication. In any case, physicians must look more closely into the possibility of treating the individual alcohol-withdrawal syndrome with an agent that is less liable to cause addiction. Carbamazipine seems to be an effective substance without dependence potential. It can be used for the treatment of severe alcohol syndromes. Other treatments are still under investigation, such as diphenylhydantion (3), gamma-hydroxybutyric acid (22) and calcium channel blockers (37).

18.2.2. Pharmacotherapy

Medication in the treatment of alcohol problems can be classified into three major strategies.
1. *Antidipsotropic medications*. These medications cause adverse reactions when alcohol is consumed. Their intended effect is to suppress drinking. Disulfiram is the antidipsotropic medicine most frequently used. It is an inhibitor of aldehyde dehydrogenase, an enzyme that catalyses the oxidation of acetaldehyde. Ethanol ingestion during disulfiram treatment results in very unpleasant symptoms (e.g. flushing, tachycardia, hypotension) which are generally believed to be consequences of the elevated concentrations of acetaldehyde (29). Disulfiram may have a long-lasting beneficial effect for certain subgroups of patients, such as elderly patients, socially stable patients, and patients who are well motivated. Treatment with disulfiram has no established effect on the long-term outcome of alcoholism (4, 73). Calcium carbimide, another dipsotropic medication, is available in Canada and Europe but has not yet been admitted for therapeutic use in the USA. Carmibide

appears to be less toxic ant to yield less side-effects than disulfiram (53). Well-controlled clinical trials with carbimide have not yet been reported.

2. *Effect-altering medications.* Effect-altering medications are designed to reduce the reinforcing properties of ethanol without producing illness rather than to induce aversive reactions to alcohol. Numerous serotonin uptake inhibitors have been shown to decrease alcohol consumption in animal models. Inhibition of serotonin, a neurohormonal compound, induces flushing. Zimelidine, citalopram, viqualine and fluoxetine appear to increase abstinent days and to lower the number of drinks on drinking days in moderately dependent drinkers (63).

3. *Psychotropics.* Psychotropic medication is designed to treat such concomitant disorders as depression, psychosis and anxiety. Their intended effect is to alleviate psychopathology that may accompany alcohol abuse, thereby diminishing the likelihood of relapse to drinking. Lithium, antidepressant and antipsychotic medications are of known therapeutic value.

18.2.3. Mutual help groups

The fellowship Alcoholics Anonymous (AA) has been founded in 1935 when two alcoholics discovered that it was possible to abstain from alcohol by mutual help. The fundamental principles of AA are embedded in its twelve traditions which have one purpose: recovering alcoholics. In the ideology of AA, alcoholism has to be regarded as a disease that has to be controlled. The members of AA have to learn to be responsible for their own behaviour. Their responsibility is not to drink (27).

The effectiveness of AA has been discussed widely. Emrick (15, 16) has reviewed evaluation studies with regard to data on AA's effectiveness as a treatment. He concludes that when alcoholics participate in AA in addition to professional treatment, results are not worse and may be better than for patients who do not engage in AA. AA involvement tends to be associated with relatively high abstinence rates but with low total improvement rates. Emrick argues that the high abstinence rates derive from a self-selection process in which those individuals who are most likely to commit themselves to and achieve abstinence are the ones most apt to join AA. According to Emrick, the effectiveness of AA as compared to other treatments of alcoholism has yet to be demonstrated. Other authors are more positive towards the effectiveness of AA (2, 30, 39, 57). Geelen (23), in his review on the effectiveness of AA, notes that AA is attractive to women, higher-educated people and people aged 35–50. This implicates that AA is self-selective. Geelen stresses that the findings of the studies reviewed were hard to interpret since most studies contained methodological weaknesses. Despite these shortcomings he concluded that the results were promising and comparable with those of professional treatments.

18.2.4. Psychotherapy

Psychotherapy means the treatment of alcoholism by psychological methods. The four basic elements of psychotherapy according to Frank (21) are: a confidential relationship between patient and therapist; a goal- and mind-mediated treatment theory; a demanding and supporting manner; and a convivial treatment institute. Several problems regarding these four elements have been reported (54).

Controlled treatment studies prior to 1980 failed to yield any evidence for the effectiveness of psychotherapy for alcoholics (43). Recent controlled studies have not substantially altered this picture (9, 50). Similarly, Annis & Chan (5) found no effect of confrontational group therapy in a random-assignment design with incarcerated alcohol-related offenders. Individuals low in self-esteem suffer detrimental effects from confrontational psychotherapy, whereas those higher in self-esteem experience some benefit. Swenson & Clay (69) observed no differences after an eight-month follow-up interval between drunk-driving offenders assigned to confrontational group therapy and those given only a home-study course.

18.2.5. Hospitalization

It is a common-sense assumption that if one kind of treatment is longer, more intensive, offered in a hospital or more expensive than another one, it should also be more effective.

Several uncontrolled studies have reported more favourable outcomes among patients receiving longer treatment in either inpatient (residential) or outpatient (non-residential) settings (19, 42, 66). Others have found no relationship (49, 51) or even a negative correlation (25). Miller & Hester conclude that 26 controlled trials have consistently failed to show an overall advantage for residential over non-residential settings, or for more intensive over less intensive interventions in treating alcohol abuse. Intensive treatment may be differentially beneficial for severely deteriorated and socially unstable (e.g. unemployed or homeless) individuals (44). In a prospective multi-centre study for inpatient treatment of female alcoholics a negative relationship between prognosis and length of treatment has been observed (i.e. the better the prognosis, the shorter the treatment). For men, no statistically confirmed correlation between rate of abstinence and length of treatment could be established (18). The value of inpatient treatment has been confirmed by Adelman & Weis (1) who found some features that appeared to be helpful in inpatient treatment. These features include a responsive intake procedure, patient participation in treatment planning, a professional staff with good interpersonal skills, interdisciplinary assessment, judicious use of medication, and a strong aftercare programme.

Inpatient hospital-based medical treatment appears to be the most common form of alcohol detoxification (62). Because of the high costs of this treatment studies have been performed in order to estimate the effectiveness of outpatient detoxification. Hayashida et al. (28) conclude that outpatient medical detoxification is an effective, safe and cost-effective treatment for patients with mild to moderate symptoms of alcohol withdrawal. These conclusions are confirmed by Collings et al. (13) who suggest that outpatient detoxification may obviate the need for many patients to be admitted, freeing psychiatric beds for other uses. Recently, in England an experiment including home detoxification has been started. Alcoholics who also participated in aftercare programmes and who scored low in an alcohol problems inventory, had better changes for a favourable progress (68).

18.2.6. Relapse prevention

Relapse prevention includes any intervention that is designed to diminish or forestall relapses after treatment. Programme components referred to as 'aftercare' have been designed to promote this same goal. The work of Vannicelli (72) and Costello (14) suggests that aftercare attendance is instrumental in fostering maintained abstinence.

One obstacle in evaluating aftercare programmes is the high drop-out rate. Ossip-Klein et al. (52) have observed that distribution of calender prompts and a contract laying down that the patient has agreed to attend aftercare increase attendance and hence the probability of long-term maintenance of alcohol treatment effects. Gilbert (24) found that meeting the alcoholics at a convenient location improved therapy attendance, but this did not correspond one-to-one with an improved treatment outcome.

Approaches of relapse prevention found to be successful vary from attendance of AA meetings (32, 71) to antidepressant agents (35) and aftercare services for adolescents rendered by clinicians (aftercare managers) (17).

Conflicting results are reported by Fitzgerald & Mulford (20) who tested frequent telephone contacts in addition to the usual aftercare treatment and found them ineffective as compared to the usual treatment without the experimental element. McClatch & Lomp (41) also did not find an advantage of extended therapy in the treatment of alcoholics. They were not able to observe a difference between mandatory aftercare, voluntary aftercare and no aftercare at all with regard to relapse rate, satisfaction with life-style and anxiety symptoms. They recommend a more individual aftercare planning for alcoholics.

Summary

Secondary prevention, i.e. early identification and treatment of developing cases can be divided into screening, recruitment and implementation. Until now, the best method of identifying alcohol abusers is the combination of the use of the gamma-glutamyltransferase level as an alcohol abuse indicator with one or more self-administered or interview tests and with the full clinical picture. The intervention to be given to early identified alcohol abusers need not to as intensive as what might be required for more damaged patients. Promising results emerge from a study in

Sweden in which a potential impact of the doctor's advice has been demonstrated. Despite the promising results of early identification there are still several reasons why early intervention programmes have not yet been implemented on a systematic basis. The reasons for this lack of action and a proposal for an early intervention strategy are given.

The approaches towards *tertiary* prevention mentioned in this chapter are detoxification, pharmacotherapies, mutual help groups, psychotherapy, hospitalization and relapse prevention. A necessary first step in the treatment of alcoholism is detoxification. Chlometiazole and benzodiazepine are established therapeutics against delirium, but these medications suffer from side-effects. Carbamazipine seems to be an effective alternative. The use of medication in the treatment of alcoholics can follow three major strategies. Antidipsotropic, effect-altering and psychotropic medication are described. Although the results concerning the effectiveness of mutual-help groups such as Alcoholics Anonymous (AA) are contradictory, results are promising. The high abstinence rates in AA may be based on a self-selection process. Evidence for the effectiveness of psychotherapy is not convincing to date. Outpatient treatment of alcoholics appears to be an effective, safe and cost-effective treatment for more socially stable and less deteriorated individuals. Conflicting results are reported concerning relapse prevention. A more individually oriented aftercare planning has been recommended.

References

1. Adelman SA, Weiss RD. What is therapeutic about inpatient alcoholism treatment? Hospital Community Psychiat 1989;40: 515-9.
2. Alford G. Alcoholics Anonymous, an empirical study. Addict Behav 1980;5: 359-70.
3. Alldredge BK, Lowenstein DH, Simon RP. Placebo-controlled trial of intravenous dephenyldanthion for short-term treatment of alcohol withdrawal seizure. Am J Med 1989;87: 645-8.
4. American College of Physicians. Disulfiram treatment of alcoholism. Ann Intern Med 1989;111: 943-5.
5. Annis HM, Chan D. The differential treatment model, empirical evidence from a personality typology of adult offenders. Crim Justice Behav 1983;110: 159-73.
6. Arnold U, Feuerlein W. Alkohol oder Psychopharmaca beim Entzugsdelir? Klinikarzt 1983;12: 203.
7. Barrison IG, Ruzek J, Murray-Lyon IM. Drinkwatchers, description of subjects and evaluation of laboratory markers of heavy drinkers. Alcohol Alcoholism 1987;22: 147-54.
8. Bernadt MW, Taylor C, Mumford J et al. Comparison of questionnaire and laboratory tests in the detection of excessive drinking and alcoholism. Lancet 1982;i: 325-8.
9. Braunstein WB, Powel BJ, McGowan JF et al. Employment factors in outpatient recovery of alcoholics, a multivariate study. Addict Behav 1983;8: 345-51.
10. Buchan IC, Buchley EG, Deacon GLS et al. Problem drinkers and their problems. J R Coll Gen Pract 1981;31: 151-3.
11. Busch H, Frings A. Pharmacotherapy of alcohol withdrawal symptoms in hospitalised patients. Pharmacopsychiatry 1988;21: 232-7.
12. Chick J, Pikkarainen J, Plant M. Serum ferritin as a marker of alcohol consumption in working men. Alcohol Alcoholism 1987;22: 75-7.
13. Collins MN, Burns T, van den Berk PAH, Tubman GF. A structured programme for out-patient alcohol detoxification. Br J Psychiat 1990;156: 871-4.

14. Costello RM. Alcoholism aftercare and outcome, cross-lagged panel analysis. Br J Addict 1980;75: 44-53.
15. Emrick CD. Alcoholics Anonymous, affiliation processes and effectiveness as treatment. Alcoholism Clin Exp Res 1987;11: 416-23.
16. Emrick CD. Alcoholics Anonymous, memberships characteristics and effectiveness as treatment. Recent Dev Alcohol 1989;7: 37-53.
17. Fertman CI, Toca OA. A drug and alcohol aftercare service, linking adolescents, families, and schools. J Alcohol Drug Educ 1989;34: 46-53.
18. Feuerlein W, Küfner H. A prospective multicentre study of in-patient treatment for alcoholics, 18- and 48-month follow up (Munich Evaluation for Alcoholism Treatment MEAT). Eur Arch Psychiatr Neurol Sci 1989;239: 144-57.
19. Finney JW, Moos RH, Chan DA. Length of stay and program component effects in the treatment of alcoholism, a comparison of two techniques for process analysis. J Consult Clin Psychol 1980;49: 120-31.
20. Fitzgerald JL, Mulford HA. An experimental test of telephone aftercare contacts with alcoholics. J Stud Alcohol 1985;46: 418-21.
21. Frank JD. Persuasion and healing. Baltimore, 1968.
22. Gallimberti L, Gentile N, Cibin M et al. Gamma-hydroxybutyric acid for treatment of alcohol withdrawal syndrome. Lancet 1989;ii: 787-9.
23. Geelen K. De effectiviteit van AA-groepen, een overzichtsstudie. Tijdschr Alcohol Drugs 1987;13: 187-93.
24. Gilbert FS. The effect of type of aftercare follow-up on treatment outcome among alcoholics. J Stud Alcohol 1988;49: 149-59.
25. Gunderson EKE, Schuckit MA. Prognostic indicators in young alcoholics. Military Med 1978;143: 168-70.
26. Hambridge DM. Gamma-glutamyl transpeptidase and mean cell volume in alcoholics. Br J Psychiat 1987;150: 568.
27. Harberden P van. Therapeutische processen in AA. T Alcohol Drugs 1987;13: 193-7.
28. Hayashida M, Alterman AI, McLellan T et al. Comparative effectiveness and costs of inpatient and outpatient detoxification of patients with mild to moderate alcohol withdrawal syndrome. N Eng J Med 1989;320: 358-65.
29. Helander A, Tottmar O, Carlsson S. Inhibition of human erythrocyte and leukocyte aldehyde dehydrogenase activity by disulfiram and cyanamide in vitro and in vivo. In: Kuriyama K, Takada A, Ishii H, eds. Biomedical and social aspects of alcohol and alcoholism. Amsterdam, New York, Oxford: Excerpta Medica, 1988: 131-4.
30. Hoffman NB, Harrison PA, Belille CA. Alcoholics Anonymous after treatment, attendance and abstinence. Int J Addict 1983;18: 311-318.
31. Institute of Medicine. Prevention and treatment of alcohol problems, research opportunities. Report of a study by a committee of the Institute of Medicine, Division of Mental Health and Behavioral Medicine. Washington, DC: National Academy Press, 1984.
32. Knouse VH, Schneider HG. Recovering alcoholics personality and aftercare factors. Psychol Report 1987;61: 595-601.
33. Kristenson H, Trell E. Indicators of alcohol consumption, comparisons between a questionnaire (Mm-MAST), interviews and serum gamma-glutamyl transferase (GGT) in a health survey of middle aged males. Br J Addict 1982;77: 297-304.
34. Kristenson H, Ohlin H, Hulten-Nosslin MJ, Trell E, Hood B. Identification and intervention of heavy drinking in middle-aged men, results and follow-up of 24–60 months of long-term study with randomized controls. Alcoholism Clin Exp Res 1983;7: 203-9.
35. Lhuintre JP, Daoust M, Moore ND et al. Ability of calcium bis acetyl homotaurine, a GABA agonist, to prevent relapse in weaned alcoholics. Lancet 1985;i:1014-6.
36. Lhuintre JP, Boismare F, Daoust M et al. Increased platelet affinity for serotonin, a marker for alcohol dependence. In: Kuriyama K, Takada A, Ishii H, eds. Biomedical and social aspects of alcohol and alcoholism. Amsterdam, New York, Oxford: Excerpta Medica, 1988: 557-60.
37. Littleton JM, Little HJ, Whittington MA. Effects of dehydropyridine calcium channel antagonists in ethanol withdrawal, doses required, stereospecificity and actions of Bay K 8644. Psychopharmacology 1990;100: 387-92.

38. Lumeng L. New diagnostic markers of alcohol abuse. Hepatology 1986;6: 742-5.
39. Machell DF. Alcoholics Anonymous, a wonderful medication with some possible side effects. J Alcohol Drug Educ 1989;34: 3.
40. Mayfield D, McLeod G, Hull P. The GAGE questionnaire, validation of a new alcoholism screening instrument. Am J Psychiatry 1974;131: 1121-3.
41. McClatchie BH, Lomp KGE. An experimental investigation of the influence of aftercare on alcoholic relapse. Br J Addict 1988;83: 1045-54.
42. McLellan AT, Luborsky L, Woody GE, O'Brien CP, Druley KA. Predicting response to alcohol and drug abuse treatments, role of psychiatric severity. Arch Gen Psychiatry 1983;40: 620-5.
43. Miller WR, Hester RK. Treating the problem drinker, modern approaches. In: Miller WR, ed. The addictive behaviors, treatment of alcoholism, drug abuse, smoking, and obesity. Oxford: Pergamon Press, 1980.
44. Miller WR, Hester RK. Inpatient alcoholism treatment, who benefits? Am Psychol 1986;41: 794-805.
45. Miller WR, Sovereign RG, Krege B. Motivational interviewing with problem drinkers, the drinkers check-up as a preventive intervention. Behav Psychother 1988;16: 251-68.
46. Nalpas B, Vassault A, Charpin S, Lacour B, Berthelot P. Serum mitochrondial aspartate aminotransferase as a marker of chronic alcoholism, diagnostoc value and interpretation in a liver unit. Hepatology 1986;6; 608-14.
47. Nalpas B, Poupon RE, Vassault A, et al. Evaluation of mAST/tAST ratio as a marker of alcohol misuse in a non-selected population. Alcohol Alcoholism 1989;24: 415-9.
48. Nederlands Huisartsen Genootschap. Problematisch alcoholgebruik, standaard M10. Utrecht, NHG, 1990.
49. Ogborne AC, Clare G. A note on the interface between a residential alcoholism rehabilitation centre and detoxification centres. Br J Addict 1979;74: 283-7.
50. Olson RP, Ganley R, Devine VT et al. Long-term effects of behavioral versus insight-oriented therapy with inpatient alcoholics. J Consult Clin Psychol 1981;49: 866-77.
51. Orford J, Hawker A. An investigation of an alcoholism rehabilitation halfway house. II. The complex question of client motivation. Br J Addict 1976;14: 409-18.
52. Ossip-Klein DJ, Vanladingham W, Prue DM, Rychtarik RG. Increasing attendance at alcohol aftercare using calender prompts and home based contracting. Addict Behav 1984;9: 85-9.
53. Peachey JE, Annis HM. New strategies for using the alcohol-sensitizing drugs. In: Naranjo CA, Sellers EM, eds. Research advances in new psychopharmacological treatments for alcoholism. Amsterdam, New York, London: Excerpta Medica, 1985: 199-216.
54. Petry J. Integrative psychologische Aspekte der Alkoholismusbehandlung. Ernährungs-Umschau 1987;34: 332-8.
55. Pokorny AD, Miller BA, Kaplan MB. The brief MAST, a shortened version of the Michigan Alcoholism Screening Test. Am J Psychiatry 1972;129: 342-5.
56. Pol S, Poynard T, Bedossa P, Aubert A, Chaput J. Diagnostic value of serum γ-glutamyl-transferase activity and mean corpuscular volume in alcohol patients with or without cirrhosis. Alcoholism Clin Exp Res 1990;14: 250-4.
57. Polich JM, Armor DJ, Braiker HB. Patterns of alcoholism over four years. J Stud Alcohol 1980;41: 397-416.
58. Reisinger PWM, Soyka M. Die Diagnose van Alkoholismus auf der Grundlage des Nachweises einer Transferrinvariante durch Polyacrylamid-Gelelektrophorese und Immunoblotting. Blutalkohol 1990;27: 427-33.
59. Rijder D. secondary prevention of alcohol related problems. Aust Drug Alcohol Rev 1987;6: 271-7.
60. Roine R, Nykänen J, Ylikahri R, Salaspuro M. Urinary dolichols as markers of alcoholism. In: Kuriyama K, Takada A, Ishii H, eds. Biomedical and social aspects of alcohol and alcoholism. Amsterdam, New York, Oxford: Excerpta Medica, 1988: 557-60.
61. Saunders JB, Aasland OG. WHO collaborative project on identification and treatment of persons with harmful alcohol consumption. Doc. WHO/MNH/DAT/86.3. Geneva: World Health Organization, 1987.

62. Sausser GJ, Fishburne SB Jr, Everett VD. Outpatient detoxification of the alcoholic. J Fam Pract 1982;14: 863-7.
63. Sellers EM, Naranjo CA, Lawrin MO. New drugs to decrease alcohol use. In: Kuriyama K, Takada A, Ishii H, eds. Biomedical and social aspects of alcohol and alcoholism. Amsterdam, New York, Oxford: Excerpta Medica, 1988: 609-12.
64. Skinner HA, Holt S, Sheu WJ, Israel Y. Clinical versus laboratory detection of alcohol abuse, the Clinical Alcohol Index. Br Med J 1986;292: 1703-8.
65. Skinner HA. Early detection of alcohol and drug problems, why? Aust Drug Alcohol Rev 1987;6: 293-301.
66. Smart RG. So some alcoholics do better in some types of treatment than others? Drug Alcohol Abuse 1978;3: 65-76.
67. Stibler H, Borg S. The value of carbohydrated-deficient transferrin as a marker of high alcohol consumption. In: Kuriyama K, Takada A, Ishii H, eds. Biomedical and social aspects of alcohol and alcoholism. Amsterdam, New York, Oxford: Excerpta Medica, 1988: 503-6.
68. Stockwell T, Bolt L, Milner I, Pugh P, Young I. Home detoxification for problem drinkers, acceptibility to clients, relatives, general practitioner and outcome after 60 days. Br J Addict 1990;85: 61-70.
69. Swenson PR, Clay TR. Effects of short-term rehabilitation on alcohol consumption and drinking related behaviors, an eight month follow-up study on drunken drivers. Int J Addict 1980;15: 821-38.
70. Takase S, Enyama K, Takada A. Changes of serum proteins in chronic alcoholics. In: Kuriyama K, Takada A, Ishii H, eds. Biomedical and social aspects of alcoholism. Amsterdam, New York, Oxford: Excerpta Medica, 1988: 507-10.
71. Thurstin AH, Alfano AM, Nerviano VJ. The efficacy of AA attendance for aftercare of inpatient alcoholics, some follow-up data. Int J Addict 1987;22: 1083-90.
72. Vannicelli M. Impact of aftercare in the treatment of alcoholics, a cross-lagged panel analysis. J Stud Alcohol 1978;39: 1875-86.
73. Wright C, Moore RD. Disulfiram treatment of alcoholism. Am J Med 1990;88: 647-55.

Subject index

absenteeism	18.1.2	--induced ketosis	5.1
absorption		--induced pancreatitis	12.1.5
– of alcohol	2.1	– intoxication	8.2.3
– of nutrients	11.1.5	– liver disease, treatment of	8.3
– phase	2.3.1	– metabolism	3
abstinent culture	17.1.2	– metabolism, modification of	3.7
abuse, measures to control	18	--related birth defects	13.1
accumulation of fat	8.2.1	– withdrawal delirium	9.2.2
acetaldehyde	2.3.1, 3.2.1, 3.3, 3.4, 4.6, 5.1, 8.2.4, 12.3.4, 13.4.1	– withdrawal syndrome	6.7, 18.2.1
		–, fusel	12.3.1
		alcoholic (definition)	1.3
		– beverages, differences between	2.1.2, 12.2
– dependence	9.2.3	– cardiomyopathy	10.7
– poisoning, acute	3.3	– cirrhosis	3.6
acetate	2.3.1, 3.4, 3.5, 5.1	– dementia	9.2.1
–, extrahepatic dissimilation of	3.5	– fibrosis	5.3
acetoacetate	5.1	– heart disease	10.7
acetyl-CoA	3.5, 3.6, 5.1, 5.3	– hepatitis	8.1
acetylsalicylic acid	11.1.2.2	– ketoacidosis	5.1
acidaemia	5.1	– ketosis	5.1
acidophilic hyalin	8.1	– liver disease	8.2
acidosis	5.1	Alcoholics Anonymous	18.2.3
–, lactic	5.1	alcoholism (definition)	1.3
–, metabolic	7.2.2, 14.6.1	–, aetiology of	15
adaptation	9.2.2	–, alpha	1.3
adenosine monophosphate, cyclic	5.2	–, beta	1.3
adenylate cyclase	15.1.4.2	–, familial history of	15.1.1
adipose tissue	5.3	–, gamma	1.3
adoption studies	15.1.3	–, paternal	13.7
adrenaline	4.7, 5.2, 10.3	–, treatment of	18.2
adrenocorticotropic hormone	4.6	aldehyde dehydrogenase	2.3,2, 3.3, 15.1.4.2, 18.2.2
advertising of alcohol	17.2.4		
–, self-censorship in	17.2.4	– activity	18.1.1
aetiology of alcoholism	15	aldorase activity, erythrocyte	8.6
aftercare	18.2.6	aldosterone	4.1
age	15.5.4.1	alkaline phosphatase	8.6
aggregation, platelet	10.2	alkalosis	
agitation, psychomotor	9.2.2	–, metabolic	5.1
alanine aminotransferase	8.6	–, respiratory	7.2.4
albumin	5.4	alpha alcoholism	1.3
alcohol		ambivalent culture	17.1.2
– abuse, measures to control	17.2, 18	amblyopia	11.3.2
– abuse, prevention of	17	amenorrhoea	4.9.1
– amnestic disorder	9.2.1	amino acids	
--associated cardiomyopathy	7.1	–, absorption of	11.1.5
--associated immunosuppression	12.3.4	–, placental transport of	13.4.2
– deficiency	2.3.1	amino groups, free	8.2.5
– dehydrogenase	3.2, 4.8, 6.1, 7.2.7, 8.4.4, 15.1.4.2	aminobutyric acid	
		aminobutyric acid, alpha-	5.4
– dehydrogenase activity	9.1.2	– /leucine ratio, alpha-	8.6
– dependence	1.3	–, gamma-	9.2.3
– education	17.2.5	aminolaevulinic acid dehydratase	8.6
--free beer	14.1	aminopyrine	2.1.3
--induced fibrosis	8.1	amylase, renal clearance of	11.1.1

anabolic steroids	8.3	biokinetics of alcohol	2
anaemia	7.2.6, 8.2.3, 11.6	biological markers of alcoholism	15.1.4
−, haemolytic	6.9, 11.6	biotransformation	3.6
−, megaloblastic	6.5, 6.6	birth	
angina		− defects, alcohol-related	7.2.7, 13.1
− pectoris	10.6	− weight	13.3.1
−, variant	10.6	black-out	9.1.1
angiography, coronary	10.4	bleeding time	10.2
anomalies, congenital	13.3.3	blindness, night	6.1
anorexia	7.2.7, 8.1	blood	11.6
antidipsotropic drugs	3.3, 18.2.2	− alcohol concentration (BAC),	
antidiuretic hormone	4.1, 7.1, 7.2.1	maximum allowable	16.2.3
antigen, human lymphocyte	15.1.4.2	− alcohol curve	2.3.1
antimicrobial drugs	3.6	− lipids	10.1
antioxidant	6.9, 8.2.4	− pressure	10.3
antitubular effect of alcohol	8.2.2	bodily constitution	2.2.2
antocholinergic drugs	2.1.3	body	
antral injury	11.1.2.2	− mass	8.4.4
anxiety	9.2.2	− temperature	2.3.2
apolipoprotein	10.1	− water	8.4.4
apprehension, increasing chance of	16.3.1	bone marrow	11.6
arachidonic acid	5.3	bourbon	4.8, 14.6.3
area under the blood alcohol curve	2.1.1, 3.2.1	brain	9
arginine vasopressin	9.2.1	− atrophy	9.2.1
aromatic hydrocarbons, polycyclic	12.3.1, 12.3.4	− damage	5.2, 6.2, 6.6, 9.2.1
arrest, increasing risk of	16.3.1	brandy	9.1.3
arrhythmia, cardiac	3.4, 10.7	breast	
arteriography	10.4	− cancer	12.1.3
articulation abilities	13.3.2	− -feeding	13.5
asbestos fibres	12.3.1	breath alcohol analysis	2.2.1
ascites	7.1		
ascorbic acid	2.3.2, 8.2.4	caffeine	2.1.1, 2.3.2
aspartate aminotransferase	8.6, 18.1.1	calcitonin	4.2
−, mitochondrial	18.1.1	calcium	7.2.3
aspirin	2.1.3, 11.1.2.2	− carbimide	18.2.2
atherosclerosis	10.4	− homeostasis	6.8
ATPase	11.1.5	calvados	12.2
atrial fibrillation	10.7	campaign, media	17.2.4
atrophic gastritis, chronic	11.1.2.2	cancer	12
atrophy		− of the breast	12.1.3
− of type IIb fibre	11.5.2	− of the digestive tract	12.1.1
−, cerebral/brain	9.2.1	− of the endolarynx	12.1.1
−, testicular	6.1	− of the epilarynx	12.1.1
autonomic regulation (control)	13.3.2	− of the hypopharynx	12.1.1
availability of alcohol	17.2.2	− of the larynx	12.1.1
		− of the liver	12.1.1, 12.1.2
barrier breaker	11.1.2.2, 12.3.4	− of the lung	12.1.4
beer	8.4.5, 12.1.6, 12.2, 13.5, 14.4, 14.5	− of the oesophagus	12.1.1, 12.3.1
		− of the oropharynx	12.1.1
behaviour	15	− of the pancreas	12.1.5
−, reflexive	13.3.2	− of the prostate	12.1.6
behavioural problems	13.3.2	− of the rectum	12.1.6, 12.2
benzodiazepines	9.2.2, 18.2.1	− of the sigmoid colon	12.1.6
beriberi, cardiac	6.2, 10.8	− of the stomach	12.1.6
beta alcoholism	1.3	carbohydrate	2.1.2, 2.3.2, 14.3
beverages, differences between	2.1.2, 12.2, 14	− intolerance	4.5
biliary pancreatic reflux	11.2.2	− metabolism	5.2, 11.2
biochemical markers		− -deficient transferrin	8.6
− for alcohol abuse	8.6	carcinogenesis	7.2.7, 12
− of alcoholism	15.1.4.2	−, mechanisms of	12.3

carcinoma, hepatocellular	12.3.3	coma	6.2
cardiac		communication skills	13.3.2
– abnormalities	13.3.3	concentration tests	9.1.1
– arrhythmia	3.4, 10.7	confrontational group therapy	18.2.4
– beriberi	6.2, 10.8	confusion	5.2
cardiomyopathy	5.3	congeners	14.6
–, alcohol-associated	7.1, 10.7	congenital anomalies	13.3.3
cardiopathy	10.3	congestion, pulmonary	6.2
cardiovascular		consciousness, disturbances of	9.2.2
– disease	10.1	constitution, bodily	2.2.2
– system	10	constitutional studies of alcoholism	15.1
– system of the unborn	13.1	control, models of	17.1
catabolism	3.2.2	coordination	9.1.1
catalase pathway	3.2.3	coronary	
cataract	11.3.2	– angiography	10.4
catechin	14.6.3	– heart disease	10.1, 10.5
catecholamine	3.4, 10.3	– stenosis	10.4
cell membrane narcosis	13.4.1	corpuscular volume, mean	8.5, 11.6, 18.1.1
cellular		corticosteroids	8.3
– mechanisms	13.4.3	cortisol	4.6, 15.1.4.2
– tolerance	9.2.2	cough, productive	11.4
central nervous system	2.3.2, 13.1, 13.3.2, 15.1.4.1	craniofacial abnormalities	13.3.3
		cream liqueur	14.3
centrilobular hypoxia	8.2.3	cultural factors of alcoholism	15.5.1
cerebral		culture	
– atrophy	9.2.1	–, abstinent	17.1.2
– dysfunction	9.2.1	–, ambivalent	17.1.2
– glucose utilization	6.2	–, overpermissive	17.1.2
– oedema	9.2.1	–, permissive	17.1.2
check-up, drinker's	18.1.2	Cushing syndrome	4.6
chest radiography	8.6	cyanide	11.3.2
chlormetiazole	9.2.2, 18.2.1	cyclic adenosine monophosphate	5.2
cholecalciferol	6.8	cytochrome P-450	3.2.2, 3.6, 6.1, 12.3.4
– metabolism	4.2		
cholecystokinin	11.1.3	– reductase	8.2.4, 12.3.4
cholesterol, HDL	8.6		
chromium	14.5	dealcoholized beverages	14.1
chronic alcohol consumption	9.2	deficiency	
cimetidine	3.6	–, alcohol	2.3.1
circadian rhythm(icity)	2.3.2	–, dietary	12.3.2
cirrhogenic activity of beverages	8.4.5	definitions of alcoholism	1
cirrhosis	7.1, 8.4.3, 10.4, 12.1.2, 12.3.3, 12.4, 15.1.4.2	delirium	
		– tremens	6.7, 9.2.2
		–, alcohol withdrawal	9.2.2
– mortality	8.4.5, 8.5	dementia, alcoholic	9.2.1
–, alcohol-induced (alcoholic)	3.6, 6.1	dependence, alcohol	1.3, 9.2.2
–, epidemiology of	8.5	depressions	9.2.2
citric acid cycle	5.1, 5.3, 6.2	deterrence, general	16.3
clearance, renal	5.1	detoxification	4.1, 18.2.1
closing time of outlets	17.2.2	diarrhoea	7.2.7
clouding, mental	9.2.2	diastolic blood pressure	10.3
cocarcinogen, ethanol as a	12.3.4	dietary deficiency	12.3.2
coenzyme A (CoA), acetyl-	3.5, 3.6	differences	
cognitive		– between beverages	14
– dysfunction	9.2.1	– in cirrhogenic activity	8.4.5
– function	6.2	digestive tract	2.1, 11.1.1
colchicine	8.3	–, cancer of the	12.1.1
collagen	5.1, 5.4, 8.2.6	dihomogammalinolenic acid	5.3
– production	8.2.2	dihydroxycholecalciferol, 1,25-	6.8
collagenase	8.3	dilatation, vascular	11.1.2.2

265

dilution effect	2.1.2
direct carcinogenic effect	12.3.1
disaccharidase	11.1.5
disaccharide	14.3
disorientation	9.2.2
dissimilation of acetate, extrahepatic	3.5
distribution	
– of alcohol	2.2
– of isozymes	3.2.1
disulfiram	3.3, 3.4, 3.6, 18.2.2
diuresis	4.1, 7.1
diurnal rhythm(icity)	2.1.3
DNA research	15.1.4.3
dolichol	8.6, 18.1.1
dopamine	3.4, 4.7, 9.2.3
dose-response relationship	13.3.1
double vision	11.3.1
drinker's check-up	18.1.2
driving while intoxicated	16
drop-out rate	18.2.6
drugs	
–, anticholinergic	2.1.3
–, antidipsotropic	3.3
–, antimicrobial	3.6
–, CNS	9.2.2
ductal plug	11.2.2
duodenopancreatic reflux	11.2.2
dysfunction, sexual	4.8
dysmorphology, facial	13.1
dyspnoea	6.2
early identification	18.1
edema, see oedema	
education programmes	16.2.1, 16.4.2, 17.2.5
–, peer-led	17.2.5
–, teacher-led	17.2.5
educational approach	16.4.2
effect-altering medication	18.2.2
eggnog	14.3
eicosanoid formation	5.3
electrolyte	
– balance	7.2.1
– excretion	10.3
electron transport	6.3
–, mitochondrial	3.2.1
electrophysiological markers	15.1.4.1
elimination	
– of alcohol	2.3
– phase	2.3.1
– rate	2.3.2
emprisonment	16.3.2
encephalopathy	
–, hepatic	4.3, 9.2.1
–, Wernicke's	6.2, 9.2.1
endogenous synthesis of fatty acids	5.3
endolarynx, cancer of the	12.1.1
endorphin, beta-	15.1.4.2
enterohepatic circulation	6.5
enterokinase	11.2.2
enzymes	
–, intestinal	11.1.5
–, pancreatic	11.2.2
epidemiology	12.1
– of liver cirrhosis	8.5
epilarynx, cancer of the	12.1.1
epinephrine, see adrenaline	
equation, Michaelis-Menten	2.3.1
erosion	
– of gastric mucosa	2.1.2
– of jejunal villi, haemorrhagic	11.1.5
–, mucosal	11.1.2.1
erythrocyte	11.6
– aldorase activity	8.6
escapism	16.1
esophagus, see oesophagus	
essential fatty acid	5.3
– metabolism	8.2.2
estrogen etc., see oestrogen etc.	
ethanol	1, 5.1, 12.3.1, 14.2
– as a cocarcinogen	12.3.4
etiol-, see aetiol-	
excessive drinking/drinker	1.1
exercise, physical	2.2, 2.3.2
expiratory volume, forced (FEV)	11.4
extracellular fluid	7.1
extrahepatic dissimilation of acetate	3.5
eye	11.3
facial dysmorphology	13.1
false neurotransmitters	3.3, 9.2.3
familial history of alcoholism	15.1.1
family	
– relations	15.5.2
– studies of alcoholism	15.1.1
fasting hypoglycaemia	5.2
fat	2.1.1
– accumulation	8.1, 8.2.1
– catabolism	5.1
– malabsorption	11.1.5
– metabolism	5.3
– synthesis	5.1
fatigue	8.1
fatty acid(s)	
– ester synthetase	3.2.3
– metabolism, essential	8.2.2
–, endogenous synthesis of	5.3
–, essential	5.3
–, polyunsaturated	5.3, 8.2.3
fatty liver	6.1, 8.1, 8.4.1
female	
– fertility	4.9.1
– sex hormones	4.9
ferritin	7.2.6
fertility	
–, female	4.9.1
–, male	4.8
fetal, see foetal	
fever	8.1
fibre, type IIb	11.5.1

fibrillation, atrial	10.7	– absorption	11.1.5
fibrinogen	10.2	– intolerance	5.2
fibrinolysis	10.2	– utilization, cerebral	6.2
fibrinolytic activity	10.2	glutamate receptor	9.1.2
fibrosis, alcohol-induced (alcoholic)	8.1	glutamic-oxaloacetic transaminase	8.6
first-order kinetics	2.1	glutamic-pyruvic transaminase	8.6
first-pass metabolism	2.3.1, 3.2.1	glutamyltransferase	
flavanoids, phenolic	14.6.3	–, alpha-	8.6
flushing syndrome	3.4	–, beta-	18.1.1
foetal		glutathione	5.4, 8.2.4
– alcohol effects	13.1	glycerophosphate, alpha-	5.3
– alcohol effects, specific	13.3	glycogen stores	5.2
– alcohol syndrome	5.3, 7.2.7, 13.1, 13.2	glycogenolysis	5.2
folate	6.5, 14.4	glycolysis	5.1
– deficiency	6.5	glycoproteins	5.4
–, placental transport of	13.4.2	gonadotropin	4.9.1
folic acid	6.5	group therapy, confrontational	18.2.4
follicle-stimulating hormone	4.8	growth	
fortification of beverages	6.2	– hormone	4.4
free amino groups	8.2.5	– hormone-releasing hormone	4.4
free radicals	8.2.4	– retardation	13.1, 13.3.1
free water	7.2.1	– retardation, intra-uterine	13.4.3
frequency distribution of alcohol users	17.1.1	–, intra-uterine (foetal)	13.3.1
fructose	5.1	gynaecomastia	4.8
functional tolerance	9.2.2	habituation	15.4
fusel alcohol	9.1.3, 12.3.1, 14.6, 14.6.2	haemochromatosis	7.2.6
		haemolysis	6.9
galactose	2.1.1, 2.3.2	haemolytic anaemia	6,9, 11.6
gamma alcoholism	1.3	haemorrhage	
gamma-glutamyltransferase	18.1.1	–, gastric	11.1.2.2
gastric		–, gastrointestinal	8.2.3
– emptying	2.1	haemorrhagic	
– emptying, delay of	2.1.1	– erosion of jejunal villi	11.1.5
– haemorrhage	11.1.2.2	– stroke	10.5
– juice	11.1.4	haemosiderosis	7.2.6
– lesions	11.1.2.2	haemostasis	10.1
– motility	2.1.1, 2.1.2, 11.1.3	hallucinations	9.2.2
– mucosa	11.1.2	hangover	9.1.3, 14.6.2
– mucosa, erosion of	2.1.2	headache	5.2
– mucosa, permeability of	11.1.2.2	heart disease, alcoholic	10.7
gastrin	11.1.3, 11.1.4	heavy drinking/drinker	1.1
gastritis	5.1	help groups, mutual	18.2.3
–, chronic atrophic	11.1.2.2	hemo-, see haemo-	
gastrointestinal		hepatic	
– haemorrhage	8.2.3	– encephalopathy	4.3, 9.2.1
– hormone	11.1.3	– lipase	10.1
– tract	11.1	hepatitis	5.1, 8.2.4
– tract bleeding	7.2.6	– B virus	12.3.3
genes involved in alcoholism	15.1.4.3	–, alcoholic	8.1
genetic control	2.1.3	hepatocellular carcinoma	12.3.3
genital system of the unborn	13.1	hepatocytes	8.1
genito-urinary malformations	13.3.3	hepatomegaly	8.1
globulin, sex hormone-binding	4.8	hereditary studies of alcoholism	15.1
glucagon	4.5, 11.2	heredity	15.1
glucocorticoids	4.6	hexosaminidase, beta-	8.6
gluconeogenesis	5.1	high-density lipoprotein	10.1
glucosamides	5.4	– cholesterol	8.6
glucose	2.3.2	histamine	11.1.2.2
		holiday heart syndrome	10.7

267

homeostasis, calcium	6.8	hypozincaemia	7.2.7
hormonal status	2.1.3	identification, early	18.1
hormone		immune dysfunction	7.2.7
– metabolism	4	immunity	8.2.5
–, adrenocorticotropic	4.6	immunosuppression, alcohol-	
–, antidiuretic	4.1, 7.1, 7.2.1	associated	12.3.4
–, female sex	4.9	infarction, myocardial	10.2
–, follicle-stimulating	4.8	inflammation of the liver	8.1
–, gastrointestinal	11.1.3	inheritability	15.1
–, growth	4.4	inpatient treatment	18.2.5
–, growth hormone-releasing	4.4	insulin	4.5, 5.1, 5.2, 11.2
–, luteinizing	4.8	– -induced hypoglycaemia	2.1.3
–, luteinizing hormone-releasing	4.8	intelligence quotient	13.3.2
–, male sex	4.8	interaction	
–, thyroid	2.3.2, 4.3	– studies	15.6
–, thyroid-stimulating	4.3	–, pharmacodynamic	3.6
–, thyrotropin-stimulating	4.3	–, pharmacokinetic	3.6
hospitalization	18.2.5	intermediate brain syndrome	9.2.1
human lymphocyte antigen	8.2.5, 15.1.4.2	interstitium, pancreatic	11.2.2
hyalin		intervention	18
– sclerosis	8.1	– programmes, implementation of	18.1.3
–, acidophilic	8.1	interview tests	18.1.1
hydrocarbons, polycyclic aromatic	12.3.1, 12.3.4	intestinal	
hydroperitoneum	7.1	– enzymes	11.1.5
hydroxybutyrate	5.1	– mucosa	11.1.5
hydroxycholecalciferol, 25-	6.8	intestine	11.1.5
hyperactivity	13.3.2	–, small	11.1.5
hyperaemia	11.1.2.1	intolerance, carbohydrate	4.5
hyperamylasaemia	11.1.1	intoxication, alcohol	8.2.3, 9.1.1
hypercortisolaemia	4.6	intra-uterine	
hyperglycaemia	4.5, 5.2	– growth	13.3.1
hyperlactacidaemia	5.1	– growth retardation	13.4.3
hyperlipaemia	5.1	intracellular water	7.1
hyperoestrogenaemia	4.3	iron	7.2.6
hyperoestrogenism	4.8	irritability	5.2
hyperphosphataemia	7.2.4	ischaemic stroke	10.5
hyperphosphaturia	7.2.4	isozymes, distribution of	3.2.1
hyperprolactinaemia	4.9.2		
hypertension	9.2.2, 10.3	jejunal villi, haemorrhagic erosion	
–, portal	8.1	of	11.1.5
hyperuricaemia	5.1	Jellinek estimation formula	8.5.4
hyperventilation	5.1	juvenile alcohol use, suppression of	17.2.3
hypoalbuminaemia	7.2.3		
hypocalcaemia	4.2, 7.2.3	katalase activity	9.1.2
hypoglycaemia	4.4, 4.5, 5.2	ketoacidosis, alcoholic	5.1
–, fasting	5.2	ketogenesis	5.1
–, insulin-induced	2.1.3	ketoglutaric acid, alpha-	3.4
–, reactive	5.2	ketone bodies	5.1
hypogonadism	7.2.7	ketosis, alcohol-induced	5.1
hypomagnesaemia	7.2.3	kinetics	
hyponatraemia	7.2.1	–, first-order	2.1
hypopharynx, cancer of the	12.1.1	–, Michaelis-Menten	2.2.2
hypophosphataemia	7.2.4	–, zero-order	2.3.1
hypothalamic		kirsch	14.6.1
– osmoreceptors	7.1	Korsakoff's psychosis	6.2, 9.2.1
– -pituitary axis	4.6		
– -pituitary-gonadal axis	4.8	lactacidosis	10.8
hypothermia	5.2	lactate	5.1, 5.4, 8.2.6, 8.6
hypoxia	8.2.2, 8.2.3	– clearance	5.1
–, centrilobular	8.2.3		

–/pyruvate ratio	3.7, 8.2.3
lactic acidosis	5.1
larynx, cancer of the	12.1.1
LD-50 dose	9.1.1
lead	14.5
learning	
– problems	13.3.2
– theories	15.4
Ledermann's theory	8.5.4, 17.1.1
leucocytes	11.6
licence revocation	16.4.1
life-style factors	8.4, 12.1.6
lipase, hepatic	10.1
lipid	8.1
– metabolism	8.2.1
– oxidation	8.2.1
– peroxidation	8.2.2
–, blood	10.1
lipoprotein	5.3, 5.4, 8.2.1
–, high-density	10.1
–, low-density	6.9, 10.1
–, very low-density	5.3
liqueur, cream	14.3
liquor-by-the-drink	17.2.2
liver	8
– cancer	12.1.1, 12.1.2
– cell mitochondria	8.2.1
– cirrhosis	7.1, 15.1.4.2
– cirrhosis, epidemiology of	8.5
– disease, alcoholic	8.2
– enlargement	8.1
– injury	6.4, 7.2.6, 12.4
lobular oxygen gradient	8.2.3
low-alcohol	
– beer	14.1
– beverages	14.1
low-density lipoprotein	6.9, 10.1
lung	11.4
– cancer	12.1.4
luteinizing hormone	4.8
– -releasing hormone	4.8
lymphocyte	
– activity	8.2.2
– antigen, human	15.1.4.2
lysine	8.2.5
macronutrients	14.3
–, metabolism of	5
magnesium	7.2.5
– deficiency	9.2.2
malabsorption, fat	11.1.5
malate	5.1
male	
– fertility	4.8
– sex hormones	4.8
malformations	13.3.3
– of the unborn child	13
–, genito-urinary	13.3.3
–, sex organ	13.3.3
Mallory bodies	8.1, 8.2.2

malnutrition	9.2.1
–, primary	11.1
–, secondary	11.1
marital state	15.5.4.3
markers of alcoholism	
–, electrophysiological	15.1.4.1
–, biochemical	15.1.4.2
–, biological	15.1.4
marrow, bone	11.6
mass media	17.2.4
maximum allowable blood alcohol concentration	16.2.3
mean corpuscular volume	8.5, 11.6, 18.1.1
measures to control alcohol abuse	17.2, 18
mechanisms of carcinogenesis	12.3
media campaigns	17.2.4
medication	
–, antidipsotropic	3.3, 18.2.2
–, detoxification	18.2.1
–, effect-altering	18.2.2
–, psychotherapeutic	18.2.4
megaloblastic anaemia	6.5, 6.6
membranes, permeability of	12.3.4
menorrhagia	4.9.1
menses, irregular	4.9.1
menstrual cycle	2.3.2
mental	
– clouding	9.2.2
– development	13.5
– function, depressed	7.2.7
– -health model	17.1.3
MEOS	3.2.2, 3.6, 4.8, 8.2.4, 8.4.4
– induction	12.3.4
metabolic	
– acidosis	7.2.2, 14.6.1
– alkalosis	5.1
– tolerance	9.2.2
metabolism	
– of macronutrients	5
–, alcohol	3
–, carbohydrate	5.2, 11.2
–, cholecalciferol	4.2
–, essential fatty acid	5.3, 8.2.2.
–, fat	5.3
–, first-pass	2.3.1, 3.2.1
–, hormone	4
–, lipid	8.2.1
–, modification of alcohol	3.7
–, prostaglandin	5.3, 8.2.2
–, protein	5.4, 7.2.7
–, vitamin	6
methanol	8.6, 14.6.1
methionine	5.4
methylgyanine transferase, O^6-	12.3.4
Michaelis-Menten kinetics	2.2.2, 2.3.1
microcephaly	13.1
microphthalmia	13.1
microsomal ethanol-oxidizing system	3.2.2, 3.6, 4.8, 8.2.4

269

microsomes	3.2.2
migraine	14.6.3
milk	2.1.1
mineral	14.5
−absorption	11.1.5
−balance	7.2
minimum	
−drinking age	17.2.3
−legal purchase age	16.2.2, 17.2.3
miscarriage	13.7
misperceptions	9.2.2
mitochondria	5.3
−, liver cell	8.2.1
mitochondrial	
−abnormalities	10.7
−aspartate aminotransferase	18.1.1
−electron transport	3.2.1
−injury	8.2.2
mitotic inhibition	11.1.5
model of control	17.1
−, Ledermann's	17.1.1
−, mental-health	17.1.3
−, single-distribution	17.1.1
−, socio-cultural	17.1.3
moderate drinking/drinker	1.1
modification of alcohol metabolism	3.7
monoamine oxidase, platelet	15.1.4.2
monokine interleukin 1	7.2.7
monosaccharide	14.3
mortality	9.2.2
−, cirrhosis	8.4.5, 8.5
motility, oesophageal	11.1.1
motives for drinking before driving	16.1
motor	
−disturbance	11.3.1
−performance	13.3.2
mucosa	
−, erosion of gastric	2.1.2
−, gastric	11.1.2
−, intestinal	11.1.5
mucosal	
−erosion	11.1.2.1
−lesions	11.1.2.1
muscle	11.5
muscular tissue	2.2
mutal help groups	18.2.3
myocardial	
−infarction	10.2
−injury	10.7
myofibrillary protein, loss of	11.5.1
myofibroblasts	8.1
myopathy, alcoholic	11.5.1, 11.5.2
NADH	5.4
NADH/NAD ratio	5.1, 5.3
narcosis, cell membrane	13.4.1
nasogastrical weight loss	3.2.2
nausea	8.1, 9.2.2, 11.1.3
necrosis of the liver	8.1
neonatal withdrawal syndrome	13.3.2

nervous system	9, 14.6.2
−, central	15.1.4.1, 13.1
neurobehavioural response	13.3.2
neurological disturbances	6.2
neuropathy	
−, peripheral	11.5.1
−, vagal	7.2.1
neurosis	15.3
neurotransmission	4.1, 9.2.3
neurotransmitters	3.4, 4.7, 11.1.3
−, 'false'	3.3, 9.2.3
niacin	3.7
nicotinamide adenine dinucleotide	3.2
night blindness	6.1, 7.2.7, 11.3.2
nitrosamines	12.2
noradrenaline	4.7, 9.2.3, 10.3
norepinephrine, see noradrenaline	
nucleic acid synthesis	13.4.1
nutrient	
−absorption	11.1.5
−transport, placental	13.4.3
nutrition-related effects of alcohol exposure	13.4.2
nystagmus	11.3.1, 14.6.2
occlusion, arterial	10.4
oculomotor system	11.3.2
Oddi, sphincter of	11.2.2
oedema	6.2
−, cerebral	9.2.1
oesophageal	
−cancer	12.1.1, 12.3.1
−motility	11.1.1
−varix	8.3
oesophagitis	11.1.1
oesophagus	11.1.1
oestradiol	4.8, 4.9.1
oestrogen	4.8, 4.9.1
oestrone	4.8
on-premise consumption	17.2.2
optic nerve damage	14.6.1
opticoneuropathy	11.3.2
organs	11
oropharynx, cancer of the	12.1.1
osmolality, plasma	7.1
osmoreceptors, hypothalamic	7.1
osmotic effect	2.1.1
osteoblastic activity	4.2
osteoporosis	6.8, 7.2.5
−, skeletal	4.2
outlets, restriction of number of	17.2.2
outpatient treatment	18.2.5
overhydration	7.1
−of the brain	4.1
overpermissive culture	17.1.2
ovulation	4.9.1
oxaloacetate	5.1
oxygen	
−demand of liver cells	8.2.3
−gradient, lobular	8.2.3

oxygenated drinking water	2.3.2	potassium	7.2.2
oxytocin	4.9.3	pregnancy	13
		prevention	
pancreas	11.2	– models	17.1
– cancer	12.1.5	– model, Ledermann's	17.1.1
pancreatic		– model, mental-health	17.1.3
– duct	11.2.2	– model, single-distribution	17.1.1
– enzymes	11.2.2	– model, socio-cultural	17.1.2
– interstitium	11.2.2	– of alcohol (ab)use	17
– reflux, biliary	11.2.2	–, primary	16.2, 17.2.5
pancreatitis	5.1, 5.2, 5.3, 7.2.6, 11.2.1, 11.2.2, 12.1.5, 15.1.4.2	–, relapse	18.2.6
		–, secondary	16.3, 18
		–, tertiary	16.4, 18
–, alcohol-induced	12.1.5	price as a control mechanism	17.2.1
–, pathogenesis of	11.2.2	primary	
parathyroid hormone	4.2	– malnutrition	11.1
paternal alcoholism	13.6	– prevention	16.2, 17.2.5
pathogenesis of pancreatitis	11.2.2	pro-oxidants	8.2.4
peer group	15.5.3	problem drinkers/drinking	1.2
peripheral neuropathy	11.5.1	–, screening for	18.1.1
perivenular zones of liver cells	8.1, 8.2.3	procyanidin	14.6.3
permeability		productive cough	11.4
– of gastric mucosa	11.1.2.2	progesterone	4.9.1
– of intestinal mucosa	11.1.5	prolactin	4.9.2, 15.1.4.2
– of membranes	12.3.4	proline	5.4
permissive culture	17.1.2	– hydroxylase	5.1, 5.4, 8.2.6
peroxidation, lipid	8.2.2	pronormoblasts, vacuolated	11.6
personal variables	15.5.4	propanolol	8.3,
personality	15.2	propylthiouracil	8.2.3, 8.3
pharmacodynamic		prostaglandin	9.1.3, 11.1.2.2, 5.3
– interactions	3.6	– E1	5.3
– tolerance	9.2.2	– metabolism	8.2.2
pharmacokinetic interactions	3.6	prostate cancer	12.1.6
pharmacotherapy	18.2.2	protein	2.1.1, 2.3.2, 14.3
phenolic flavanoids	14.6.3	– metabolism	5.4, 7.2.7
phenothiazines	9.2.2	– synthesis	10.7, 13.4.1, 13.4.3
phenylalanine	5.4	pseudo-Cushing syndrome	4.6, 13.5
phosphate	7.2.4	psychoanalytic studies	15.3
phospholipids	12.3.4	psychological studies	15.2
physical exercise	2.2, 2.3.2	psychomotor agitation	9.2.2
phytoestrogen	14.6.3	psychosis, Korsakoff's	6.2
placental transport		psychotherapy	18.2.4
– of amino acids	13.4.2	pulmonary	
– of folate	13.4.2	– congestion	6.2
– of zinc	13.4.2	– function	11.4
plasminogen	10.2	punishment	
platelet	8.6	–, increasing risk of	16.3.1
– affinity to serotonin	18.1.1	–, severity of	16.3.2
– aggregation	10.2	purchase age, minimum legal	17.2.3, 17.2.3
– function	11.6	purine degradation	8.2.4
– monoamine oxidase	15.1.4.2	pylorospasm	2.1.2
poisoning, methanol	14.6.1	pyrazole	8.2.1
polycyclic aromatic hydrocarbons	12.3.1, 12.3.4	pyridoxal-5'-phosphate	6.4
polyneuropathy	6.2	pyridoxine	6.4
polysaccharide	14.3	pyruvate	3.4, 5.1
polyunsaturated fatty acids	5.3, 8.2.3		
portal		racial differences in distribution of isozymes	3.2.1, 3.4
– hypertension	8.1	radicals, free	8.2.4
– -systemic encephalopathy	5.4	radiography, chest	8.6
– venous pressure	8.3		

reaction tests	9.1.1	sexual dysfunction	4.8
reactive hypoglycaemia	5.2	shuttle system	3.2.1
recruitment of clients	18.1.2	sigmoid colon, cancer of the	12.1.6
rectal cancer	12.1.6, 12.2	silicium	14.5
reference man/woman	2.2.2	single-distribution model	17.1.1
reflex control	9.1.1	sister chromatid exchange	12.3.4
reflexive behaviour	13.3.2	skeletal	
reflux		– abnormalities	13.3.3
–, biliary pancreatic	11.2.2	– osteoporosis	4.2
–, duodenopancreatic	11.2.2	– system of the unborn	13.1
regional variation	8.5.2	skin lesions	7.2.7
relapse prevention	18.2.6	sleepiness	14.6.2
religion	15.5.4.4	small intestine	11.1.5
renal		smoking	12.1.1, 11.3.2, 12.1.4
– abnormalities	13.3.3		
– clearance	5.1	sobering agents	2.3.2
renin	4.1	sociability	16.1
respiratory		social drinking/drinker	1.1
– alkalosis	7.2.4	socio-cultural	
– depression	8.2.3	– model	17.1.2
– disease	11.4	– studies	15.5
– equilibrium	2.2.1	socio-economic status	15.5.4.2
restriction of alcohol use	17.2	sodium	7.2.1
retardation, growth	13.1, 13.3.1	somatomedin	4.4
retinal	4.8, 6.1	somatotropin	4.4
retinol	4.8, 6.1	sorbitol	2.1.1, 5.1
– -binding protein	6.1	spermatogenesis	4.8, 6.1
reverse cholesterol transport	10.1	spermatozoal motility	4.8
revocation, licence	16.4.1	sphincter of Oddi	11.2.2.
rhabdomyolysis	7.2.3	spirit	8.4.5
rhythm(icity)		steatosis	8.2.2, 15.1.4.2
–, circadian	2.3.2	stenosis, coronary	10.4
–, diurnal	2.1.3	steroids, anabolic	8.3
riboflavin	3.7, 6.3, 12.3.2, 14.4	stomach cancer	12.1.6
		stroke	
rum	9.1.3	–, haemorrhagic	10.5
		–, ischaemic	10.5
safety limits	12.4	studies	
salicylic acid	11.1.2.2	–, adoption	15.1.3
saliva, determination of alcohol in	2.2	–, constitutional	15.1
salvary glands	11.1.1	–, family	15.1.1
sclerosis, hyalin	8.1	–, hereditary	15.1
screening for problem drinkers	18.1.1	–, interaction	15.6
secondary		–, psychoanalytic	15.3
– malnutrition	11.1	–, psychological	15.2
– prevention	16.3, 18	–, socio-cultural	15.5
selenium	8.2.4	–, twin	15.1.2
self-censorship in advertising	17.2.4	stupor	6.2
self-control	2.3.2	sugar	2.3.2, 14.3
serotonin	3.4, 9.1.2, 9.2.3	supplementation, vitamin	6.1, 6.2
–, platelet affinity to	18.1.1	suppression of juvenile alcohol use	17.2.3
severity of punishment	16.3.2	sweating	5.2
sex	15.5.4.1	synaptic membranes	3.3
– differences	2.2.2	syndrome	
– differences in liver disease	8.4.4	–, alcohol withdrawal	6.7, 18.2.1
– hormone	8.4.4	–, Cushing	4.6
– hormone, female	4.9.	–, flushing	3.4
– hormone, male	4.8	–, intermediate brain	9.2.1
– hormone-binding globulin	4.8	–, neonatal withdrawal	13.3.2
– organ malformations	13.3.3	–, pseudo-Cushing	4.6

−, Wernicke-Korsakoff's	9.2.1, 10.8
−, Zieve's	6.9
synthetase, fatty acid ester	3.2.3
systolic blood pressure	10.3
tachycardia	5.2, 9.2.2
tannin	12.3.1, 14.6.3
tear fluid, alcohol in	2.2
temperature, body	2.3.2
temporal relations	8.5.1
teratogenicity	13.4.1
tertiary prevention	16.4, 18
testicular atrophy	6.1
testosterone	4.8
therapeutic approach	16.4.2
therapy, confrontational group	18.2.4
thermoregulation	2.3.2
thiamin	6.2, 10.8
−deficiency	6.2, 9.2.1
−pyrophosphate	6.2
threshold values	8.4
thrombocytopenia	11.6
thromboxane	10.2
thyroid	
−activity	4.3
−hormone	2.3.2, 4.3
−inhibitor	8.3
−-stimulating hormone	4.3
thyrotropin	4.3
−-stimulating hormone	4.3
thyroxine	4.3
tissue-plasminogen activator	10.2
tobacco use	12.1.1
tocopherol, alpha-	8.2.4
tolerance	
−, functional	9.2.2
−, metabolic	9.2.2
trace elements	14.5
traffic	16
transferrin	5.4, 7.2.6
−, carbohydrate-deficient	8.6
treatment	
−of alcohol liver disease	8.3
−of alcoholism	18.1, 18.2
tremor	9.2.2
tremulousness	5.2
triglycerides	5.3, 8.2.1, 8.6
tryptophan	5.4
twin studies	15.1.2
tyrosine	5.4, 8.2.5
U-shaped relationship	10.5
unborn child, malformations of the	13
unconsciousness	9.1.1
underreporting of alcohol use	8.4.4
urate	8.6
uric acid	5.1
urokinase-plasminogen activator	10.2
vacuolation of pronormoblasts	11.6

vagal neuropathy	7.2.1
valine	8.2.5
variant angina	10.6
varix, oesophageal	8.3
vascular dilatation	11.1.2.2
vasoconstriction	10.4
vasodilatation	10.4
vasopressin	4.1, 7.1, 9.2.1
venous pressure, portal	8.3
villi, jejunal	11.1.5
vision	11.3
vitamin(s)	2.3.2, 14.3
−A	6.1, 11.3.2, 12.3.2
−A deficiency	6.1, 11.3.2
−absorption	11.1.5
−B-1, see thiamin	
−B-2, see riboflavin	
−B-6	6.4, 12.3.2, 14.4
−B-6 deficiency	6.4
−B-12	6.6, 11.3.2
−B-12 deficiency	6.6
−C	3.7, 6.7, 14.4
−D	6.8
−E	3.7, 6.9, 8.2.4, 12.3.2
−deficiency	9.2.1, 11.3.2
−metabolism	6
−supplementation	6.1, 6.2
vomiting	8.1
water balance	7.1, 7.2.1
−, disturbance of	9.1.3
weight loss	8.1
−, nasogastrical	3.2.2
Wernicke-Korsakoff's syndrome	6.2, 9.2.1, 10.8
Wernicke's encephalopathy	6.2, 9.2.1
whisky	9.1.3, 12.2
wine	8.4.5, 10.5, 12.1.6, 14.4, 14.5
withdrawal	4.1, 9.2.2
−delirium, alcohol	9.2.2
−syndrome, alcohol	6.7, 18.2.1
−syndrome, neonatal	13.3.2
wodka	9.1.3
wound healing	7.2.7
xanthine dehydrogenase	8.2.4
xanthine oxidase	8.2.4
xenobiotics	3.2.2, 3.6
xylose	2.3.2
young drivers	16.2
zero-order kinetics	2.3.1
Zieve's syndrome	6.9
zinc	7.2.7
−metabolism	7.2.7
−, placental transport of	13.4.2